D1264353

AN INTRODUCTION TO NONLINEAR PARTIAL DIFFERENTIAL EQUATIONS

PURE AND APPLIED MATHEMATICS

A Wiley-Interscience Series of Texts, Monographs, and Tracts

Founded by RICHARD COURANT

A complete list of the titles in this series appears at the end of this volume.

AN INTRODUCTION TO NONLINEAR PARTIAL DIFFERENTIAL EQUATIONS

J. DAVID LOGAN
Department of Mathematics and Statistics
University of Nebraska–Lincoln
Lincoln, Nebraska

A Wiley-Interscience Publication
JOHN WILEY & SONS, INC.
New York • Chichester • Brisbane • Toronto • Singapore

A NOTE TO THE READER:
This book has been electronically reproduced from digital information stored at John Wiley & Sons, Inc. We are pleased that the use of this new technology will enable us to keep works of enduring scholarly value in print as long as there is a reasonable demand for them. The content of this book is identical to previous printings.

This text is printed on acid-free paper.

Library of Congress Cataloging in Publication Data:

Logan, J. David (John David)
An introduction to nonlinear partial differential equations / J. David Logan.
p. cm.—(Pure and applied mathematics)
Includes bibliographical references and index.
ISBN 0-471-59916-6
1. Differential equations, Nonlinear. 2. Differential equations, Partial. I. Title. II. Series: Pure and applied mathematics (John Wiley & Sons: Unnumbered)
QA377.L58 1994
515′.353—dc20 93-29705

Printed in the United States of America

10 9 8 7 6 5

To Tess and David for all the affection and joy they bring to my life

CONTENTS

PREFACE

This introductory text on nonlinear partial differential equations evolved from a graduate course that the author has taught at the University of Nebraska on numerous occasions during the last ten years. The audience has typically been a mixture of graduate students from mathematics, physics, and engineering, with an occasional student from chemistry and biology. The prerequisites were a standard, elementary course in partial differential equations (PDEs) emphasizing Fourier analysis and separation of variables, and an elementary course in ordinary differential equations where the phase plane was discussed.

There are outstanding treatises on this topic, notably the books by G. B. Whitham (*Linear and Nonlinear Waves*) and J. Smoller (*Shock Waves and Reaction–Diffusion Equations*), both of which are benchmarks in nonlinear partial differential equations. But beginning students seemed to struggle when I used these books as texts, so I resorted to distributing exercises and supplementary notes of a more elementary nature to build a foundation for reading those two standard, authoritative treatments. Therefore, this book came about as a pedagogical effort to introduce, at a fairly elementary level, nonlinear PDEs in a style that is accessible to students with diverse backgrounds and interests, and possibly with a desire to move more deeply into the subject later by studying Whitham or Smoller, for example. The text also includes numerous exercises.

There is enough independence among the chapters to allow the instructor considerable flexibility in choosing topics for a course. The text may be used for a second course in partial differential equations, a course in nonlinear PDEs, or an advanced course in applied mathematics emphasizing modeling. The range of applications include biological problems, chemistry, porous media, combustion, detonation, traffic flow, water waves, plug flow reactors, heat transfer, and other topics of interest in applied mathematics.

Nonlinear partial differential equations is a vast area and practitioners include applied mathematicians, analysts, and others in science and engineering. It is one of the most popular areas in modern analysis. Any limited selection of topics, like the ones herein, will not be satisfying to all workers in the area. I have chosen topics that reflect my own interests in wave propagation and reaction–diffusion equations. Thus the scope of the book is limited, and many interesting and important topics are not covered. For example, there is little about dispersive waves or scattering theory; and nothing has been said about some of the topological concepts, such as degree theory, that have had a significant impact on the subject in recent times. I wanted to present a concise introduction to the subject, emphasizing time-evolution problems, that struck a balance between the mathematical and physical aspects of the subject and was accessible to students in a wide range of engineering and physical fields. Thus, by necessity and design, both the scope and the breadth of the text are limited. The approach is analytical and technique oriented rather than numerical.

The goal of the first chapter is to develop a point of view about how to understand problems involving PDEs and how the subject interrelates with physical phenomena. The subject of nonlinear PDEs is developed from the basic conservation law, which, when appended to constitutive relations, gives rise to the fundamental models of diffusion, convection (or transport), and reaction. There is emphasis on understanding that nonlinear hyperbolic and parabolic PDEs describe evolutionary processes; a solution is a signal that is propagated into a spacetime domain from the boundaries of that domain. There is emphasis on the structure of the various equations and what the terms in the equations describe physically.

Chapters 2 and 3 deal with wave propagation and hyperbolic problems. In Chapter 2 we assume that the equations have smooth solutions and we develop algorithms to solve the equations analytically. In Chapter 3 we study discontinuous solutions and shock formation, and we introduce the concept of a weak solution. In keeping with our strategy of thinking about initial waveforms evolving in time, we focus on the initial value problem rather than the general Cauchy problem. The idea of characteristics is central and forms the thread that weaves through these two chapters.

In Chapter 4 we investigate equations that model basic diffusion processes. After establishing a probabilistic basis for diffusion, we examine various methods that are useful in studying the solution structure of diffusion problems. These methods include phase-plane analysis, similarity methods, and asymptotic expansions. The prototype equations for reaction–diffusion and convection–diffusion, Fisher's equation and Burgers' equation, respectively, are studied in detail with emphasis on the existence of traveling wave solutions, the stability of those solutions, and the asymptotic behavior of solutions in general. An appendix to Chapter 4 presents the basics of phase-plane analysis.

In Chapter 5 we use the shallow water equations as the prototype of a hyperbolic system, and those equations are taken to illustrate the basic concepts associated with hyperbolic systems: characteristics, Riemann's method, the hodograph transformation, and asymptotic behavior. Also, the general classification of systems of first order PDEs is developed, and weakly nonlinear methods of analysis are described, the latter being illustrated by a derivation of Burgers' equation.

In Chapter 6 we discuss systems of reaction-diffusion equations, emphasizing applications and model building. We also expend some effort addressing theoretical concepts such as existence, uniqueness, and comparison and maximum principles, energy estimates, blowup, and invariant sets.

Chapter 7 contains a discussion of a specialized topic, chemically reacting fluids. The choice of this topic reflects the interests of the author in combustion and detonation phenomena. But it is reasonable to include these topics in an introductory treatment because the methods and techniques developed to understand reactive flow phenomena have been applied in many other areas modeled by reaction–diffusion–convection processes, and a good understanding of this area of application is important if one is eventually to study nonlinear processes.

Elliptic equations are introduced in Chapter 8. Nonlinear eigenvalue problems and stability and bifurcation phenomena form the core of the study. The reader should be alerted to the fact that elliptic equations are slighted in this treatment and that the small amount of material on elliptic equations in no way reflects its relative importance in the field.

The references, collected at the end of each chapter, are selected to suit the needs of an introductory text, pointing the reader to parallel treatments and suggesting resources for further study. Thus, most references are to texts or monographs, with an occasional citation of a survey paper.

The exercises at the end of each section have been collected from a variety of sources over many years: other books, research papers, our department's master's degree examinations, and so on. Only in some cases have I identified the source.

I would like to acknowlege the support of my colleagues with whom I have had the privilege of working during the past few years. In particular, my sabbatical at Rensselaer Polytechnic Institute with Ash Kapila spurred my interests in combustion and asymptotics; at the University of Nebraska, Professors Steve Cohn, Steven Dunbar, Glenn Ledder, and Thomas Shores have helped make our weekly seminars a tremendous learning experience. My own view of the subject has also been strongly influenced by the works of Whitham and Smoller, the classic treatise of R. Courant and K. O. Friedrichs (*Supersonic Flow and Shock Waves*), and the recent treatment of mathematical biology by J. D. Murray (*Mathematical Biology*). Finally, my appreciation goes to Elizabeth Murphy, the editor at Wiley, who has done a skillful job of managing this project.

1

PARTIAL DIFFERENTIAL EQUATIONS

The subject of partial differential equations is one of the basic areas of applied analysis and it is difficult to imagine any area of applications where its impact is not felt. In the last few decades there has been tremendous emphasis on understanding and modeling nonlinear processes; such processes are often governed by partial differential equations, and the subject of nonlinear partial differential equations has become one of the most active areas in applied mathematics and analysis.

In this initial chapter, which is introductory in nature, the focus is on developing in the student a point of view about how to understand problems involving partial differential equations and how the subject interrelates with physical phenomena. The purpose is to provide a transition from an elementary course emphasizing eigenfunction expansions and linear problems to a more sophisticated way of thinking about problems that is suggestive of and consistent with the methods to be developed in nonlinear analysis.

Section 1.1 is a summary of some of the basic terminology of elementary partial differential equations, including a review of the problem of classification. In Section 1.2 we begin the study of the origins of partial differential equations in physical problems. This interdependence is developed from the basic, one-dimensional conservation law. In Section 1.3 we show how constitutive relations can be appended to the conservation law to obtain equations that model the fundamental processes of diffusion, convection or transport, and reaction. Some of the common equations, such as the diffusion equation, Burgers' equation, Fisher's equation, the porous media equation, and so on, are obtained as models of these processes. In Section 1.4 we introduce initial and boundary value problems to observe how auxiliary data specialize the problems. Finally, in Section 1.5 we discuss wave propagation in order to fix the notion of how evolution equations carry boundary and initial signals into the domain of interest. We also introduce some common techniques for determining solutions of a certain form (e.g., traveling wave solutions). The ideas presented in this chapter are meant to build an understanding of evolutionary processes so that the fundamental concepts of hyperbolic problems and characteristics, as well as diffusion problems, can be examined in later chapters with a firmer base. Some of the material will be review for many readers, and those portions of the text can be skimmed quickly.

1.1 PARTIAL DIFFERENTIAL EQUATIONS

PDEs and Solutions

A partial differential equation (hereafter abbreviated PDE) is an equation involving an unknown function of several variables and its partial derivatives. To fix the notion, a *second order PDE in two independent variables* is an equation of the form

$$G(x, t, u, u_x, u_t, u_{xx}, u_{tt}, u_{xt}) = 0, \qquad (x, t) \in D \tag{1}$$

where, as indicated, the independent variables x and t lie in some given domain D in R^2. By a *solution* to (1) we mean a twice continuously differentiable function $u = u(x, t)$ defined on D which, when substituted into (1), reduces (1) to an identity on the domain D. The function $u(x, t)$ is assumed to be twice continuously differentiable, so that it makes sense to calculate its first and second derivatives and substitute them into (1); a solution like this is also called a *classical solution* or a *genuine solution*. Later we extend the notion of solution to include functions that may have discontinuities, or discontinuities in their derivatives; such functions are called *weak solutions*. The xt-domain D in R^2, where the problem is defined, is referred to as a *spacetime domain*, and PDEs that include time t as one of the independent variables are called *evolution* equations. When the two independent variables are both spatial variables, say x and y rather than x and t, the PDE will be called an *equilibrium* or *steady-state* equation. Evolution equations govern time-dependent processes, and equilibrium equations often govern physical processes after the transients caused by initial conditions or boundary conditions die away.

Graphically, a solution $u = u(x, t)$ of (1) can be represented as a smooth surface in three-dimensional xtu-space lying over the domain D in the xt-plane, as shown in Figure 1.1. An alternative representation that is used more frequently is a graph in the xu-plane of the function $u = u(x, t_0)$ for some fixed time $t = t_0$ (see Figures 1.1 and 1.2). Such representations are called *time snapshots* or *profiles* of the solution; time snapshots are profiles in space of the solution $u = u(x, t)$ frozen at a fixed time t_0, or, stated differently, slices of the solution surface at a fixed time t_0. Occasionally, several time snapshots are shown simultaneously on the same set of xu axes to indicate how the profiles are changing in time. It is also helpful on

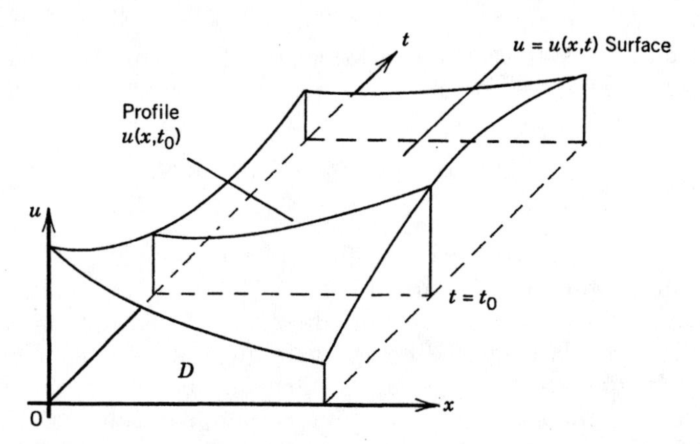

Figure 1.1. Solution surface $u = u(x, t)$ in xtu-space over the domain D, also showing a time snapshot or wave profile $u(x, t_0)$ at time t_0.

occasion to think of a solution in abstract terms. For example, suppose that $u = u(x, t)$ is a solution of a PDE for $x \in R$ and $0 \le t \le T$. Then for each t, $u(x, t)$ is a function of x (a profile), and it generally belongs to some space of functions $\mathbf{X}$. To fix the idea, suppose that $\mathbf{X}$ is the set of all twice continuously differentiable functions on R that vanish at infinity. Then the solution can be regarded as a mapping from the time interval $[0, T]$ into the function space $\mathbf{X}$. That is, to each t in $[0, T]$ we associate a function $u(\cdot, t)$, which is the wave profile at time t. The dot represents the spatial variable; often the dot is just deleted and we write $u = u(t)$ to denote the solution profile at time t.

A PDE of the type (1) has infinitely many solutions, depending on arbitrary functions. For example, the *wave equation*

$$u_{tt} - c^2 u_{xx} = 0 \tag{2}$$

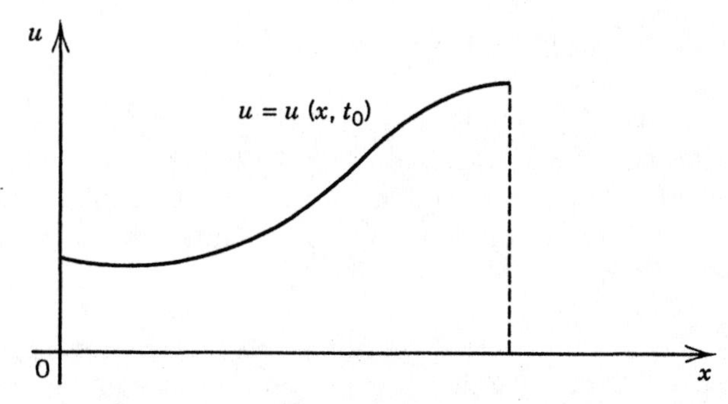

Figure 1.2. Time snapshot $u(x, t_0)$ at $t = t_0$ graphed in xu-space. Often several snapshots for different times t are graphed on the same set of xu coordinates to indicate how the wave profiles are evolving in time.

has a general solution that is the superposition (sum) of a right traveling wave $F(x - ct)$ of speed c and a left traveling wave $G(x + ct)$ of speed c; that is,

$$u(x,t) = F(x - ct) + G(x + ct) \tag{3}$$

for any twice continuously differentiable functions F and G. One may contrast the situation in ordinary differential equations, where solutions depend on arbitrary constants; there, initial or boundary conditions fix the arbitrary constants and often, therefore, pick out a unique solution. For PDEs the situation is similar; initial and boundary conditions are usually imposed and select out one of the infinitude of solutions. These auxiliary or subsidiary conditions are usually suggested by the underlying physical problem from which the PDE arises, or by the type of PDE. A condition on u or its derivatives given at $t = 0$ along some segment of the x-axis is called an *initial condition*, while a condition along any other curve in the xt-plane is called a *boundary condition*. PDEs with auxiliary conditions are called *initial value problems, boundary value problems, or initial–boundary value problems*, depending on the type of subsidiary conditions that are specified.

Example 1. The initial value problem for the wave equation is

$$u_{tt} - c^2 u_{xx} = 0, \qquad x \in R, \quad t > 0 \tag{4}$$

$$u(x,0) = f(x), \qquad u_t(x,0) = g(x), \quad x \in R \tag{5}$$

where f and g are given twice continuously differentiable functions on R. The unique solution is given by (see Exercise 3)

$$u(x,t) = \frac{1}{2}[f(x - ct) + f(x + ct)] + \frac{1}{2c}\int_{x-ct}^{x+ct} g(s)\, ds \tag{6}$$

which is called *d'Alembert's formula*. So, in this example, we can think of the auxiliary data (5) as selecting out one of the infinitude of solutions given by (3). Note that the solution at (x, t) depends only on the initial data (5) in the interval $[x - ct, x + ct]$.

The statements in the preceding paragraphs regarding the single second order PDE (1) can be generalized in various directions. Higher order equations (as well as first order equations), several independent variables, and several unknown functions (governed by systems of PDEs) are all possibilities.

Classification

In the study of PDEs the strategy is to classify them into different types, depending either on the types of physical phenomena from which they arise, or on some mathematical basis. As the reader has learned from previous experience, there are three fundamental types of equations: those that govern diffusion processes, those that govern wave propagation, and those that govern equilibrium phenomena. Equations of mixed type also occur. We consider a single, second order PDE of the form

$$au_{xx} + 2bu_{xt} + cu_{tt} = d(x,t,u,u_x,u_t), \qquad (x,t) \in D \tag{7}$$

where a, b and c are continuous functions on D, and not all of a, b, and c vanish simultaneously at some point of D. The function d on the right side is assumed to be continuous as well. Classification is made on the basis of the combination of the second order derivatives in the equation. If we define the *discriminant* Δ by $\Delta = b^2 - ac$, then (7) is *hyperbolic* if $\Delta > 0$; (7) is *parabolic* if $\Delta = 0$; (7) is *elliptic* if $\Delta < 0$.

Hyperbolic and parabolic equations are typically evolution equations that usually govern wave propagation and diffusion processes, respectively, and elliptic equations are often associated with equilibrium or steady-state processes. In the latter case, we usually use x and y as independent variables rather than x and t. There is also a close relationship between the classification type and the kinds of initial and boundary conditions that may be imposed on a PDE to obtain a well-posed mathematical problem or one that is physically relevant. Again we point out that the classification is made based on the highest order derivatives in (7), or the principal part of the equation, and because Δ depends on x and t, equations may change type as x and t vary throughout the domain.

Now we demonstrate that equation (7) can be transformed to certain simpler, or *canonical* forms, depending on the classification, by a change of independent variables

$$\xi = \xi(x,t), \qquad \eta = \eta(x,t) \tag{8}$$

We now perform this calculation, with the view of actually trying to determine (8) such that (7) reduces to a simpler form in the new $\xi\eta$-coordinate system. The transformation (8) is assumed to be invertible, which requires that the Jacobian $J = \xi_x\eta_t - \xi_t\eta_x$ not vanish in any region where the transformation is to be applied. A straightforward application of the chain rule, which the reader can verify, shows that the left side of (7) becomes, under the change of independent variables (8),

$$au_{xx} + 2bu_{xt} + cu_{tt} + \cdots = Au_{\xi\xi} + 2Bu_{\xi\eta} + Cu_{\eta\eta} + \cdots \tag{9}$$

where the three dots denote terms of lower order, and where

$$A = a\xi_x^2 + 2b\xi_x\xi_t + c\xi_t^2$$
$$B = a\xi_x\eta_x + b(\xi_x\eta_t + \xi_t\eta_x) + c\xi_t\eta_t$$
$$C = a\eta_x^2 + 2b\eta_x\eta_t + c\eta_t^2$$

We notice that the expressions for A and C have the same form, namely

$$a\phi_x^2 + 2b\phi_x\phi_t + c\phi_t^2$$

and they are independent.

As we shall now demonstrate, in the *hyperbolic case* it is possible to choose ξ and η such that $A = C = 0$. To this end let us set

$$a\phi_x^2 + 2b\phi_x\phi_t + c\phi_t^2 = 0 \tag{10}$$

If the discriminant Δ is positive, we can write (10) as (assume that a is not zero)

$$\frac{\phi_x}{\phi_t} = -\frac{b \pm \sqrt{b^2 - ac}}{a}$$

To determine ϕ we regard it as defining loci (curves) in the xt-plane via the equation $\phi(x, t) = \text{const}$. The differentials dx and dt along one of these curves satisfy the relation $\phi_x\, dx + \phi_t\, dt = 0$ or $dt/dx = -\phi_x/\phi_t$. Therefore,

$$\frac{dt}{dx} = \frac{b \pm \sqrt{b^2 - ac}}{a} \tag{11}$$

is a differential equation whose solutions determine the curves $\phi(x, t) = \text{const}$. Thus, upon choosing the + and − sign in (11), respectively, we obtain $\xi(x, t)$ and $\eta(x, t)$ as integral curves of (11), making $A = C = 0$. Consequently, if (7) is hyperbolic, it can be reduced to the *canonical hyperbolic form*

$$u_{\xi\eta} + \cdots = 0$$

where the three dots denote lower order derivatives (we leave it as an exercise to show that B is nonzero in this case).

The differential equations (11) are called the *characteristic equations* associated with (7), and the two sets of solution curves $\xi(x, t) = \text{const}$ and $\eta(x, t) = \text{const}$ are called the *characteristic curves*, or just the *characteristics*; ξ and η are called *characteristic coordinates*. In summary, in the hyperbolic case there are two real families of characteristics that provide a coordinate system where the equation reduces to a simpler form. As we shall observe in the sequel, characteristics are the fundamental concept in the analysis of hyperbolic problems because characteristic coordinates form a natural curvilinear coordinate system in which to examine these problems.

In the *parabolic case* ($b^2 - ac = 0$) there is just one family of characteristic curves, defined by

$$\frac{dt}{dx} = \frac{b}{a}$$

Thus we may choose $\xi = \xi(x, t)$ as an integral curve of this equation to make $A = 0$. Then, if $\eta = \eta(x, t)$ is chosen as any smooth function independent of ξ (i.e., so that the Jacobian is nonzero), one can easily determine $B = 0$ automatically, giving the *parabolic canonical form*

$$u_{\xi\xi} + \cdots = 0$$

Characteristics do not play an important role in parabolic problems.

In the *elliptic case* ($b^2 - ac < 0$) there are no real characteristics. However, it is still possible to eliminate the mixed derivative term in (7) to obtain the *elliptic canonical form*

$$u_{\xi\xi} + u_{\eta\eta} + \cdots = 0$$

The procedure is to go ahead and determine complex characteristics by solving (11), and then take real and imaginary parts to determine a transformation (8) that makes $A = C$ and $B = 0$ in (9).

***Example* 2.** It is easy to see that the characteristic curves for the wave equation (2), which is hyperbolic, are the straight lines $x - ct =$ const and $x + ct =$ const. These are shown in Figure 1.3. In this case the characteristic coordinates are given by $\xi = x - ct$ and $\eta = x + ct$. In these coordinates the wave equation becomes $u_{\xi\eta} = 0$; verification is requested in the Exercises. We regard the characteristics as curves in spacetime moving with speeds c and $-c$, and from the general solution (3) we observe that signals are propagated along these curves. In hyperbolic problems, in general, the characteristics are curves in spacetime along which signals are transmitted in the system.

If the coefficients a, b, and c of the second order derivatives in equation (7) depend on x, t, u, u_x, and u_t, (7) is called a *quasilinear* equation. In this case we make the same classification as above, depending on the sign of the discriminant Δ; now the type of the equation not only depends on the spacetime domain, but also on the solution u itself. The canonical forms listed above are no longer valid in this case, and the characteristics defined by

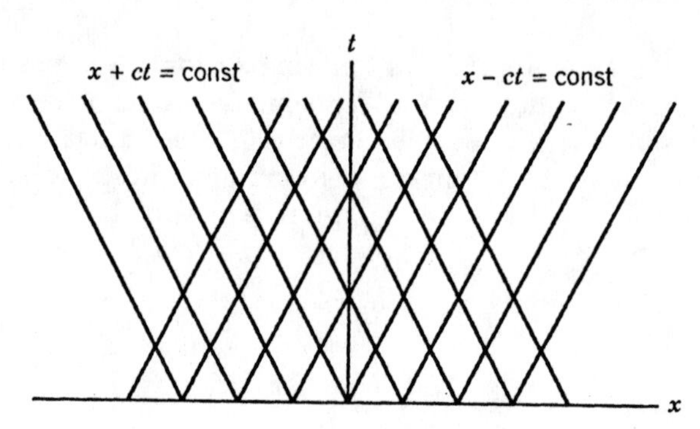

Figure 1.3. Characteristic diagram for the wave equation showing the forward and backward characteristics $x - ct = \text{const}$ and $x + ct = \text{const}$.

(11) cannot be determined a priori since a, b, and c depend on u, the unknown solution itself. Therefore, a significant increase in difficulty arises when the principal part of the equation is nonlinear.

There are other ways to approach the classification problem. In the preceding discussion our focus was on determining transformations under which a simplification occurs. In Section 8.1 we take a different perspective when we ask if it is possible to determine the solution u near a curve where the values of u and its first derivatives are known. That discussion is accessible to the reader at the present juncture, if desired. Yet another view of classification is presented in Chapter 5, where systems of hyperbolic equations are discussed.

Nonlinearity

It is also important to distinguish PDEs as *linear* or *nonlinear*. Roughly, a PDE is linear if the sum of two solutions is again a solution, and constant multiples of solutions are solutions. Otherwise, a PDE is said to be nonlinear. The division of PDEs into these two categories is a significant one; as we shall observe, the mathematical methods devised in dealing with these two types of equations are often entirely different, and the behavior of solutions differs substantially. An underlying reason for these differences is the fact that the solution space to a linear, homogeneous PDE is a vector space, and the linear structure of that space can be used with advantage in constructing solutions with desired properties that can meet diverse boundary and initial conditions. Such is not the case for nonlinear equations.

More formally, linearity and nonlinearity are usually defined in terms of the properties of the operator that defines the PDE itself. Let us assume that the PDE (1) can be written in the form

$$Lu = F \tag{12}$$

where $F = F(x, t)$ and L is an operator that contains all the operations (differentiation, multiplication, composition, etc.), that act on $u = u(x, t)$. For example, the wave equation $u_{tt} - u_{xx} = 0$ can be written $Lu = 0$, where L is the partial differential operator $\partial_t^2 - \partial_x^2$. In (12) we reiterate that all terms involving the unknown function u are on the left side of the equation and are contained in the expression Lu; the right side of (12) contains in F only expressions involving the independent variables x and t. If $F = 0$, (12) is said to be *homogeneous*; otherwise, it is *nonhomogeneous*. We say that an operator L is *linear* if it is additive and if constants factor out of the operator, that is, (i) $L(u + v) = Lu + Lv$, and (ii) $L(cu) = cLu$, where u and v are functions (in the domain of the operator) and c is any constant. The PDE (12) is *linear* if L is a linear operator; otherwise, the PDE is *nonlinear*.

Example 3. The equation $Lu = u_t + uu_x = 0$ is nonlinear since, for example, $L(cu) = cu_t + c^2uu_x$, which does not equal $cLu = c(u_t + uu_x)$.

Conditions (i) and (ii) stated above imply that a linear homogeneous equation $Lu = 0$ has the property that if $u_1, u_2, \ldots, u_n$ are n solutions, the linear combination

$$u = c_1u_1 + c_2u_2 + \cdots + c_nu_n$$

is also a solution for any choice of the constants $c_1, c_2, \ldots, c_n$. This fact is called the *superposition principle* for linear equations. For nonlinear equations, one cannot superimpose solutions in this manner. The superposition principle can often be extended to infinite sums for linear problems, provided that certain convergence requirements are satisfied. Superposition for linear equations allows one to construct, from a given set of solutions, another solution that meets initial or boundary requirements by choosing the constants $c_1, c_2, \ldots$ judiciously. This observation is the basis for the Fourier method, or eigenfunction expansion method, for linear, homogeneous boundary value problems. In addition, superposition can often be extended to a family of solutions depending on a continuum of values of a parameter. More precisely, if $u = u(x, t; k)$ is a family of solutions of a linear homogeneous PDE for all values of k in some interval of real numbers I, one can superimpose these solutions formally by integration, that is, by defining

$$u(x, t) = \int_I c(k)u(x, t; k)\, dk$$

where $c = c(k)$ is now a function of the parameter k. Under certain conditions that must be established, the superposition $u(x, t)$ may again be a solution. As in the discrete case, there may be flexibility in selecting $c(k)$ to meet boundary or initial conditions. In fact, this procedure is the vehicle for so-called transform methods for solving PDEs (Laplace transforms, Fourier transforms, etc.). None of these methods based on superposition are applicable to nonlinear problems, and other methods must be sought. In summary, there is a profound difference between solution methods for linear and nonlinear PDEs.

If most of the solution methods used for linear problems are inapplicable to nonlinear equations, what methods can be developed in the nonlinear case? We mention a few.

1. *Perturbation Methods.* Perturbation methods are applicable to most problems where a small or large parameter can be identified. In this case an approximate solution is sought as a series expansion in the parameter.
2. *Similarity Methods.* The similarity method is based on the PDE and its auxiliary conditions being invariant under a local Lie group of transformations. The invariance transformation allows one to identify a canonical change of variables that reduces the PDE to an ordinary differential equation or reduces the order of the PDE.
3. *Characteristic Methods.* Nonlinear hyperbolic equations, which are usually associated with wave propagation, can often be analyzed with success in characteristic coordinates (i.e., coordinates in spacetime along which the waves or signals propagate).
4. *Transformations.* Sometimes it is possible to identify transformations that map a given nonlinear equation into a simpler equation that can be solved.
5. *Numerical Methods.* The advent of fast, large-scale computers in the past few decades has given tremendous impetus to the development and analysis of numerical algorithms to solve nonlinear problems.
6. *Ad Hoc Methods.* The mathematical and applied science literature is full of articles illustrating special methods that analyze a certain type of nonlinear PDE, or restricted classes of nonlinear PDEs.

Of course, the methods in the list above are primarily solution methods, which represent only one aspect of the subject of nonlinear PDEs. There are other basic issues in the subject as well: namely, questions of existence and uniqueness of solutions, and the investigation of stability properties of solutions. These and other theoretical questions have spawned investigations based on modern topological and algebraic concepts, and the subject of nonlinear PDEs has evolved into one of the most diverse, active areas of applied analysis. We shall address these issues in due course.

Terminology. We introduce some notation and terminology for some function spaces that commonly occur in analysis. Let D be a domain in either one or several dimensions. By the set of all continuous functions on D that have n continuous derivatives in D (partial derivatives if D is of dimension greater than 1). The space of continuous functions on D is denoted by $C(D)$, and $C^\infty(D)$ denotes the space of continuous functions on D that have derivatives of all orders. When a function u belongs to one of these sets [e.g., $C^n(D)$], we sometimes say that u is of class C^n on D. We will usually assume that D is an open domain (i.e., a set that does not contain any of its boundary), so that we do not have to deal with the question of existence of derivatives at boundary points; also, we shall always assume that D is simply connected.

EXERCISES

1. Find a formula for the solution of the problem

$$u_{xt} = h(x,t), \qquad x > 0, \quad t > 0$$
$$u(x,0) = f(x), \quad x > 0; \qquad u(0,t) = g(t), \quad t > 0$$

where $f(0) = g(0)$. Assume sufficient differentiability.

2. Show that the wave equation $u_{tt} - c^2u_{xx} = 0$ reduces to the canonical form $u_{\xi\eta} = 0$ under the change of variables $\xi = x - ct$, $\eta = x + ct$. Use this information to show that the general solution of the wave equation is $u(x,t) = F(x - ct) + G(x + ct)$, where F and G are arbitrary $C^2(R)$ functions.
3. Derive d'Alembert's formula (6) for the initial value problem (4)–(5). (Use Exercise 2 and determine F and G from the initial conditions.)
4. Let u be of class C^3. Prove that $u = u(x,t)$ is a solution of the wave equation $u_{tt} - c^2u_{xx} = 0$ iff u satisfies the difference equation

$$u(x - ck, t - h) + u(x + ck, t + h)$$
$$= u(x - ch, t - k) + u(x + ch, t + k)$$

for all constants $h, k > 0$. Interpret this result geometrically in the xt-plane by observing that the difference equation relates the value of u at the vertices of a characteristic parallelogram whose sides are the characteristic straight lines $x + ct = \text{const}$ and $x - ct = \text{const}$ (see Figure 1.4).

5. If the initial data f and g for the initial value problem for the wave equation in Exercise 3 have compact support (i.e., f and g vanish for $|x|$ sufficiently large), prove that the solution $u(x,t)$ has compact support in x for each fixed t.
6. Find the general solution of the PDE

$$x^2u_{xx} + 2xtu_{xt} + t^2u_{tt} = 0$$

by transforming the equation to canonical form using characteristic coordinates $\xi = t/x$, $\eta = t$.

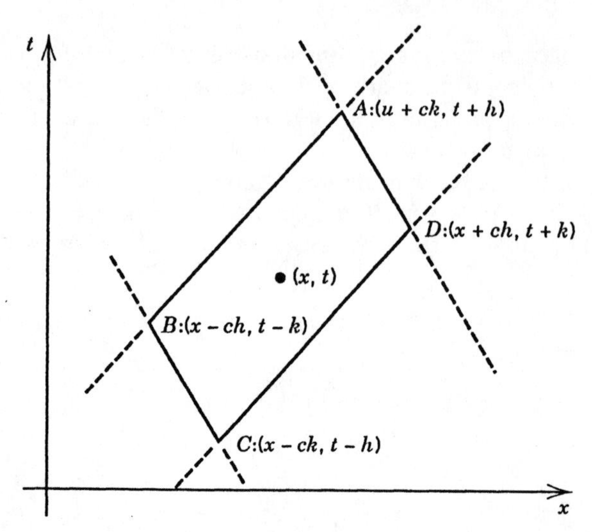

Figure 1.4. Characteristic parallelogram *ABCD* whose sides are characteristic straight lines $x - ct =$ const and $x + ct =$ const. The numbers h and k are positive constants that define the size of *ABCD*.

7. Let $u = u(x,t)$ be a solution of the nonlinear equation

$$a(u_x, u_t)u_{xx} + 2b(u_x, u_t)u_{xt} + c(u_x, u_t)u_{tt} = 0$$

Introduce new independent variables via $\xi = \xi(x,t)$ and $\eta = \eta(x,t)$, and a new function $\phi = \phi(\xi, \eta)$ defined by $\phi = xu_x + tu_t - u$. Prove that $\phi_\xi = x$, $\phi_\eta = t$, and ϕ satisfies the *linear* PDE

$$a(\xi, \eta)\phi_{\eta\eta} - 2b(\xi, \eta)\phi_{\xi\eta} + c(\xi, \eta)\phi_{\xi\xi} = 0$$

(This transformation, known as a *hodograph* transformation or a *Legendre* transformation, transforms a nonlinear equation of the given form to a linear equation by reversing the roles of the dependent and independent variables.)

8. Find a particular solution to the initial value problem

$$u_{tt} - c^2 u_{xx} = \phi(x,t), \qquad x \in R, \quad t > 0$$

$$u(x,0) = u_t(x,0) = 0, \qquad x \in R$$

Suggestion: Integrate the PDE over a characteristic triangle having vertices (x,t), $(x - ct, 0)$, and $(x + ct, 0)$, and use Green's theorem in the plane. Derive a solution in the case $\phi(x,t) = x^2$.

9. Find a formula for the solution of the initial–boundary value problem

$$u_{tt} = c^2 u_{xx}, \qquad x > 0, \quad t > 0$$
$$u(x,0) = f(x), \qquad u_t(x,0) = g(x), \quad x > 0$$
$$u(0,t) = h(t), \qquad t > 0$$

Suggestion: Use D'Alembert's formula for $x > ct$ and the difference equation in Exercise 4 for $0 < x < ct$. Assume sufficient differentiability.

10. Consider the PDE

$$4u_{xx} + 5u_{xt} + u_{tt} = 2 - u_t - u_x$$

Find the characteristic coordinates and graph the characteristic curves in the *xt*-plane. Reduce the equation to canonical form and find the general solution of the given equation.

1.2 CONSERVATION LAWS

Many of the fundamental equations occurring in the natural and physical sciences are obtained from *conservation laws*. Conservation laws are just balance laws, equations expressing the fact that some quantity is balanced throughout a process. In thermodynamics, for example, the first law states that the change in internal energy in a given system is equal to, or is balanced by, the total heat added to the system plus the work done on the system. Thus the first law of thermodynamics is really an energy balance law, or conservation law. As another example, consider a fluid flowing in some region of space that consists of chemical species undergoing chemical reaction. For a given chemical species, the time rate of change of the total amount of that species in the region must equal the rate at which the species flows into the region, minus the rate at which the species flows out, plus the rate at which the species is created, or consumed, by the chemical reactions. This is a verbal statement of a conservation law for the amount of the given chemical species. Similar balance or conservation laws occur in all branches of science. In the biosciences, for example, the rate of change of a given animal population in a certain region must equal the birth rate, minus the death rate, plus the migration rate into or out of the region.

Mathematically, conservation laws usually translate into differential equations, which are then regarded as the *governing equations* or *equations of motion* of the process. These equations dictate how the process evolves in time. Here we are interested in processes governed by partial differential equations. We now formulate the basic one-dimensional conservation law, out of which will evolve some of the basic models and concepts in nonlinear PDEs.

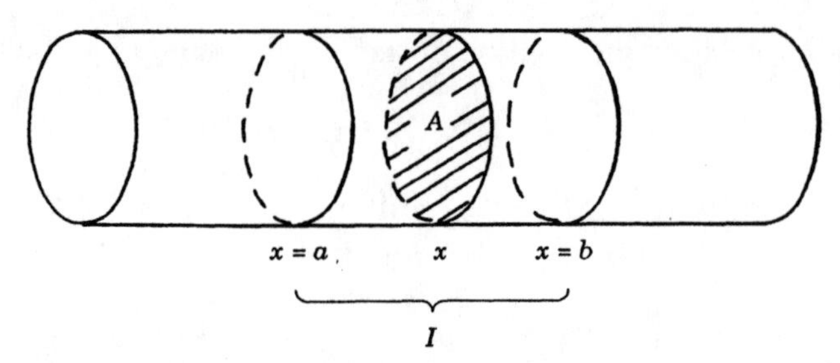

Figure 1.5. Cylindrical tube of cross-sectional area A showing a cross section at x and a finite section I: $a \leq x \leq b$.

Let us consider a quantity $u = u(x, t)$ that depends on a single spatial variable $x \in R$ and time $t > 0$. We assume that u is a density or concentration measured in an amount per unit volume, where the amount may refer to population, mass, energy, or any quantity. By definition, u varies in only one spatial direction, the direction denoted by x. We imagine further that the quantity is distributed in a tube of cross-sectional area A (see Figure 1.5). Again, by assumption, u is constant in any cross section of the tube, and the variation is only in the x direction. Now consider an arbitrary segment of the tube denoted by the interval $I = [a, b]$. The total amount of the quantity u inside I at time t is

$$\text{total amount of quantity in } I = \int_a^b u(x, t) A \, dx$$

Now assume that there is motion of the quantity in the tube in the axial direction. We define the *flux* of u at x at time t to be the scalar function $\phi(x, t)$; that is, $\phi(x, t)$ is the amount of the quantity u flowing through the cross section at x at time t, per unit area, per unit time. Thus the dimensions of ϕ are $[\phi] = \text{amount}/(\text{area} \cdot \text{time})$, where the bracket notation denotes *dimensions of*. By convention, we take ϕ to be positive if the flow at x is in the positive x direction, and ϕ is negative at x if the flow is in the negative x direction. Therefore, at time t the net rate that the quantity is flowing into the interval I is the rate that it is flowing in at $x = a$ minus the rate that it is flowing out at $x = b$. That is,

$$\text{net rate that the quantity flows into } I = A\phi(a, t) - A\phi(b, t)$$

Finally, the quantity u may be created or destroyed inside I by some external or internal source (e.g., by a chemical reaction if u were a species concentration, or by birth or death if u were a population density). We denote this *source function*, which is a local function acting at each x, by $f(x,t,u)$ and its dimensions are given by $[f] = \text{amount}/(\text{volume} \cdot \text{time})$. Consequently, f is the rate that u is created (or destroyed) at x at time t, per unit volume. Note that the source function f may depend on u itself, as well as space and time. If f is positive, we say that it is a *source*, and if f is negative, we say that it is a *sink*. Now, given f, we may calculate the total rate that u is created in I by integration. We have

$$\text{rate that quantity is produced in } I \text{ by sources} = \int_a^b f(x,t,u(x,t))\,A\,dx$$

The fundamental conservation law may now be formulated for the quantity u: For any interval I, we have

time rate of change of the total amount in I

= net rate that the quantity flows into I

+ rate that the quantity is produced in I

In terms of the mathematical symbols and expressions that we introduced above, we have, after canceling the constant cross-sectional area A,

$$\frac{d}{dt}\int_a^b u(x,t)\,dx = \phi(a,t) - \phi(b,t) + \int_a^b f(x,t,u)\,dx \tag{1}$$

In summary, (1) states that the rate that u changes in I must equal the net rate at which u flows into I plus the rate that u is produced in I by sources. Equation (1) is called a *conservation law in integral form*, and it holds even if u, ϕ, or f are not smooth (continuously differentiable) functions. The latter remark is important when we consider in subsequent chapters physical processes giving rise to shock waves, or discontinuous solutions.

If some restrictive conditions are place on the triad u, ϕ, and f, (1) may be transformed into a single partial differential equation. Two results from elementary integration theory are required to make this transformation: (i) the fundamental theorem of calculus, and (ii) the result on differentiating an integral with respect to a parameter in the integrand. Precisely,

(i) $\int_a^b \phi_x(x,t)\,dx = \phi(b,t) - \phi(a,t)$.
(ii) $d/dt \int_a^b u(x,t)\,dx = \int_a^b u_t(x,t)\,dx$.

These two results are valid if ϕ and u are continuously differentiable functions on R^2. Of course, (i) and (ii) remain correct under less stringent conditions, but our assumption of smoothness is all that is required in the subsequent discussion. Therefore, assuming smoothness of u and ϕ, as well as continuity of f, equations (i) and (ii) imply that the conservation law (1) may be written

$$\int_a^b [u_t(x,t) + \phi_x(x,t) - f(x,t,u)]\, dx = 0 \qquad \text{for all intervals } I = [a,b] \tag{2}$$

Because the integrand is a continuous function of x, and because (2) holds for all intervals of integration I, it follows that the integrand must vanish identically; that is,

$$u_t + \phi_x = f(x,t,u), \qquad x \in R, \quad t > 0 \tag{3}$$

Equation (3) is a partial differential equation relating the density $u = u(x,t)$ and the flux $\phi = \phi(x,t)$. Both are regarded as unknowns, whereas the source function f is assumed to be given. Equation (3) is called a *conservation law in differential form*, in contrast to the integral form (1). The ϕ_x term is called the *flux term* since it arises from the movement, or transport, of u through the cross section at x. The source term f is sometimes called a *reaction term* (especially in chemical contexts) or a *growth* or *interaction* term (in biological contexts). To be consistent with standard usage of terminology, (3) is frequently called a conservation law when the source term f is absent; in that case we should call (3) a conservation law with sources. We shall always make clear to what context we are referring. Finally, we have defined the flux ϕ as a function of x and t; it may happen that this dependence on space and time may occur through dependence on u or its derivatives. For example, a physical assumption may require us to posit $\phi(x,t) = \phi(x,t,u(x,t))$, where the flux is dependent on u itself.

Conservation in Higher Dimensions

It is straightforward to formulate conservation laws in higher dimensions. In this section we limit the discussion to three-dimensional Euclidean space R^3. For notation, we let $x = (x_1, x_2, x_3)$ denote a point in R^3, and we assume that $u = u(x,t)$ is a scalar density function representing the amount per unit volume of some quantity of interest distributed throughout some domain in R^3. In this domain let V be an arbitrary region, and assume that V has a smooth boundary, which is denoted by ∂V. It follows that, similar to the one-dimensional case, the total amount of the quantity in V is given by the volume integral

$$\text{total amount in } V = \int_V u(x,t)\, dx$$

where $dx = dx_1\, dx_2\, dx_3$ represents a volume element in R^3. We prefer to write the volume integral over V with a single integral sign rather than the usual triple integral. Now, we know that the time rate of change of the total

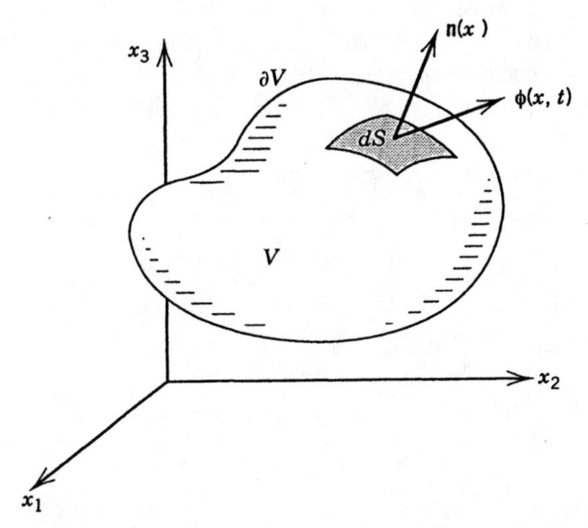

Figure 1.6. Volume V with boundary ∂V showing a surface element dS with outward normal $\mathbf{n}$ and flux vector $\boldsymbol{\phi}$.

amount in V must be balanced by the rate that the quantity is produced in V by sources, plus the net rate that the quantity flows through the boundary of V. We let $f(x, t, u)$ denote the source term, so that the rate that the quantity is produced in V is given by

$$\text{rate that } u \text{ is produced by sources} = \int_V f(x, t, u)\, dx$$

In three dimensions the flow can be in any direction, and therefore the flux is given by a vector $\boldsymbol{\phi}(x, t)$. If $\mathbf{n}(x)$ denotes the outward unit normal vector to the region V (see Figure 1.6), the net outward flux of the quantity u through the boundary ∂V is given by the surface integral

$$\text{net outward flux through } \partial V = \int_{\partial V} \boldsymbol{\phi}(x, t) \cdot \mathbf{n}(x)\, dS$$

where dS denotes a surface element on ∂V. Finally, therefore, the conservation law, or balance law for u, is given by

$$\frac{d}{dt}\int_V u\, dx = -\int_{\partial V} \boldsymbol{\phi} \cdot \mathbf{n}\, dS + \int_V f\, dx \tag{4}$$

The minus sign on the flux term occurs because outward flux decreases the rate that u changes in V.

The integral form of the conservation law (4) can be reformulated as a local condition, that is, a PDE, provided that u and ϕ are sufficiently smooth functions. In this case the surface integral can be written as a volume integral over V using the *divergence theorem* (the divergence theorem is the fundamental theorem of calculus in three dimensions). The divergence theorem is embodied in the expression

$$\int_V \operatorname{div} \phi \, dx = \int_{\partial V} \phi \cdot \mathbf{n} \, dS \tag{5}$$

where div is the divergence operator. Using (5) and bringing the derivative under the integral on the left side of (4) yields

$$\int_V u_t \, dx = -\int_V \operatorname{div} \phi \, dx + \int_V f \, dx$$

Using the arbitrariness of V then gives the differential form of the balance law as

$$u_t + \operatorname{div} \phi = f(x, t, u), \qquad x \in V, \quad t > 0 \tag{6}$$

Equation (6) is the three-dimensional version of equation (3), the conservation law in one dimension.

EXERCISES

1. In the derivation of the fundamental conservation law [(1) or (3)] we assumed that the cross-sectional area A of the tube was constant. Derive integral and differential forms of the conservation law in the case that the area is a slowly varying function of x, that is, $A = A(x)$. [Note that $A(x)$ cannot change significantly over small changes in x; otherwise, the one-dimensional assumption of the state functions u and ϕ being constant in any cross section would be violated; see Figure 1.7.

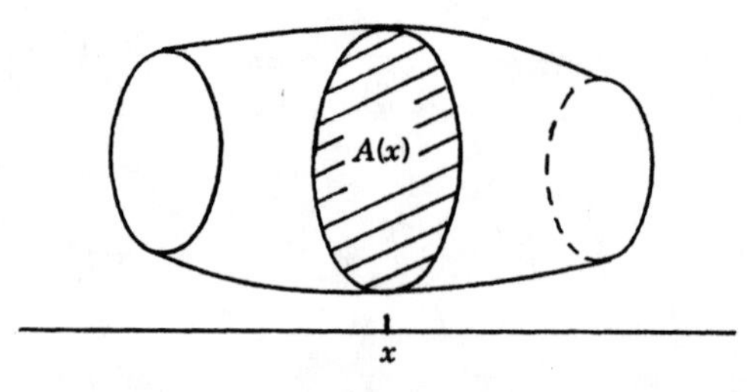

Figure 1.7

2. In the case that there are no sources and $\phi = \phi(u)$, show that the conservation law (1) is equivalent to

$$\int_a^b u(x, t_2)\, dx = \int_a^b u(x, t_1)\, dx + \int_{t_1}^{t_2} \phi(u(a, t))\, dt - \int_{t_1}^{t_2} \phi(u(b, t))\, dt$$

for all t_1 and t_2.

1.3 CONSTITUTIVE RELATIONS AND EXAMPLES

Because equation (3) [or (6)] in Section 1.2 is a single PDE for two unknown quantities (the density u and the flux ϕ), intuition indicates that another equation is required in order to have a well-determined system. This additional equation is often an equation that is based on an assumption about the physical properties of the medium, which in turn is based on empirical reasoning. Equations expressing these assumptions are called *constitutive relations* or *equations of state*. Thus constitutive equations are on a different level from the basic conservation law; the latter is a fundamental law of nature connecting the density u to the flux ϕ, whereas a constitutive relation is often an approximate equation whose origin is in empirics.

Example 1 (*Diffusion Equation*). At the outset assume that no sources are present ($f = 0$) and the process is governed by the basic conservation law in one dimension

$$u_t + \phi_x = 0, \qquad x \in R, \quad t > 0 \tag{1}$$

In many physical problems it is observed that the amount of the substance represented by the density u that flows through a cross section at x at time t is proportional to the density gradient u_x, [i.e., $\phi(x, t) \propto u_x(x, t)$]. If $u_x > 0$, then $\phi < 0$ (the substance flows to the left), and if $u_x < 0$, then $\phi > 0$ (the substance flows to the right). Figure 1.8 illustrates the situation. For example, by the second law of thermodynamics, heat behaves in this manner; heat flows from hotter regions to colder regions, and the steeper the temperature distribution curve, the more rapid the flow of heat. As another example, if u represents a concentration of insects, one might observe that insects move from high concentrations to low concentrations with a rate proportional to the concentration gradient. Therefore, we assume the basic constitutive law

$$\phi(x, t) = -Du_x(x, t) \tag{2}$$

which is known as *Fick's law*. The positive proportionality constant D, called the *diffusion constant*, has dimensions given by $[D] = \text{length}^2/\text{time}$. Fick's law accurately describes the behavior of many physical and biological systems,

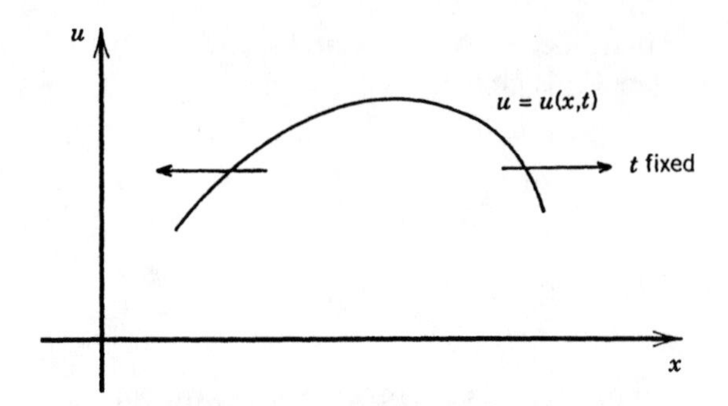

Figure 1.8. Time snapshot of the density distribution $u(x, t)$ illustrating Fick's law. The arrows indicate the direction of the flow, from higher concentrations to lower concentrations. The flow is said to be *down the gradient*.

and in Chapter 4 we give a supporting argument for it based on a probability model and random walk.

Equations (1) and (2) give a pair of PDEs for the two unknowns u and ϕ. They combine easily to form a single second order linear PDE for the unknown density $u = u(x, t)$ given by

$$u_t - Du_{xx} = 0 \tag{3}$$

Equation (3), called the *diffusion equation*, governs conservative processes when the flux is specified by Fick's law. It may not be clear to the novice at this time why (3) should be termed the diffusion equation; suffice it to note for the moment that Fick's law seems to imply that a substance leaks (diffuses) into adjacent regions because of concentration gradients. We refer to the Du_{xx} term in (3) as the *diffusion term*.

We remark that the diffusion constant D defines a characteristic time (or time scale) T for the process. If L is a length scale (e.g., the length of the container), the quantity

$$T = \frac{L^2}{D} \tag{4}$$

is the only constant in the process with dimensions of time, and T gives a measure of the time required for discernible changes in concentration to occur.

Example 2 (*Reaction–Diffusion Equation*). If sources are present ($f \neq 0$), the conservation law

$$u_t + \phi_x = f(x, t, u) \tag{5}$$

and Fick's law (2) combine to give

$$u_t - Du_{xx} = f(x,t,u) \tag{6}$$

which is called a *reaction–diffusion equation*. Reaction–diffusion equations may be nonlinear if the source term f (or, as it is also called, the reaction term) is nonlinear in u. These equations are of great interest in nonlinear analysis and applications, particularly in combustion processes and in biological systematics. The following example introduces Fisher's equation, which is an important reaction–diffusion equation.

Example 3 (*Fisher Equation*). In studies of elementary population dynamics it is often proposed that a population is governed by the logistics law, which states that the rate of change of a population $u = u(t)$ is given by

$$\frac{du}{dt} = ru\left(1 - \frac{u}{K}\right) \tag{7}$$

where $r > 0$ is the *growth rate* and $K > 0$ is the *carrying capacity*. The integral curves of (7) are shown in Figure 1.9. Initially, if u is small, the linear growth term ru in (7) dominates and rapid population growth results; as u becomes large, the quadratic competition term $-ru^2/K$ kicks in to inhibit the growth. For large times t, the population equilibrates toward the asymptotically stable state $u = K$, the carrying capacity.

Now suppose that the population u is a population density (population per unit volume) and depends on a spatial variable x as well as time t [i.e., $u = u(x,t)$]. Then, as in the preceding discussion, a conservation law may be formulated as

$$u_t + \phi_x = ru\left(1 - \frac{u}{K}\right) \tag{8}$$

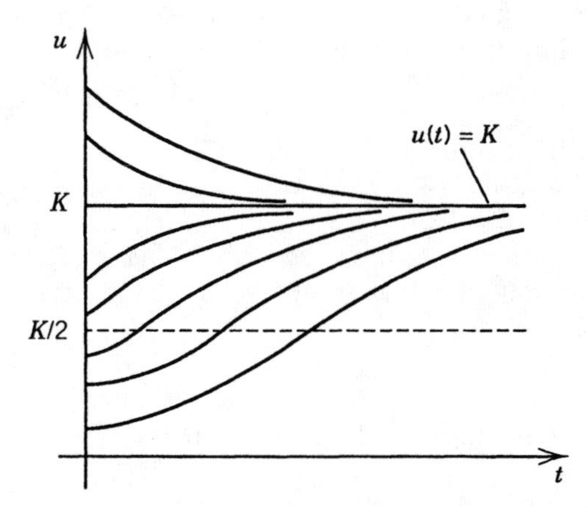

Figure 1.9. Integral curves (i.e., solution curves) of the logistics equation (7). The limiting or asymptotic population K is called the carrying capacity.

where $f = f(u) = ru(1 - u/K)$ is the assumed local source term given by the logistics growth law, and ϕ is the population flux. Assuming Fick's law for the flux, we have

$$u_t - Du_{xx} = ru\left(1 - \frac{u}{K}\right) \tag{9}$$

The reaction–diffusion equation (9) is known as *Fisher's equation*, after R. A. Fisher, who studied the equation in the context of investigating the distribution of an advantageous gene in a given population. A complete analysis of this equation is given in a subsequent chapter.

Example 4 (*Burgers' Equation*). To the basic conservation law (1) with no sources we now append the constitutive relation

$$\phi = -Du_x + Q(u) \tag{10}$$

It follows immediately that the density u satisfies the second order PDE

$$u_t - Du_{xx} + Q(u)_x = 0 \tag{11}$$

Now there are two terms contributing to the flux, a Fick's law type of term $-Du_x$, which introduces a diffusion effect, and a flux term $Q(u)$, depending only on u itself, that leads to what later will be described as transport, or convection. In the special case that $Q(u) = u^2/2$, equation (11) can be written

$$u_t + uu_x = Du_{xx} \tag{12}$$

which is known as *Burgers' equation*. This equation is one of the fundamental model equations in fluid mechanics that illustrates the coupling between convection and diffusion; we derive (12) in a later section using a weakly nonlinear approximation of the equations of gas dynamics. When $D = 0$ (no diffusion),

$$u_t + uu_x = 0 \tag{13}$$

which is called the *inviscid Burgers' equation*. This is the prototype equation for nonlinear convection. We shall observe that (13) is hyperbolic, whereas (12) is parabolic.

Example 5 (*Advection Equation*). If the flux is given by the simple linear relationship

$$\phi = cu \tag{14}$$

where c is a positive constant, the basic conservation law becomes

$$u_t + cu_x = 0 \tag{15}$$

which is the *advection equation*. *Advection* refers to the horizontal movement of a physical property (e.g., a density wave); the term is similar to *convection* or *transport*, which have nearly the same meaning, and we shall use these terms interchangeably. This linear first order PDE (15) is a simple model of transport, and as we shall observe in the sequel, is the simplest wave equation.

Example 6 (*Diffusion in R^3*). In Section 1.2 we obtained the basic conservation law,

$$u_t + \operatorname{div} \phi = f(x, t, u)$$

In R^3, Fick's law takes the form $\phi = -D \operatorname{grad} u$; that is, the flux is in the direction of the negative gradient of u. Using the vector identity $\operatorname{div}(\operatorname{grad} u) = \Delta u$, where $\Delta = \partial^2/\partial x_1^2 + \partial^2/\partial x_2^2 + \partial^2/\partial x_3^2$ is the *Laplacian* operator, and assuming that the diffusion coefficient D is constant, the conservation law becomes

$$u_t - D\Delta u = f(x, t, u)$$

which is a reaction–diffusion equation in R^3. If there are no sources ($f = 0$), we obtain the diffusion equation $u_t - D\Delta u = 0$.

In summary, we have introduced a few of the fundamental PDEs that have been examined extensively by practitioners of the subject. An understanding of these equations and a development of intuition regarding the fundamental processes that they, and other equations, describe is one of the goals of this treatment. We end this section with a specific example of modeling flow through a porous medium.

Porous Media Equation

Consider a fluid (e.g., water) seeping downward through the soil, and let $\rho = \rho(x, t)$ be the density of the fluid, with positive x measured downward. In a given volume of soil, only a fraction of the space is available to the fluid, the remaining being reserved for the soil itself. In this sense, the soil is a porous medium through which the fluid flows. If the fraction of the volume that is available to the fluid is denoted by κ, called the *porosity*, the fluid mass balance law in a given section between $x = a$ and $x = b$ is given by

$$\frac{d}{dt}\int_a^b \kappa\rho(x,t)\,dx = \phi(a,t) - \phi(b,t) \tag{16}$$

where ϕ is the mass flux. We can write the mass flux ϕ as $\phi = \rho v$, where v is the volumetric flow rate. Then, assuming the requisite smoothness, the integral balance law (16) can be written as a PDE:

$$\kappa\rho_t + (\rho v)_x = 0 \tag{17}$$

Here we are assuming that the porosity is constant, but in general it could depend on x or even ρ. The conservation law (17) contains two unknowns, the density and the flow rate, and therefore we expect a constitutive equation that also relates these two quantities. For reasonably slow flows it is observed experimentally that the volumetric flow rate is given by

$$v = -\frac{\mu}{\nu}(p_x + g\rho) \tag{18}$$

where $p = p(x,t)$ is the pressure, and the positive constants g, μ, and ν are the acceleration due to gravity, the permeability, and the viscosity of the fluid, respectively. Equation (18) is known as *Darcy's law*, and it is a basic assumption in many groundwater problems. It is physically plausible because it confirms our intuition that the flow rate should depend on the pressure gradient as well as gravity. Darcy's law is actually a statement replacing a momentum balance law. When (18) is substituted into (17), we obtain what is called the *groundwater equation*,

$$\kappa\rho_t - \frac{\mu}{\nu}(\rho p_x + g\rho^2)_x = 0 \tag{19}$$

We remark that equation (19) contains two unknowns, ρ and p, because the constitutive relation (18) introduced yet another unknown. Consequently, we require another equation to obtain a determined system.

When the fluid is a gas it is common to neglect gravity and assume an equation of state for the gas of the form $p = p(\rho)$. In particular, we assume a *γ-law gas* having the equation of state

$$\frac{p}{p_0} = \left(\frac{\rho}{\rho_0}\right)^\gamma \tag{20}$$

where p_0 and ρ_0 are positive constants and $\gamma > 1$. Substituting (20) into (19) and setting $g = 0$ yields a single PDE for the density ρ, which has the form

$$\rho_t - \frac{\alpha}{\kappa}(\rho^\gamma \rho_x)_x = 0 \tag{21}$$

where α is a constant given by $\alpha = \gamma\mu p_0/\nu\rho_0^\gamma$. Equation (21) is a nonlinear diffusion equation called the *porous media equation*; it governs flows subject to three laws: mass conservation, Darcy's law, and the gas equation of state. (A detailed review of some of the important results for the porous media equation is contained in an article by Aronson [1986].)

EXERCISES

1. Write the Korteweg–deVries (KdV) equation

$$u_t + uu_x + u_{xxx} = 0$$

in the form of a conservation law, identifying the flux ϕ.

2. Show that Burgers' equation (12) can be transformed into the diffusion equation (3) by the transformation

$$u = -\frac{2Dv_x}{v}$$

This transformation is called the *Cole–Hopf transformation*.

3. Show that the equation

$$u_t + kuu_x + q(t)u = 0$$

can be reduced to the inviscid Burgers' equation

$$v_s + vv_x = 0$$

using the transformation

$$v = u \exp \int q(t)\, dt \qquad s = \int k \exp\left(-\int q(y)\, dy\right) dt$$

Show that the same transformation takes

$$u_t + kuu_x + q(t)u - Du_{xx} = 0$$

into

$$v_s + vv_x - g^{-1}(s)v_{xx} = 0$$

where $g = (k/D)\exp[-\int q(t)\, dt]$.

4. By rescaling, show that the porous media equation can be written in the form

$$u_\tau + (u^m)_{\xi\xi} = 0 \qquad (m > 2)$$

for appropriately chosen dimensionless variables τ, ξ, and u.

5. In a chain reaction the number of neutrons in any volume V increases at a rate proportional to the number present. If $u = u(x,t)$ denotes the neutron density and $\phi = \phi(x,t)$ is the neutron flux, conclude that

$$\frac{d}{dt}\int_V u\,dx = -\int_{\partial V} \boldsymbol{\phi}\cdot\mathbf{n}\,dS + \alpha\int_V u\,dx$$

where α is a constant. If $\boldsymbol{\phi} = -D\,\text{grad}\,u$, with D constant, and if the density is sufficiently smooth, show that

$$u_t - D\,\Delta u = \alpha u$$

where Δ is the Laplacian operator.

6. Consider a porous medium where the fluid is water, and assume that the density ρ is constant. What equation must the pressure p satisfy? Describe the pressure distribution.

1.4 INITIAL AND BOUNDARY VALUE PROBLEMS

Heretofore we have encountered four basic types of PDEs governing four different types of physical processes:

$$\begin{aligned}
u_t - Du_{xx} &= 0 && \text{(diffusion)}\\
u_t - Du_{xx} &= f(x,t,u) && \text{(reaction–diffusion)}\\
u_t + uu_x &= 0 && \text{(convection)}\\
u_t + uu_x &= Du_{xx} && \text{(convection–diffusion)}
\end{aligned}$$

As we remarked in Section 1.1, we are seldom interested in the general solution to a PDE, which contains arbitrary functions. Rather, we are usually interested in solving the PDE subject to auxiliary conditions such as initial conditions, boundary conditions, or both.

One of the fundamental problems of PDEs is a *pure initial value problem* (or *Cauchy problem*) on R having the form

$$u_t + F(x,t,u,u_x,u_{xx}) = 0, \qquad x \in R, \quad t > 0 \tag{1}$$

$$u(x,0) = u_0(x), \qquad x \in R \tag{2}$$

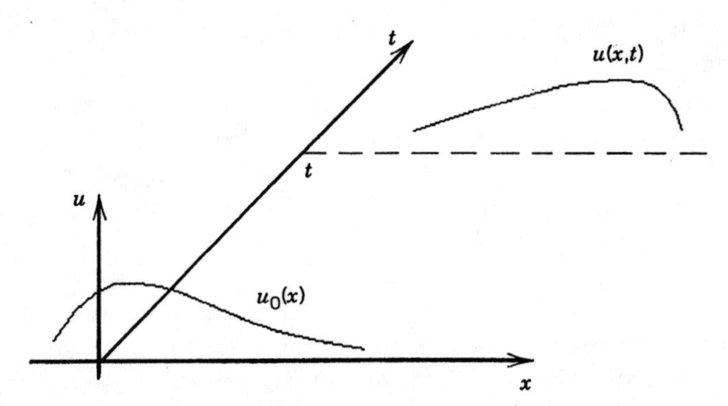

Figure 1.10. Schematic indicating the time evolution, or propagation, of an initial signal or waveform $u_0(x)$.

where $u_0(x)$ is a given function defined on the real line. Interpreted physically, the function $u_0(x)$ represents a *signal* at time $t = 0$, and the PDE is the equation that propagates the signal in time. Figure 1.10 depicts this interpretation. In wave propagation problems, the signal is usually called a *wave* or wavelet. There are several fundamental questions associated with the pure initial value problem (1)–(2).

1. *Existence of Solutions*. Given an initial signal $u_0(x)$ satisfying specified regularity conditions (e.g., continuous, bounded, integrable, or whatever), does a solution $u = u(x, t)$ exist for all $x \in R$ and $t > 0$? If a solution exists for all $t > 0$, it is called a *global solution*. Sometimes solutions are only *local*; that is, they exist for only finite times. As we shall observe, for nonlinear hyperbolic problems, for example, the signal can sometimes only be propagated up to a finite time and blowup occurs; the signal experiences a gradient catastrophe where u_x becomes infinite and the solution ceases to be smooth. In some problems the solution u itself may blow up. This occurs in some nonlinear reaction-diffusion equations.

2. *Uniqueness*. If a solution of (1)–(2) exists, is the solution unique? For a properly posed physical problem we expect an affirmative answer, and therefore we expect the governing initial value problem, which is regarded as a mathematical model for the physical system, to mirror the properties of the system when considering uniqueness and existence questions.

3. *Continuous Dependence on Data*. Another requirement of a physical problem is that of stability; that is, if the initial condition is changed by only a small amount, the system should behave in nearly the same way. Mathematically, this is translated into the statement that the solution should depend continuously on the initial data. In PDEs, if the initial value problem has a unique solution that depends continuously upon the initial conditions, we say that the problem is *well-posed*. A similar statement is made for

boundary value problems. A basic question in PDEs is the problem of well-posedness.

4. *Asymptotic Behavior*. If an initial signal can be propagated for all times $t > 0$, one may inquire about its asymptotic behavior (i.e., the form of the signal for long times). If the signal decays, for example, what is the decay rate? Does the signal disperse, or does the signal remain coherent for large times?

These are just a few of the issues in the study of PDEs. The primary issue, however, from the point of view of the applied scientist may be methods of solution. If a physical problem leads to an initial value problem as a mathematical model, what methods are available or can be developed to obtain a solution, either exact or approximate? Or, even if no solution can be obtained (say, other than numerical), what properties can be inferred from the governing PDEs themselves? For example, what would the speed of propagation be? Are the solutions wavelike, diffusionlike, or dispersive in nature? These questions are all addressed in subsequent chapters.

Now we investigate several initial value problems, to illustrate the diversity of solutions.

Example 1 (*Diffusion Equation*). The initial value problem for the diffusion equation is

$$u_t - Du_{xx} = 0, \qquad x \in R, \quad t > 0$$

$$u(x,0) = u_0(x), \qquad x \in R$$

This is a linear problem and an integral representation of the solution can be found exactly in the form

$$u(x,t) = (4\pi Dt)^{-1/2} \int_R u_0(\xi) \exp\left[-\frac{(x-\xi)^2}{4Dt}\right] d\xi$$

In the next section this solution will be derived. As it turns out, the solution is valid for all $t > 0$ and $x \in R$ under rather mild restrictions on the initial signal $u_0(x)$, and the solution has a high degree of smoothness even if the initial data are discontinuous.

Example 2 (*Advection Equation*). Consider the linear initial value problem

$$u_t + cu_x = 0, \qquad x \in R, \quad t > 0$$

$$u(x,0) = u_0(x), \qquad x \in R$$

where c is a positive constant. It is easy to check that $u(x,t) = f(x - ct)$ is a

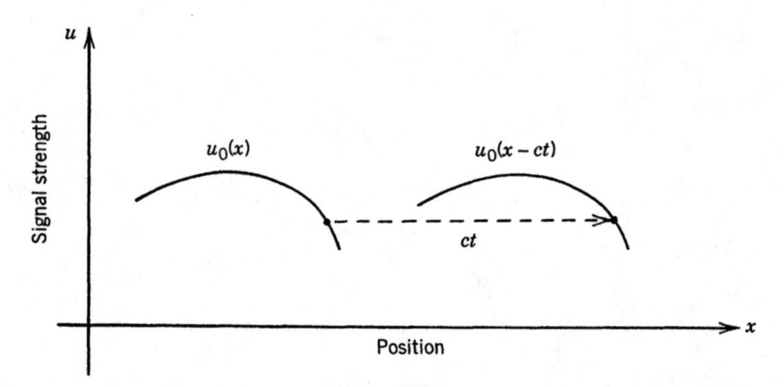

Figure 1.11. Right traveling wave, which represents the solution to the advection equation, shown in xu-space.

solution of the PDE for differentiable function f. Therefore, applying the initial condition to determine f, we have $u(x,0) = f(x) = u_0(x)$. So the global solution to the initial value problem is

$$u(x,t) = u_0(x - ct), \qquad x \in R, \quad t > 0$$

Graphically, the solution is the initial signal $u_0(x)$ shifted to the right by the amount ct, as shown in Figures 1.11 and 1.12. Therefore, the initial signal moves forward undistorted in spacetime at speed c.

Example 3 (*Inviscid Burgers' Equation*). Now we consider a more complicated example involving a nonlinear PDE:

$$u_t + uu_x = 0, \qquad x \in R, \quad t > 0$$

$$u(x,0) = (1 + x^2)^{-1}, \qquad x \in R$$

Note here that the initial condition has been fixed. In this case, in contrast to the two preceding examples, the solution does not exist for all $t > 0$. The

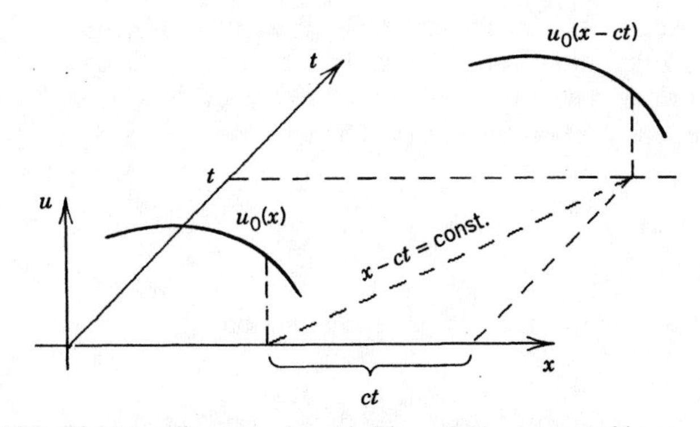

Figure 1.12. Right traveling wave shown in Figure 1.11 represented in xtu-space.

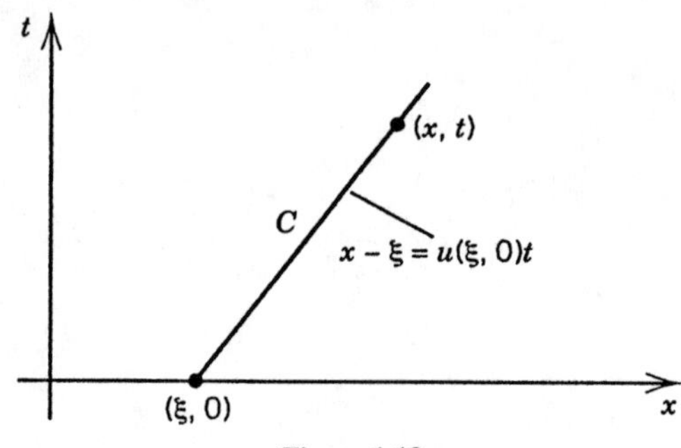

Figure 1.13

initial waveform distorts during propagation and a gradient catastrophe occurs at a finite time. The argument we present to show this nonexistence of a global solution is typical of the kinds of general arguments that are developed later to study nonlinear hyperbolic problems. Assume that the problem has a solution $u = u(x, t)$ and consider the family of curves in xt space defined by the differential equation

$$\frac{dx}{dt} = u(x, t)$$

Denote a curve C in this family by $x = x(t)$. Along this curve we have $du/dt = u_t(x(t), t) + u_x(x(t), t)\, dx/dt = 0$, and therefore $u =$ constant on C. The curve C must be a straight line because $d^2x/dt^2 = du/dt = 0$. The curve C, which is called a *characteristic curve*, is shown in Figure 1.13 emanating from a point $(\xi, 0)$ on the initial time line (x-axis) to an arbitrary point (x, t) in spacetime. The equation of C is given by

$$x - \xi = u(\xi, 0)t$$

where its speed (the reciprocal of its slope), $u(\xi, 0) = (1 + \xi^2)^{-1}$, is determined by the initial condition and the fact that u is constant on C. Now let us determine how the gradient u_x of u evolves along curve C. For simplicity denote $g(t) = u_x(x(t), t)$. Then

$$g'(t) = u_{xx}\frac{dx}{dt} + u_{xt} = (u_t + uu_x)_x - u_x^2 = -u_x^2 = -g(t)^2$$

The general solution of the differential equation $g' = -g^2$ is

$$g(t) = \frac{1}{t + c}, \qquad c \text{ constant}$$

But $g(0) = 1/c = -2\xi/(1+\xi^2)^2$ is the initial gradient, and therefore g is given by

$$g(t) = \frac{1}{t - (1+\xi^2)^2/2\xi}$$

Note that ξ may be chosen positive. Therefore, along the straight line C the gradient u_x becomes infinite at a finite time $t = (1+\xi^2)^2/2\xi$. Thus a smooth solution cannot exist for all $t > 0$.

The nonexistence of a global solution to the initial value problem is a typically nonlinear phenomenon. Since physical processes are often governed by nonlinear equations, we may well ask what happens after the gradient catastrophe. Actually, the distortion of the wave profile and development of an infinite gradient is the witnessing of the formation of a shock wave (i.e., a discontinuous solution that propagates thereafter). However, the idea of a nonsmooth solution to a PDE is a concept that must be formulated precisely, and we carry out this program in Chapter 3.

Another type of problem associated with PDEs is a signaling problem. In this case the domain of the problem is the first quadrant $x > 0$, $t > 0$ in spacetime and auxiliary data are given along the positive x-axis as initial data as well as along the positive t-axis as boundary or signaling data (see Figure 1.14). The form of the *signaling problem* is

$$u_t + F(x, t, u, u_x, u_{xx}) = 0, \qquad x > 0, \quad t > 0 \tag{3}$$

$$u(x, 0) = u_0(x), \qquad x > 0 \tag{4}$$

$$u(0, t) = u_1(t), \qquad t > 0 \tag{5}$$

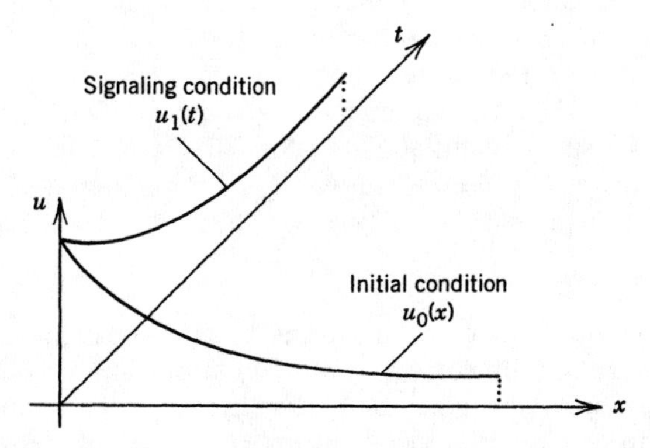

Figure 1.14. Schematic representing a signaling problem where signaling data are prescribed at $x = 0$ along the time axis, and initial data are prescribed at $t = 0$ along the spatial axis.

where $u_0(x)$ is the given initial state and $u_1(t)$ is a specified signal imposed at $x = 0$ for all times $t > 0$. As in the case of the initial value problem, the signaling problem may or may not have a solution that exists for all times t.

In lieu of the condition (5) given at $x = 0$, one may impose a condition on the derivative u of the form

$$u_x(0,t) = u_2(t), \qquad t > 0 \tag{6}$$

If the PDE is a conservation law and Fick's law $\phi(x,t) = -Du_x(x,t)$ holds, condition (6) translates into a condition on the flux ϕ. The condition that the flux be zero at $x = 0$ is the physical condition that the amount of u that passes through $x = 0$ is zero; in heat flow problems this condition is often called the *insulated boundary condition*.

In the case that the spatial domain is finite, that is, $a < x < b$, one may expect to impose boundary data along both $x = a$ and $x = b$, and therefore we consider the *initial–boundary value problem*

$$u_t + F(x,t,u,u_x,u_{xx}) = 0, \qquad a < x < b, \quad t > 0 \tag{7}$$

$$u(x,0) = u_0(x), \qquad a < x < b \tag{8}$$

$$u(a,t) = u_1(t), \qquad u(b,0) = u_2(t), \quad t > 0 \tag{9}$$

where u_0, u_1, and u_2 are given functions. If (7) is the diffusion equation, this problem has a solution under mild restrictions on the data. However, if we consider the advection equation, the problem does not in general have a solution for arbitrary boundary data, as the following example shows.

Example 4. Consider the initial–boundary value problem

$$u_t + cu_x = 0, \qquad 0 < x < 1, \quad t > 0$$

with initial and boundary conditions given by (8) and (9) with $a = 0$ and $b = 1$. We have noted already that the general solution of the advection equation is $u(x,t) = f(x - ct)$, for an arbitrary function f. Consequently, u must be constant on the straight lines $x - ct =$ constant (see Figure 1.15). Clearly, therefore, data cannot be independently specified along the boundary $x = b$. In this case, the initial data along $0 < x < 1$ is carried along the straight lines to the segment A on $x = 1$; the boundary data along $x = 0$ is carried to the segment B on $x = 1$. Thus $u(1,t)$ cannot be specified arbitrarily.

Example 4 shows that we must be careful in defining a well-posed problem. Mathematically correct initial conditions and boundary conditions are often associated with types of PDEs (hyperbolic, parabolic, elliptic), as well as their order. Conditions that ensure well-posedness are often suggested by the underlying physical problem.

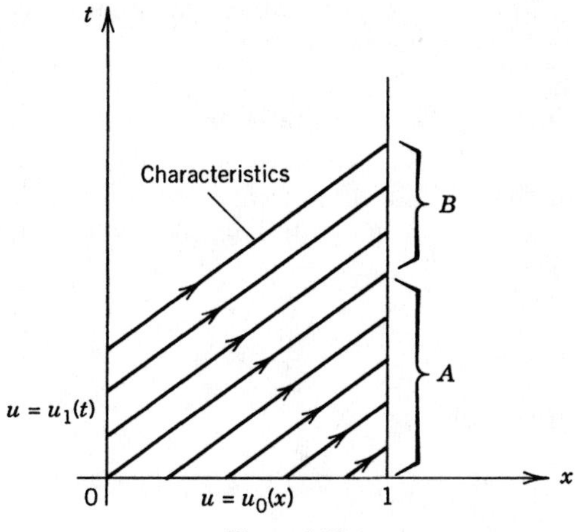

Figure 1.15

Example 5 (*Wave Equation*). The wave equation

$$u_{tt} - c^2 u_{xx} = 0, \qquad x \in R, \quad t > 0 \tag{10}$$

is second order in t, and therefore it does not fit into the category of equations defined by (1). Because it is second order in t, we are guided by our experiences with ordinary differential equations to impose two conditions at $t = 0$, a condition on u and a condition on u_t. Therefore, the pure initial value problem for the wave equation consists of (10) subject to the initial conditions

$$u(x,0) = u_0(x), \qquad u_t(x,0) = u_1(x), \quad x \in R \tag{11}$$

where u_0 and u_1 are given functions. If $u_0 \in C^2(R)$ and $u_1 \in C^1(R)$, the unique, global solution to (10)–(11) is given by D'Alembert's formula (see Exercise 3 of Section 1.1) with $f = u_0$ and $g = u_1$.

Example 6. The initial value problem

$$u_{tt} + u_{xx} = 0, \qquad x \in R, \quad t > 0 \tag{12}$$

$$u(x,0) = u_0(x), \qquad u_t(x,0) = u_1(x), \quad x \in R \tag{13}$$

is not well posed because small changes, or perturbations, in the initial data can lead to arbitrarily large changes in the solution (see Exercise 6). Equation (12) is Laplace's equation, which is elliptic. In general, initial conditions are not correct for elliptic equations, which are naturally associated with equilibrium phenomena and boundary data.

To summarize, the auxiliary conditions that are imposed on a PDE must be considered carefully. In the sequel, as the subject of nonlinear PDEs is developed, the reader should become aware of what conditions go with what equations in order to have, in the end, a well-formulated problem.

EXERCISES

1. Solve the signaling problem

$$u_t + cu_x = 0, \qquad x > 0, \quad t > 0$$
$$u(x,0) = 1, \qquad x > 0$$
$$u(0,t) = \frac{1+t^2}{1+2t^2}, \qquad t > 0$$

2. Obtain the solution to the initial value problem

$$u_t - u_{xx} = 0, \qquad x \in R, \quad t > 0$$
$$u(x,0) = u_0 \quad \text{if } |x| < L \quad \text{and} \quad u(x,0) = 0 \quad \text{if } |x| > L$$

where u_0 is a constant, in the form

$$u(x,t) = -\frac{u_0}{2}\left[\operatorname{erf}\left(\frac{x-L}{\sqrt{4t}}\right) - \operatorname{erf}\left(\frac{x+L}{\sqrt{4t}}\right)\right]$$

where erf is the *error function*

$$\operatorname{erf}(x) = \frac{2}{\sqrt{\pi}}\int_0^x \exp(-s^2)\,ds$$

Show that for x fixed and for large t,

$$u(x,t) \sim \frac{u_0 L}{\sqrt{\pi t}}$$

3. Find a formula for the solution to the initial–boundary value problem

$$u_t - u_x = 0, \qquad 0 < x < 1, \quad t > 0$$
$$u(x,0) = 2, \qquad 0 < x < 1$$
$$u(1,t) = \frac{2}{1+t^2}, \qquad t > 0$$

4. Use direct integration to solve the signaling problem

$$u_{xt} = h(x,t), \qquad x > 0, \quad t > 0$$

$$u(x,0) = u_0(x), \qquad x > 0$$

$$u(0,t) = u_1(t), \qquad t > 0$$

where u_0 and u_1 are class $C^2(R)$ and $u_0(0) = u_1(0)$.

5. Let $u \in C^1$ be a solution to the initial value problem

$$u_t + Q(u)_x = 0, \qquad x \in R, \quad t > 0$$

$$u(x,0) = u_0(x), \qquad x \in R$$

where $Q(0) = 0$, $Q''(u) > 0$, and u_0 is integrable on R with $u_0(x) = 0$ for $x < x_0$ (for some x_0), and $u_0(x) > 0$ otherwise. Define

$$U(x,t) = \int_{-\infty}^{x} u(s,t)\, ds$$

(a) Show that U satisfies the equation $U_t + Q(U_x) = 0$.
(b) Prove that $Q(u) \geq Q(v) + c(v)(u - v)$, where $c(u) = Q'(u)$.
(c) Prove that $U_t + c(v)U_x \leq c(v)v - Q(v)$.
(d) Show that $U(x,t) \leq U(\xi,0) + t[c(v)v - Q(v)]$, where ξ is the point where the line defined by $dx/dt = c(v)$ through the point (x,t) intersects the x axis.
(e) In the inequalities above, show that equality holds if $v = u$.
(f) Give a geometrical–physical interpretation of the results above.
(This exercise was adapted from Lax [1973].)

6. (a) Show that the initial value problem (12)–(13) is not well posed. *Hint:* Consider the solutions $u_n(x,t) = n^{-1}\cos nx \cosh nt$.
(b) Consider the initial value problem for the *backward* diffusion equation:

$$u_t + u_{xx} = 0, \qquad x \in R, \quad t \geq 0$$

$$u(x,0) = 1, \qquad x \in R$$

Show that the solution does not depend continuously on the initial condition by considering the functions

$$u_n(x,t) = 1 + n^{-1}\exp(n^2 t)\sin nx$$

1.5 WAVES

One of the cornerstones in the study of nonlinear PDEs (and, for that matter, linear PDEs) is wave propagation. A *wave* is a recognizable signal that is transferred from one part of the medium to another part with a recognizable speed of propagation. Energy is often transferred as the wave propagates, but matter may not be. There is hardly any area of science or engineering where wave phenomena are not important. We mention a few areas where wave propagation is of fundamental importance.

Fluid mechanics (water waves, aerodynamics)
Acoustics (sound waves in air and liquids)
Elasticity (stress waves, earthquakes)
Electromagnetic theory (optics, electromagnetic waves)
Biology (epizootic waves)
Chemistry (combustion and detonation waves)

Traveling Waves

The simplest form of a mathematical wave is a function of the form

$$u(x,t) = f(x - ct) \tag{1}$$

We interpret the density u as the strength of the signal. At $t = 0$ the wave has the form $f(x)$, which is the initial wave profile. Then $f(x - ct)$ represents the profile at time t, which is just the initial profile translated to the right ct spatial units. The constant c therefore represents the speed of the wave (see Figures 1.11 and 1.12). Evidently, therefore, (1) represents a right traveling wave of speed c. Similarly, $u(x,t) = f(x + ct)$ represents a left traveling wave of speed c. These types of waves propagate undistorted along the straight lines $x - ct = \text{constant}$ (or $x + ct = \text{constant}$) in spacetime.

One of the fundamental questions in the theory of nonlinear PDEs is: Can a given PDE propagate such a traveling wave? In different words, does a traveling wave solution exist for a given PDE? This question is generally asked without regard to initial conditions (i.e., time is usually regarded as varying from $-\infty$ to $+\infty$), so that the wave is assumed to have existed for all times. However, boundary conditions of the form

$$u(-\infty,t) = \text{constant} \qquad u(+\infty,t) = \text{constant} \tag{2}$$

are usually imposed. A *wavefront type-solution* to a PDE is a solution of the form $u(x,t) = f(x - ct)$ [or $f(x + ct)$] subject to the conditions (2) of constancy at plus and minus infinity (not necessarily the same constant); the function f is assumed to have the requisite degree of smoothness defined by

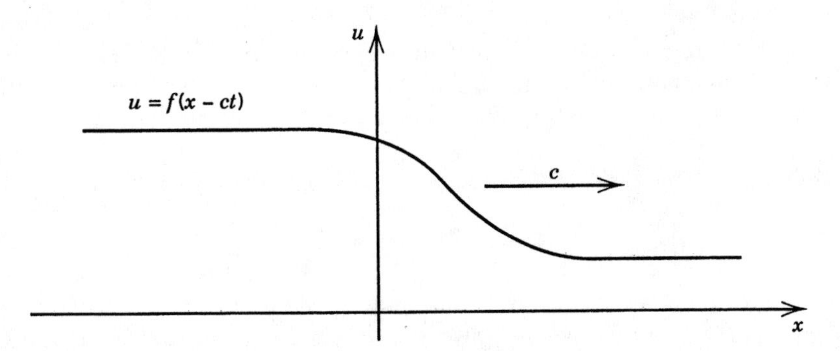

Figure 1.16. Traveling wavefront with speed c.

the PDE [e.g., $C^1(R), C^2(R), \ldots$]. Figure 1.16 shows a typical wavefront. If u approaches the same constant at both plus and minus infinity, the wavefront solution is called a pulse.

Example 1 (*Advection Equation*). It has already been observed that

$$u(x,t) = f(x - ct)$$

where f is a differentiable function, is a solution to the PDE

$$u_t + cu_x = 0 \tag{3}$$

Therefore, the advection equation (3) admits wavefront-type solutions, and it is the simplest example of a wave equation.

Example 2. We have already noted that the equation $u_{tt} - c^2 u_{xx} = 0$, which is called the wave equation, has solutions that are the superposition of both right and left traveling waves (see Section 1.1).

Example 3 (*Diffusion Equation*). The diffusion equation cannot propagate nonconstant wavefronts. To observe this fact, we substitute $u = f(z)$, where $z = x - ct$, into the diffusion equation $u_t - Du_{xx} = 0$ to obtain the ordinary differential equation

$$-cf'(z) - Df''(z) = 0, \qquad z \in R$$

for the wave profile f. This linear differential equation has general solution

$$f(z) = a + b\exp\left(-\frac{cz}{D}\right)$$

where a and b are arbitrary constants. The only possibility for f to be constant at both plus and minus infinity is to require $b = 0$. Thus there are no nonconstant traveling wavefront solutions to the diffusion equation.

Example 4 (*Korteweg–deVries Equation*). The KdV equation is given by

$$u_t + uu_x + u_{xxx} = 0 \tag{4}$$

This nonlinear equation admits traveling wave solutions of different types. One particular type of traveling wave that arises from the KdV equation is the soliton, or solitary wave, which is now derived. Let us assume that $u = f(z)$, where $z = x - ct$. Substituting into (4) gives

$$-cf' + ff' + f''' = 0 \tag{5}$$

where the prime denotes d/dz. Integrating (5) two times yields, after rearrangement,

$$\frac{df}{dz} = 3^{-1/2}\left(-f^3 + 3cf^2 + 6af + 6b\right)^{1/2} \tag{6}$$

where a and b are constants of integration. Here we took the plus sign on the square root; later we observe that this can be done without loss of generality. The first order differential equation (6) for the form of the wave $f(z)$ is separable. However, the expression under the radical on the right side is a cubic in f, and therefore different cases must be considered depending on the number of real roots, double roots, and so on. We examine only one case, namely when the cubic has three real roots, one being a double root. So let us assume that

$$-f^3 + 3cf^2 + 6af + 6b = (f - \gamma)^2(\alpha - f), \qquad 0 < \gamma < \alpha$$

Then the differential equation (6) becomes

$$(f - \gamma)^{-1}(\alpha - f)^{-1/2}\, df = 3^{-1/2}\, dz \tag{7}$$

The substitution $f = \gamma + (\alpha - \gamma)\operatorname{sech}^2 w$ easily reduces (7) to

$$-2(\alpha - \gamma)^{-1/2}\, dw = 3^{-1/2}\, dz$$

which immediately integrates to $w = [(\alpha - \gamma)/12]^{1/2} z$. Consequently, traveling waves in the case we are considering have the form

$$u(x,t) = \gamma + (\alpha - \gamma)\operatorname{sech}^2[\kappa(x - ct)], \qquad \kappa = \left(\frac{\alpha - \gamma}{12}\right)^{1/2} \tag{8}$$

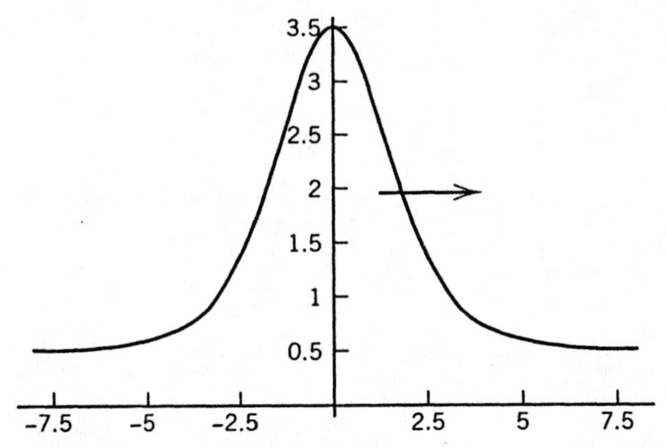

Figure 1.17. MATHEMATICA plot of the soliton (8) at $t = 0$ with $\gamma = 0.5$ and $\alpha - \gamma = 3.0$. A soliton is an example of a pulse.

A graph of the waveform, which is a pulse, is shown in Figure 1.17. It is instructive to write the roots α and γ in terms of the original parameters. To this end we have

$$\begin{aligned} -f^3 + 3cf^2 + 6af + 6b &= (f - \gamma)^2(\alpha - f) \\ &= -f^3 + (\alpha + 2\gamma)f^2 - (2\gamma\alpha + \gamma^2)f + \gamma^2\alpha \end{aligned}$$

Therefore, the wave speed c is given by

$$c = \frac{\alpha + 2\gamma}{3} = \frac{\alpha - \gamma}{3} + \gamma \tag{9}$$

We note that the speed of the wave relative to the state γ ahead of the wave is proportional to the amplitude $A = \alpha - \gamma$. For linear waves, say governed by the wave equation, the speed of propagation is independent of the amplitude of the wave. Such a wave form (8) is known as a *soliton*, or *solitary wave*, and many of the important equations of mathematical physics exhibit soliton-type solutions (e.g., the Boussinesq equation, the Sine–Gordon equation, the Born–Infeld equation, and nonlinear Schrödinger-type equations). In applications, the value of such solutions is that if a pulse or signal travels as a soliton, the information contained in the pulse can be carried over long distances with no distortion or loss of intensity. Solitons occur in fluid mechanics, nonlinear optics, and other nonlinear phenomena.

The problem of determining whether or not a given PDE admits wavefront-type solutions is a fundamental one, and it will occupy our attention in subsequent chapters. The search for these types of solutions is one of the basic methods in the analysis of nonlinear problems.

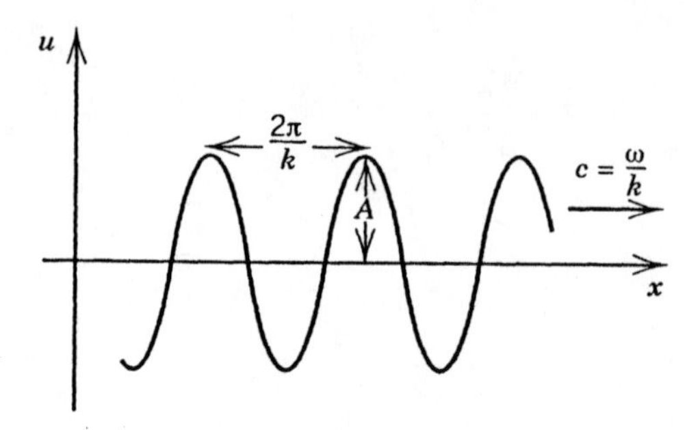

Figure 1.18. Plane wave.

Plane Waves

Another type of wave that is of interest in many contexts is a so-called *plane wave*, or *wave train*. These waves are traveling, periodic waves of the form

$$u(x, t) = A\cos(kx - \omega t) \tag{10}$$

where A is the amplitude of the wave, k the wave number, and ω the frequency. The *wave number* k is a measure of the number of spatial oscillations (per 2π units) observed at a fixed time, and the *frequency* ω is a measure of the number of oscillations in time (per 2π units) observed at a fixed spatial location. The number $\lambda = 2\pi/k$ is the *wavelength*, and $P = 2\pi/\omega$ is the *time period*. The wavelength measures the distances between successive crests, and the period measures the time for an observer located at a fixed position x to see a repeat pattern (see Figure 1.18). We remark that (10) may be written

$$u(x, t) = A\cos k\left(x - \frac{\omega}{k}t\right)$$

so (10) represents a traveling wave moving to the right with speed $c = \omega/k$. This number is called the *phase velocity*, and it is the speed that one would have to move to remain on the crest of the wave. For calculations the complex form

$$u(x, t) = A\exp[i(kx - \omega t)] \tag{11}$$

is often preferred because differentiations with the exponential function are simpler. After the calculations are completed, one can use Euler's formula $\exp(i\theta) = \cos\theta + i\sin\theta$ and then take real or imaginary parts to recover a real solution. We remind the reader that the real and imaginary parts of a complex solution to a linear differential equation are both real solutions. We emphasize that the technique of looking for plane wave solutions of the form (11) is applicable only to linear equations; however, a modification of this method does apply to some nonlinear problems, as we shall observe in the sequel. Finding plane wave solutions is a powerful method for determining properties of linear problems.

Example 5 (*Diffusion Equation*). We seek solutions of the form of equation (11) to the diffusion equation

$$u_t - Du_{xx} = 0 \tag{12}$$

Substituting (11) into (12) gives

$$\omega = -iDk^2 \tag{13}$$

Equation (13) is a condition, called a *dispersion relation*, between the frequency ω and the wave number k that must be satisfied in order that (12) admit plane wave solutions. Thus we have determined a class of solutions

$$u(x, t; k) = A \exp(-Dk^2 t)\exp(ikx), \qquad k \in R \tag{14}$$

for the diffusion equation depending on an arbitrary parameter k, the wave number. The factor $\exp(ikx)$ represents a spatial oscillation, and the factor $A \exp(-Dk^2 t)$ represents a decaying amplitude. Note that the rate of decay depends on the wave number k; waves of shorter wavelength decay more rapidly that waves of longer wavelength.

Example 6 (*Wave Equation*). Substituting (11) into the wave equation

$$u_{tt} - c^2 u_{xx} = 0 \tag{15}$$

implies that

$$\omega = \pm ck \tag{16}$$

Therefore, the wave equation admits solutions of the form

$$u(x, t) = A \exp[ik(x \pm ct)]$$

which are right and left sinusoidal traveling waves of speed c.

The technique of looking for solutions of the form (11) for linear, homogeneous PDEs of the form

$$Lu = 0 \tag{17}$$

where L is a linear constant-coefficient operator, always leads to a relation connecting the frequency ω and the wave number k of the form

$$G(k, \omega) = 0 \tag{18}$$

This condition is called the *dispersion relation* corresponding to the PDE (17), and it characterizes sinusoidal wavetrain solutions entirely. Generally, linear PDEs (17) are classified according to their dispersion relation in the following way. Let us assume that (18) may be solved for ω in the form

$$\omega = \omega(k) \tag{19}$$

We say that the PDE is *dispersive* if $\omega(k)$ is real and if $\omega''(k) \neq 0$. When $\omega(k)$ is complex, the PDE is said to be *diffusive*. The diffusion equation, from the example above, has dispersion relation $\omega = -ik^2D$, which is complex; thus the diffusion equation is classified as diffusive. The classical wave equation (15) is neither diffusive nor dispersive under the foregoing classification. Even though its solutions are wavelike, it is not classified as dispersive since its dispersion relation (16) satisfies $\omega'' = 0$. For the wave equation, the speed of propagation is c and is independent of the wave number k. For dispersive equations, the speed of propagation, or the phase velocity ω/k, depends on the wave number (or equivalently, the wavelength). The wave equation is generally regarded as the prototype of a hyperbolic equation, and the term *dispersive* is reserved for equations where the phase velocity depends on k.

***Example** 7 (Schrödinger Equation)*. The Schrödinger equation for a free particle is

$$u_t = iu_{xx}$$

It is easy to see that the dispersion relation is $\omega = k^2$, so that the Schrödinger equation is dispersive. Note that the Schrödinger equation is neither parabolic nor hyperbolic.

Fourier Method

We now introduce a common technique, using superposition and the Fourier integral theorem, for constructing more general types of solutions to linear constant-coefficient PDEs than those given by equation (11). If it is possible to obtain a one-parameter family of solutions $u = u(x, t; k)$ to a linear, homogeneous PDE (17), where k is a parameter belonging to a set S, one can form the function

$$u(x, t) = \int_S A(k)u(x, t; k)\, dk \tag{20}$$

by integrating these solutions against some unknown continuum of coefficients $A(k)$ over the set S. For example, the $u(x, t; k)$ could be solutions of the form (11). Under certain conditions that must be established, we can write

$$Lu(x,t) = L\left(\int_S A(k)u(x,t;k)\,dk\right) = \int_S A(k)Lu(x,t;k)\,dk$$

$$= \int_S A(k)0\,dk = 0$$

and hence $u(x, t)$ given by (20) is a solution of (17). Equation (20) represents the superposition of the continuum of solutions $u(x, t; k)$ over the set S. To show that $u(x, t)$ is a solution, the calculation requires bringing the differential operator L under the integral sign; the validity of this step must be checked in order to have a rigorous argument. In practice, one just *formally* carries out the calculations, and then at the end verifies that the function obtained by this kind of superposition is indeed a solution (in applied mathematics, a *formal argument* is a calculation that can usually be rigorously verified under certain conditions and assumptions).

Example 8. We have already constructed a class of solutions $u(x, t; k)$, $k \in R$, given by (14), to the diffusion equation. Formally superimposing these solutions, we obtain

$$u(x,t) = \int_R A(k)\exp(-k^2Dt)\exp(ikx)\,dk \tag{21}$$

which can be verified to be a solution to the diffusion equation, provided that $A(k)$ is a well-behaved function (e.g., continuous, bounded, and integrable on R).

Having (21) as a solution to the diffusion equation now opens up the possibility of selecting the function $A(k)$ so that other conditions (e.g., an initial condition) can be met. Therefore, let us impose the initial condition

$$u(x,0) = u_0(x), \qquad x \in R$$

to the diffusion equation. From (21) it follows that

$$u_0(x) = \int_R A(k)e^{ikx}\,dk \tag{22}$$

We recognize $u_0(x)$ as the inverse Fourier transform of $A(k)$, up to a constant multiple, and therefore $A(k)$ must be the Fourier transform of the function $u_0(x)$. We remind the reader of these facts by recalling the *Fourier integral theorem* (see, e.g., Stakgold [1979]).

Theorem 1. Let f be a continuous, bounded, integrable function on R, and let

$$\hat{f}(k) = (2\pi)^{-1/2} \int_R f(x) e^{-ikx}\, dx$$

be the Fourier transform of f. Then for all $x \in R$,

$$f(x) = (2\pi)^{-1/2} \int_R \hat{f}(k) e^{ikx}\, dk$$

Applying this theorem to (22) leads us to conclude that

$$A(k) = \frac{1}{2\pi} \int_R u_0(x) e^{-ikx}\, dx$$

Consequently, we have obtained the solution to the initial value problem for the diffusion equation

$$u_t - Du_{xx} = 0, \qquad x \in R, \quad t > 0 \tag{23}$$

$$u(x,0) = u_0(x), \qquad x \in R \tag{24}$$

in the form

$$u(x,t) = \frac{1}{2\pi} \int_R \left(\int_R u_0(\xi) e^{-ik\xi}\, d\xi \right) \exp(-k^2 Dt) \exp(ikx)\, dk \tag{25}$$

By interchanging the order of integration we can formally write the solution (25) as

$$u(x,t) = \frac{1}{2\pi} \int_R u_0(\xi) \left(\int_R \exp[ik(x-\xi) - k^2 Dt]\, dk \right) d\xi \tag{26}$$

The inner iterated integral may be calculated outright by noticing

$$\int_R \exp[ik(x-\xi) - k^2 Dt]\, dk = 2\int_0^\infty \exp(-k^2 Dt) \cos[k(x-\xi)]\, dk$$

$$= \left(\frac{\pi}{Dt} \right)^{1/2} \exp\left[-\frac{(x-\xi)^2}{4Dt} \right]$$

Therefore, the solution to the initial value problem (23)–(24) for the diffusion equation is given by

$$u(x,t) = \frac{1}{(4\pi Dt)^{1/2}} \int_R u_0(\xi) \exp\left[-\frac{(x-\xi)^2}{4Dt} \right] d\xi \tag{27}$$

Note that we have derived this solution formally; we have not proved that it is indeed a solution to the initial value problem (23)–(24). A proof would consist of a rigorous argument that (26) satisfies the diffusion equation (23) and the initial condition (24). To show well-posedness, a uniqueness argument would have to be supplied as well as a proof that the solution is stable to small perturbations of the initial data. We shall not carry out these proofs here, but rather, refer the reader to the references for the details. See also Section 4.1 for a review of some of the results.

The method presented in the preceding paragraphs (finding plane wave solutions followed by superposition and use of the Fourier integral theorem) is equivalent to the classical Fourier transform method learned in elementary courses where one takes Fourier transforms of the partial differential equation and initial conditions to reduce the problem to an ordinary differential equation in the transform domain, which is then solved; then the inverse Fourier transform is applied to return the solution in the original domain. This method is generally applicable to the pure initial value problem for linear equations with constant coefficients.

Nonlinear Dispersion

For nonlinear equations we do not expect plane wave solutions, that is, sinusoidal traveling wave solutions of the form (11), and therefore a dispersion relation will not exist as it does for linear equations. Moreover, superposition for nonlinear equations is invalid. However, in some nonlinear problems, there may exist traveling periodic wavetrains of the form

$$u(x,t) = U(\theta), \qquad \theta = kx - \omega t \tag{28}$$

where U is a periodic function. For example, consider the nonlinear PDE

$$u_{tt} - u_{xx} + f'(u) = 0 \tag{29}$$

where $f(u)$ is some function of u, yet to be specified. If (28) is substituted into (29), we obtain the ordinary differential equation

$$(\omega^2 - k^2)U_{\theta\theta} + f'(U) = 0$$

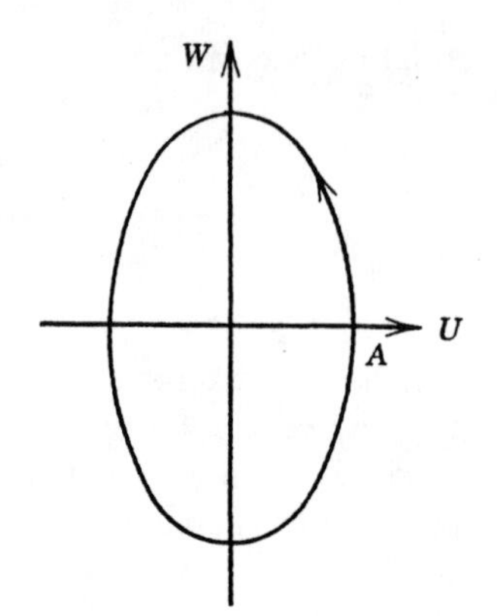

Figure 1.19. Phase space orbit representing the periodic solution (32) with amplitude A.

Multiplication by U_θ and subsequent integration gives

$$\tfrac{1}{2}(\omega^2 - k^2)U_\theta^2 + f(U) = A \tag{30}$$

where A is a constant of integration. Our goal is to determine U as a periodic function of θ. Equation (30) has the same form as an energy conservation law, where f is a potential function, which suggests introducing the coordinate W defined by $W = U_\theta$. Upon solving for W^2, this gives

$$W^2 = \frac{2}{\omega^2 - k^2}[A - f(U)] \tag{31}$$

which are the integral curves of (30). Assuming that $\omega^2 > k^2$, we have

$$W = \pm\left(\frac{2}{\omega^2 - k^2}\right)^{1/2}[A - f(U)]^{1/2} \tag{32}$$

which defines a locus of points in the UW-plane. For example, let us choose $f(U) = U^4$. Then the locus (32) is a closed path, representing a periodic solution of (30), as shown in Figure 1.19. Notice that A, which has been taken positive, is the amplitude of the oscillation represented by the closed path. To find U as a function of θ [in the case $f(U) = U^4$] we write (30) as

$$\theta = \pm\left(\frac{\omega^2 - k^2}{2}\right)^{1/2}\int_0^U (A - s^4)^{-1/2}\,ds \tag{33}$$

This formula defines the periodic function $U = U(\theta)$ implicitly; in this case $U(\theta)$ can be determined explicitly as an elliptic function. We leave this exercise to the reader. The period of the oscillation can be determined by integrating over one period in the integral on the right side of (33), taking care to choose the appropriate sign. If P denotes the period, we will obtain upon integration an equation of the form

$$P = P(\omega, k, A) \tag{34}$$

In other words, the frequency ω will depend on the amplitude A as well as the wave number k. Consequently, the wave speed $c = \omega/k$ will be amplitude dependent. This amplitude dependence in the nonlinear dispersion relation (34) is one important distinguishing aspect of nonlinear phenomena.

These calculations can be carried out for nonlinear equations of the form (29) for various potential functions $f(u)$. Periodic solutions are obtained when U oscillates between two simple zeros of $A - f(U)$. A thorough discussion of nonlinear dispersion can be found in Whitham [1974].

EXERCISES

1. Find the dispersion relation for the linearized Burgers' equation

$$u_t + au_x = Du_{xx} \qquad (a, D > 0)$$

and show that the equation is diffusive. Use superposition and Fourier integral theorem to find an integral representation of the solution of the pure initial value problem for this equation.

2. Consider the KdV equation in the form $u_t - 6uu_x + u_{xxx} = 0$, $x \in R$, $t > 0$. Let $u = u(x, t)$ be a solution that decays, along with its derivatives, very rapidly to zero as $|x| \to \infty$. Show that $\int_R u\,dx$ and $\int_R u^2 dx$ are both constant in time.
3. Examine the form of traveling wave solutions of the KdV equation (4) in the case that the cubic expression on the right side of (6) has a triple real root.
4. Determine and sketch traveling wave solutions of the equation

$$u_{tt} - u_{xx} = -\sin u$$

in the form

$$u(x,t) = 4\arctan\left\{\exp \pm\left[\frac{x - ct}{(1 - c^2)^{1/2}}\right]\right\}, \qquad 0 < c < 1$$

5. For the following PDEs find the dispersion relation and classify the equations as diffusive, dispersive, or neither:
   (a) $u_{tt} + a^2 u_{xxxx} = 0$ (the beam equation)
   (b) $u_t + au_x + bu_{xxx} = 0$ (linearized KdV equation)
   (c) $u_{tt} - c^2 u_{xx} + bu = 0$ (Klein–Gordon equation)

6. Find solutions of the outgoing signaling problem for the wave equation:

$$u_{tt} - c^2 u_{xx} = 0, \qquad x > 0, \quad t \in R$$

$$u_x(0, t) = s(t), \qquad t \in R$$

7. In equation (29) take $f(u) = u^2/2$ and determine periodic solutions of the form $u = U(\theta)$, where $\theta = kx - \omega t$. What is the period of oscillation? Does ω depend on the amplitude of oscillation in this case?
8. In equation (29) assume that the potential function $f(u)$ has the expansion

$$f(u) = \frac{u^2}{2} + \sigma u^4 + \cdots$$

where σ is a small known parameter. Assuming that the amplitude is small, show that periodic wavetrains are given by

$$U(\theta) = a\cos\theta + \tfrac{1}{8}\sigma a^3 \cos 3\theta + \cdots$$

where the frequency and amplitude are

$$\omega^2 = 1 + k^2 + 3\sigma a^2 + \cdots$$

$$A = \frac{a^2}{2} + \frac{9}{8}\sigma a^4 + \cdots$$

REFERENCES

The subject of partial differential equations, being such a vast area, has spawned a large number of outstanding texts and monographs, too numerous to cite. Therefore, we mention only a few references, most of which are books rather than specialized articles and which represent a cross section of the subject. In later chapters, as the material becomes more specialized, additional references are cited.

There are a large number of books on elementary PDEs, mostly emphasizing Fourier series, integral transforms, and boundary value problems. Myint-U [1987] is representative of these texts. A list of more advanced texts covering the general subject is: Colton [1988] Garabedian [1986], Guenther and Lee [1988], John [1982], Kevorkian [1990], and Zauderer [1989]. More specialized books include Friedman [1964] for parabolic equations, Gilbarg and Trudinger [1983] for elliptic equations, and Protter and Weinberger [1967] for maximum principles. The classic treatises by Courant and Hilbert [1953; 1962] still provide a wealth of information, particularly on elliptic and hyperbolic equations. Both Kevorkian and Cole [1981] and Zauderer [1989] discuss perturbation techniques. LeVeque [1992] contains an excellent introductory treatment of conservation laws and numerics.

For general works on nonlinear PDEs we cite Ames [1965, 1972], the recent book by Pao [1992], Smoller [1983], and Whitham [1974]. Also see Kreiss and Lorenz [1989].

The origin of partial differential equations as mathematical models of physical phenomena is discussed in many places. For general texts we mention Lin and Segal [1974] and Logan [1987]. Biological applications can be found in Murray [1989], and the basic equations of fluid mechanics and gas dynamics are given in Courant and Friedrichs [1948] and Whitham [1974]. Problems in fluid mechanics have been one of the motivating forces for the theoretical development of nonlinear PDEs.

Ames, W. F., 1965. *Nonlinear Partial Differential Equations in Engineering*, Vol. 1, Academic Press, New York.

Ames, W. F., 1972. *Nonlinear Partial Differential Equations in Engineering*, Vol. 2, Academic Press, New York.

Aronson, D. G., 1986. The porous medium equation, in *Nonlinear Diffusion Problems*, Lecture Notes in Mathematics, No. 1224, Springer-Verlag, New York, pp. 1–46.

Colton, D., 1988. *Partial Differential Equations*, Random House, New York.

Courant, R., and K. O. Friedrichs, 1948. *Supersonic Flow and Shock Waves*, Wiley-Interscience, New York. Reprinted by Springer-Verlag, New York, 1976.

Courant, R., and D. Hilbert, 1953. *Methods of Mathematical Physics*, Vol. 1, Wiley-Interscience, New York.

Courant, R., and D. Hilbert, 1962. *Methods of Mathematical Physics*, Vol. 2, Wiley-Interscience, New York.

Friedman, A., 1964. *Partial Differential Equations of Parabolic Type*, Prentice Hall, Englewood Cliffs, N.J.

Garabedian, P. R., 1986. *Partial Differential Equations*, 2nd ed., Chelsea, New York.

Gilbarg, D., and N. S. Trudinger, 1983. *Elliptic Partial Differential Equations of Second Order*, 2nd ed., Springer-Verlag, New York.

Guenther, R. B., and J. W. Lee, 1988. *Partial Differential Equations of Mathematical Physics and Integral Equations*, Prentice Hall, Englewood Cliffs, N.J.

John, F., 1982. *Partial Differential Equations*, 4th ed., Springer-Verlag, New York.

Kevorkian, J., 1990. *Partial Differential Equations*, Brooks/Cole, Pacific Grove, Calif.

Kevorkian, J., and J. D. Cole, 1981. *Perturbation Methods in Applied Mathematics*, Springer-Verlag, New York.

Kreiss, H. O., and J. Lorenz, 1989. *Initial-Boundary Value Problems and the Navier–Stokes Equations*, Academic Press, New York.

Lax, P., 1973, *Hyperbolic Systems of Conservation Laws and the Mathematical Theory of Shock Waves*, Society for Industrial and Applied Mathematics, Philadelphia.

LeVeque, R. J., 1992. *Numerical Methods for Conservation Laws*, Birkhäuser, Boston.

Lin, C. C., and L. A. Segel, 1974. *Mathematics Applied to Deterministic Problems in the Natural sciences*, Macmillan, New York. Reprinted by the Society for Industrial and Applied Mathematics, Philadelphia.

Logan, J. D., 1987. *Applied Mathematics: A Contemporary Approach*, Wiley-Interscience, New York.

Murray, J. D., 1989. *Mathematical Biology*, Springer-Verlag, New York.

Myint-U, T., 1987. *Partial Differential Equations for Scientists and Engineers*, 3rd ed., North-Holland, New York.

Pao, C. V., 1992. *Nonlinear Elliptic and Parabolic Equations*, Plenum Press, New York.

Protter, M. H., and H. Weinberger, 1967. *Maximum Principles in Differential Equations*, Prentice Hall, Englewood Cliffs, N.J.

Smoller, J., 1983. *Shock Waves and Reaction–Diffusion Equations*, Springer-Verlag, New York.

Stakgold, I., 1979. *Green's Functions and Boundary Value Problems*, Wiley-Interscience, New York.

Whitham, G. B., 1974. *Linear and Nonlinear Waves*, Wiley-Interscience, New York.

Zauderer, E., 1989. *Partial Differential Equations of Applied Mathematics*, 2nd ed., Wiley-Interscience, New York.

2

FIRST ORDER EQUATIONS AND CHARACTERISTICS

A single first order PDE is a hyperbolic-type equation that is wavelike, that is, associated with the propagation of signals at finite speeds. The fundamental idea associated with hyperbolic equations is the notion of a characteristic, a curve in spacetime (a hypersurface in higher dimensions) along which these signals are propagated. There are several ways of looking at the concept of a characteristic; one definition, and this forms the basis of our definition of a characteristic, is that it is a curve along which the PDE can be reduced to a simpler form, for example, an ordinary differential equation. But ultimately, the characteristics are the curves along which information is transmitted in the system.

In this chapter the aim is to build a solid base of understanding of characteristics by examining a first order PDE in one spatial variable and time. We focus on the initial value problem and the signaling problem, and the analysis is carried out under the assumption that a continuous smooth solution exists. In Chapter 3 the concept of a weak solution is taken up, and the smoothness assumptions are relaxed. In Sections 2.1 and 2.2 we study linear and nonlinear equations, respectively, and in Section 2.3 we examine the general quasilinear equation. Then in Section 2.4 we take up the subject of wavefront expansions and show that discontinuities in the derivatives also propagate along the characteristics. In Section 2.5 we discuss the general nonlinear equation of first order.

2.1 LINEAR FIRST ORDER EQUATIONS

Advection Equation

We have already noted that the initial value problem for the advection equation, namely

$$u_t + cu_x = 0, \qquad x \in R, \quad t > 0 \tag{1}$$

$$u(x,0) = u_0(x), \qquad x \in R \tag{2}$$

has a unique solution given by $u(x,t) = u_0(x - ct)$, which is a right traveling wave of speed c. We now present an alternative derivation of this solution that is in the spirit of the analysis of hyperbolic PDEs. In this entire section we assume that the initial signal u_0 is a smooth (continuously differentiable) function.

First we recall a basic fact from elementary calculus. If $u = u(x,t)$ is a function of two variables and $x = x(t)$ defines a smooth curve C in the xt-plane, the total derivative of u along the curve C is given, according to the chain rule, by

$$\frac{d}{dt}u(x(t),t) = u_t(x(t),t) + u_x(x(t),t)\frac{dx}{dt}$$

This expression defines how u is changing along C. Therefore, by observation, the left side of equation (1) is a total derivative of u along the curves defined by the equation $dx/dt = c$. We may therefore recast (1) into the statement

$$\frac{du}{dt} = 0 \quad \text{along the curves defined by } \frac{dx}{dt} = c$$

or, equivalently,

$$u = \text{constant} \quad \text{on} \quad x - ct = \xi$$

where ξ is a constant. [Note that in these expressions we have surpressed the arguments of u in order to have concise notation; all expressions involving u are assumed to be evaluated along the curve, i.e., at $(x(t),t)$.] Consequently, the PDE (1) reduced to an ordinary differential equation, which was subsequently integrated, along the family of curves $x - ct = \xi$, which are solutions to $dx/dt = c$. If we draw one of these curves in the xt-plane that passes through an arbitrary point (x,t), it intersects the x axis at $(\xi, 0)$; its speed is c and its slope is $-1/c$ (because we are graphing t versus x rather than x versus t). It is common to refer to the speed of a curve in spacetime, rather than its slope (see Figure 2.1). Now, since u is constant on this curve,

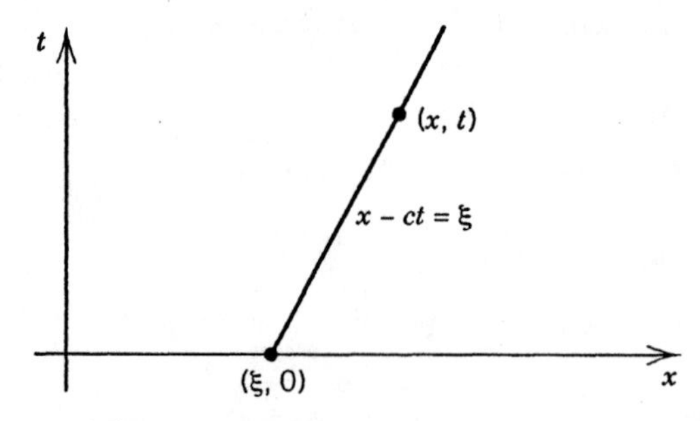

Figure 2.1. Characteristic curve with speed c passing through the arbitrary point (x,t).

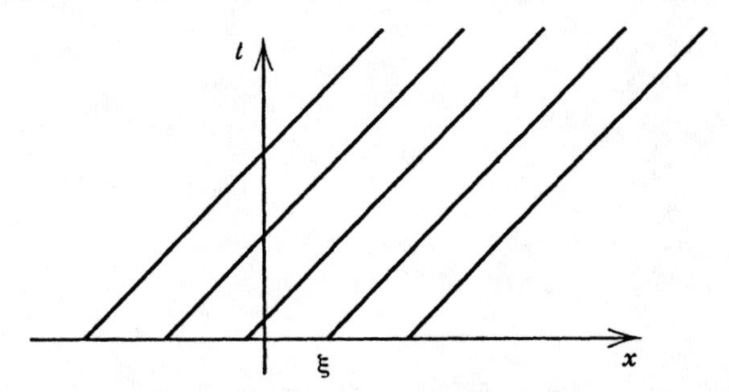

Figure 2.2. Characteristic diagram for the PDE (1) showing the family of characteristic curves $x - ct = \xi$. The curves have speed c and are parameterized by their intersection ξ with the x-axis.

we have

$$u(x,t) = u(\xi,0) = u_0(\xi) = u_0(x - ct)$$

which is the solution to the initial value problem (1)–(2). The totality of all the curves $x - ct = \xi$, where ξ is constant, is called the set of *characteristics curves* for this problem. The constant ξ acts as a parameter and distinguishes the curves. A graph of the set of characteristic curves on a spacetime diagram (i.e., in the xt-plane) is called the *characteristic diagram* for the problem. Figure 2.2 shows the characteristic diagram for the problem (1)–(2). The characteristic curves, or just characteristics, are curves in spacetime along which signals are propagated. In the present case, the signal is the *constancy of u* that is carried along the characteristics. Further, and this will turn out to be the defining feature, the PDE reduces to an ordinary differential equation along the characteristics.

Variable Coefficients

We may extend the preceding notions to more complicated problems. First consider the linear initial value problem

$$u_t + c(x,t)u_x = 0, \qquad x \in R, \quad t > 0 \tag{3}$$

$$u(x,0) = u_0(x), \qquad x \in R \tag{4}$$

where $c = c(x,t)$ is a given continuous function. The left side of the PDE (3) is a total derivative along the curves in the xt-plane defined by the differential equation

$$\frac{dx}{dt} = c(x,t) \tag{5}$$

Along these curves

$$\frac{du}{dt} = u_t + u_x \frac{dx}{dt} = u_t + c(x,t)u_x = 0$$

or, in other words, u is constant. Therefore,

$$u = \text{constant} \quad \text{on} \quad \frac{dx}{dt} = c(x,t)$$

Again in this case, the PDE (3) has reduced to an ordinary differential equation (which was integrated to a constant) along a special family of curves, the characteristics defined by (5). The function $c(x,t)$ gives the speed of these characteristic curves, which varies in spacetime. We consider an example.

Example 1. Consider the initial value problem

$$u_t - xtu_x = 0, \qquad x \in R, \quad t > 0 \tag{6}$$

$$u(x,0) = u_0(x), \qquad x \in R \tag{7}$$

The characteristics are defined by the equation

$$\frac{dx}{dt} = -xt$$

which, upon quadrature, gives

$$x = \xi \exp\left(-\frac{t^2}{2}\right) \tag{8}$$

where ξ is a constant. On these curves the PDE becomes

$$u_t - xtu_x = u_t - \frac{dx}{dt}u_x = \frac{du}{dt} = 0$$

or

$$u = \text{constant} \quad \text{on} \quad x = \xi \exp\left(-\frac{t^2}{2}\right)$$

The characteristic diagram is shown in Figure 2.3, with an arbitrary point (x,t) labeled on a given characteristic that emanates from a point $(\xi, 0)$ on

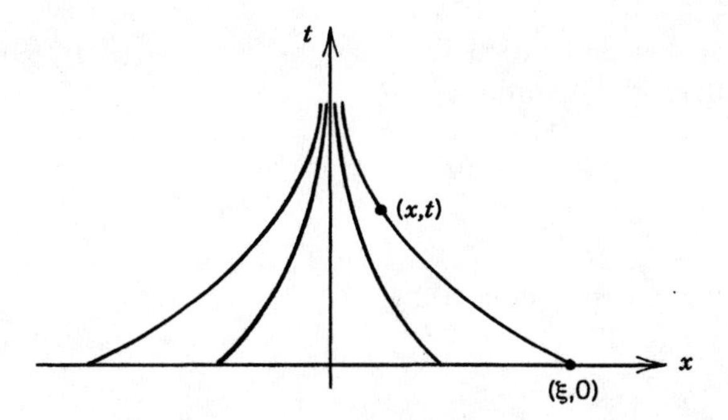

Figure 2.3. Characteristic diagram for (4) showing the family of characteristics given by (6).

the x-axis. From the constancy of u along the characteristics, we have

$$u(x,t) = u(\xi,0) = u_0(\xi) = u_0\left(x \exp\left(\frac{t^2}{2}\right)\right)$$

which one can easily check to be a solution to the initial value problem (6)–(7), valid for all $t > 0$. Figure 2.4 shows how the initial signal $u_0(x)$ is focused along the characteristics to a region near $x = 0$ as time increases.

In principle, this method, called the *method of characteristics*, can be extended to nonhomogeneous initial value problems of the form

$$u_t + c(x,t)u_x = f(x,t), \qquad x \in R, \quad t > 0 \tag{9}$$

$$u(x,0) = u_0(x), \qquad x \in R \tag{10}$$

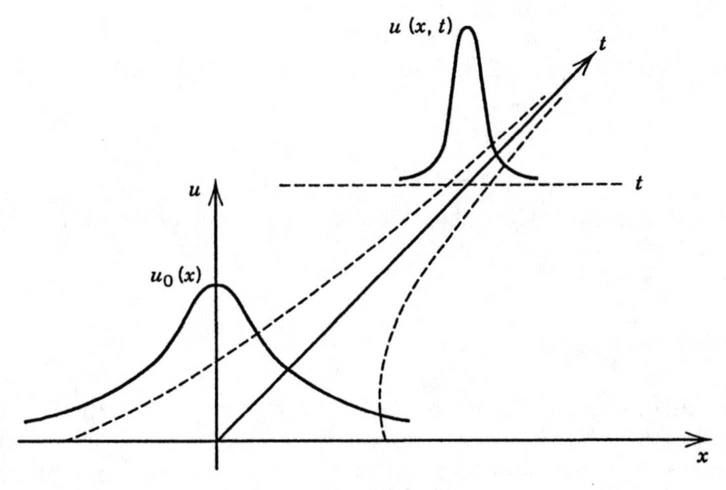

Figure 2.4

where c and f are given continuous functions. Now the PDE (9) reduces to the ordinary differential equation

$$\frac{du}{dt} = f(x,t) \tag{11}$$

on the characteristic curves defined by

$$\frac{dx}{dt} = c(x,t) \tag{12}$$

The pair of differential equations (11)–(12) can, in theory, be solved subject to the initial conditions

$$x = \xi, \qquad u = u_0(\xi) \quad \text{at} \quad t = 0$$

to find the solution. The constant ξ is again a parameter that distinguishes each characteristic by defining its intersection with the x-axis. In practice it may be impossible to solve the characteristic system (11)–(12) in closed form, and one may have to adopt numerical quadrature methods. We remark that it follows from the basic existence and uniqueness results from ordinary differential equations that a unique solution of (11)–(12) exists in some neighborhood of each point $(x_0, 0)$; however, a global solution for all $t > 0$ may not exist.

EXERCISE

1. Solve the initial value problem

$$u_t + 2u_x = 0, \qquad x \in R, \quad t > 0$$

$$u(x,0) = \frac{1}{1+x^2}, \qquad x \in R$$

Sketch the characteristics. Sketch the wave profiles at $t = 0$, $t = 1$, and $t = 4$.

2.2 NONLINEAR EQUATIONS

When nonlinear terms are introduced into PDEs the situation changes dramatically from the linear case discussed in Section 2.1. The method of characteristics, however, still provides the vehicle for obtaining information about how initial signals are propagated by the PDE. To begin, let us consider the simple nonlinear initial value problem

$$u_t + c(u)u_x = 0, \qquad x \in R, \quad t > 0 \tag{1}$$

$$u(x,0) = u_0(x), \qquad x \in R \tag{2}$$

where $c = c(u)$ is a given smooth function of u; here and in this entire section, the initial signal u_0 is assumed to be smooth as well. We recognize (1) as the basic conservation law

$$u_t + \phi(u)_x = 0, \qquad c(u) = \phi'(u)$$

where $\phi = \phi(u)$ is the flux. The nonlinearity in (1) occurs in the convective term $c(u)u_x$. Equation (1), sometimes called a *kinematic wave equation*, arises in nonlinear wave phenomena when dissipative effects such as viscosity and diffusion are ignored. It is a special case of the more general quasilinear equation that is discussed in the next section.

To analyze (1) let us assume for a moment that a C^1 solution $u = u(x,t)$ exists for $t > 0$. Motivated by the approach used for linear equations in Section 2.1, we define characteristic curves by the differential equation

$$\frac{dx}{dt} = c(u) \tag{3}$$

where $u = u(x,t)$. Of course, contrary to the situation for linear equations, the right side of this equation is not known a priori because the solution is not yet determined. Thus the characteristics cannot be determined in advance. In any case, along the curves defined by (3) the PDE (1) becomes

$$u_t + c(u)u_x = u_t + u_x\frac{dx}{dt} = \frac{du}{dt} = 0$$

or $u =$ constant. Thus u is constant on the characteristic curves. It is easy to observe that the characteristic curves defined by (3) are straight lines, for

$$\frac{d^2x}{dt^2} = \frac{d(dx/dt)}{dt} = \frac{du}{dt} = 0$$

Therefore, let us draw a characteristic back in time from an arbitrary point (x,t) in spacetime to a point $(\xi, 0)$ on the x-axis (see Figure 2.5). The equation of this characteristic is given by

$$x - \xi = c(u_0(\xi))t \tag{4}$$

where we have noted that its speed (reciprocal slope) is given by dx/dt or $c(u(0,\xi))$, since u is constant on the entire characteristic line. Further, it

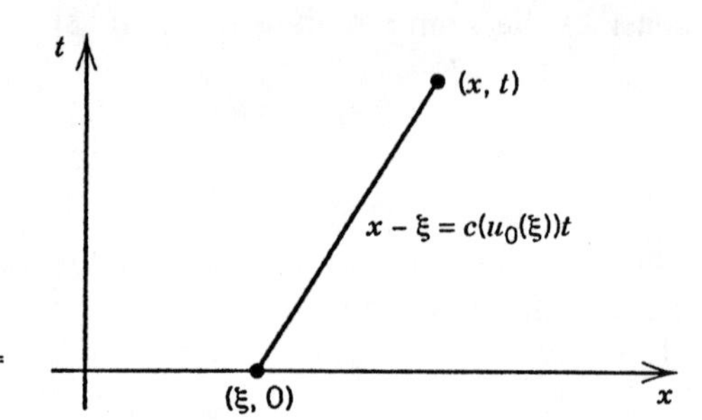

Figure 2.5. Characteristic given by $x - \xi = c(u_0(\xi))t$ and having speed $c(u_0(\xi))$.

follows that

$$u(x,t) = u(\xi,0) = u_0(\xi)$$

Therefore, if a solution to the initial value problem exists for $t > 0$, then necessarily it must be given by

$$u(x,t) = u_0(\xi) \tag{5}$$

where $\xi = \xi(x,t)$ is given implicitly by (4).

The solution of (1)–(2) determined implicitly by (4) and (5) was obtained under the assumption that a C^1 solution exists for all $t > 0$. Let us now verify that we indeed have a solution by substituting (5) into the PDE (1). We have

$$u_t = u_0'(\xi)\xi_t, \qquad u_x = u_0'(\xi)\xi_x$$

The partials ξ_t and ξ_x can be calculated by implicit differentiation of (4) to obtain

$$-\xi_t = c'(u_0(\xi))u_0'(\xi)\xi_t t + c(u_0(\xi)), \qquad 1 - \xi_x = c'(u_0(\xi))u_0'(\xi)\xi_x t$$

and therefore

$$u_t = -\frac{c(u_0(\xi))u_0'(\xi))}{D}, \qquad u_x = \frac{u_0'(\xi)}{D} \tag{6}$$

where $D = 1 + c'(u_0(\xi))u_0'(\xi)t$. Consequently, $u_t + c(u)u_x = 0$.

The preceding equations were obtained under the assumption that a unique solution $u = u(x,t)$ exists for all $t > 0$. Implicit, therefore, is the assumption that each point (x,t) in the xt-plane uniquely determines a value of ξ, and the partial derivatives (6) exist and are finite for all x, t. It seems clear that the derivatives in (6), however, will not be finite unless some restrictions are placed upon the initial signal u_0 and the function $c(u)$ in order to guarantee that the denominator D will not vanish. One such set of conditions is that the derivatives u' and c' both have the same sign; then the denominator D in (6) is always positive and it can be verified that (4) and (5) provide a solution to (1)–(2). We record this result as a theorem.

Theorem 1. If the functions c and u_0 are $C^1(R)$ and if u_0 and c are either both nondecreasing or both nonincreasing on R, the nonlinear initial value problem (1)–(2) has a unique solution defined implicitly by the parametric equations

$$u(x,t) = u_0(\xi)$$
$$x - \xi = c(u_0(\xi))t$$

Example 1. Consider the initial value problem

$$u_t + uu_x = 0, \qquad x \in R, \quad t > 0 \tag{7}$$

$$u(x,0) = 0 \quad \text{if } x \leq 0; \qquad u(x,0) = e^{-1/x} \quad \text{if } x > 0 \tag{8}$$

Here $c(u) = u$ and $u_0(x)$ are smooth nondecreasing functions on R. The initial signal $u_0(x)$ is shown in Figure 2.6. The characteristics emanating from a point $(\xi, 0)$ on the x-axis have speed $c(u_0(\xi)) = u_0(\xi)$. Hence $c = 0$ if $\xi \leq 0$ and the characteristics are vertical lines; $c = \exp(-1/\xi)$ if $\xi > 0$ and the characteristics fan out and approach speed unity as ξ gets large. Figure 2.7 shows the characteristic diagram. The wave in the region where the

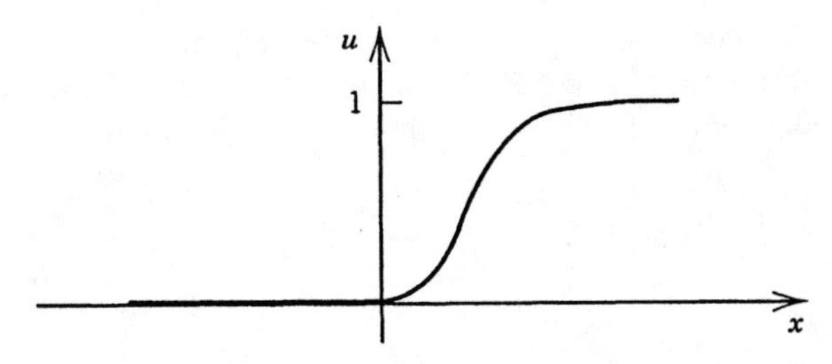

Figure 2.6. Graph of the initial waveform (8).

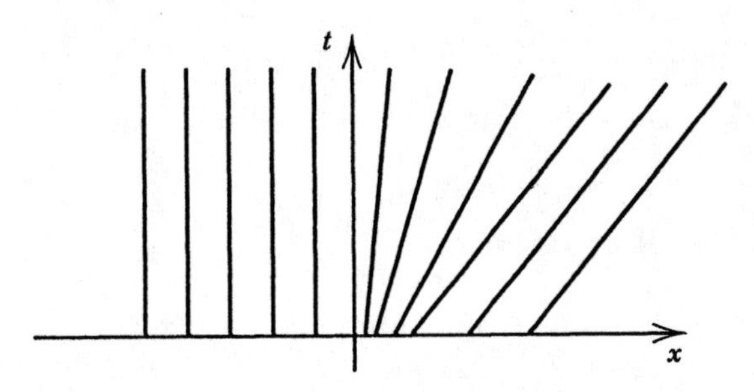

Figure 2.7. Characteristic diagram associated with (7)–(8).

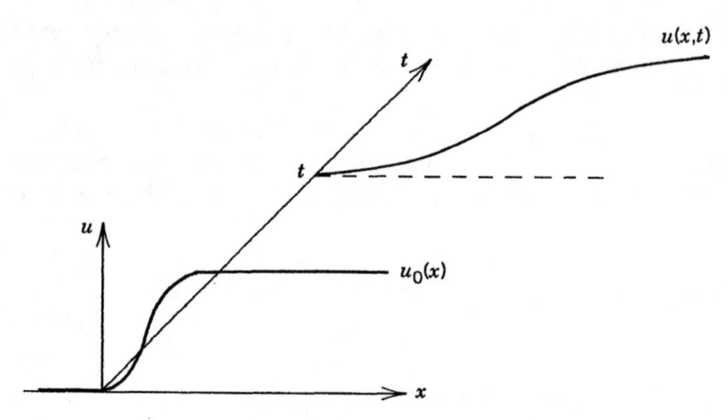

Figure 2.8. Rarefaction wave.

characteristics fan out is called a *rarefaction wave* or *release wave*. Figure 2.8 shows how the initial wave is propagated along the characteristics (recall that u is constant on the characteristics). The solution to the problem (7)–(8) is given implicitly by

$$u(x,t) = 0 \qquad \text{for } x \leq 0$$

$$u(x,t) = \exp\left(-\frac{1}{\xi}\right), \qquad \text{where} \quad x - \xi = t\exp\left(-\frac{1}{\xi}\right) \quad \text{for } x > 0$$

Remark. In Chapter 3 we address the important question of what happens when the hypotheses of Theorem 1 are not satisfied and the derivatives u_x and u_t in (6) below up. In this case the wave develops a discontinuity known as a *shock*.

EXERCISE

1. Consider the initial value problem

$$u_t + uu_x = 0, \qquad x \in R, \quad t > 0$$

$$u(x,0) = -x, \qquad x \in R$$

Sketch the characteristic diagram and find the solution.

2.3 QUASILINEAR EQUATIONS

In this section we discuss the nonlinear PDE

$$u_t + c(x,t,u)u_x = f(x,t,u), \qquad x \in R, \quad t > 0 \tag{1}$$

where the coefficients c and f are continuous functions. Such equations are called *quasilinear* because of the way the nonlinearity occurs—that is, the equation is linear in the derivatives and the nonlinearity occurs through multiplication by the coefficients which depend on u. We shall consider the initial value problem and append to (1) the initial condition

$$u(x,0) = u_0(x), \qquad x \in R \tag{2}$$

where $u_0(x)$ is continuously differentiable on R. Our approach in this section will be the same as in Section 2.2; we examine (1) and (2) under the assumption that a smooth solution exists, postponing a discussion of discontinuous solutions to Chapter 3. The goal at present is to build further on the method of characteristics and show how an algorithm can be constructed to solve (1)–(2).

Therefore, let $u = u(x,t)$ be a smooth solution to (1)–(2). From the prior discussion we observe that the PDE (1) reduces to the ordinary differential equation

$$\frac{du}{dt} = f(x,t,u) \tag{3}$$

along the family of curves (characteristics) defined by

$$\frac{dx}{dt} = c(x,t,u) \tag{4}$$

We may regard (3) and (4) as a system of two ordinary differential equations (called the *characteristic system*) for u and x, and we may solve them in principle, subject to the initial conditions

$$u = u_0(\xi), \qquad x = \xi \quad \text{at} \quad t = 0 \tag{5}$$

to obtain the solution. Here, as before, we regard ξ as a number that parameterizes the characteristic curves, giving the intersection of the curves with the x-axis.

Example 1. Consider the initial value problem

$$u_t + uu_x = -u, \qquad x \in R, \quad t > 0 \tag{6}$$

$$u(x,0) = -\frac{x}{2}, \qquad x \in R \tag{7}$$

Then the characteristic system is

$$\frac{du}{dt} = -u \quad \text{on} \quad \frac{dx}{dt} = u \tag{8}$$

with initial data

$$u = -\frac{\xi}{2}, \qquad x = \xi \quad \text{at} \quad t = 0 \tag{9}$$

It is straightforward to solve the characteristic system (8) to obtain

$$u = ae^{-t}, \qquad x = b - ae^{-t} \tag{10}$$

where a and b are constants of integration. Applying the initial conditions (9) gives a and b in terms of the parameter ξ according to

$$a = -\frac{\xi}{2}, \qquad b = \frac{\xi}{2}$$

Then the solution of the characteristic system (8) is, in parametric form,

$$u = -\frac{\xi}{2}e^{-t}, \qquad x = \frac{\xi}{2}(1 + e^{-t}), \qquad \xi \in R$$

In the present case we may eliminate the parameter ξ from the second equation to obtain

$$\xi = \frac{2x}{1 + e^{-t}}$$

Thus

$$u(x,t) = -\frac{xe^{-t}}{1 + e^{-t}}$$

which is a smooth solution valid for all $t > 0$ and $x \in R$.

We emphasize that it is unlikely in the general case that the calculations can be performed as in the example and a smooth, closed-form solution derived.

The signaling problem, as well as other boundary value problems, may be handled in a similar manner using the method of characteristics.

Example 2. Consider the initial boundary value problem

$$u_t + uu_x = 0, \qquad x > 0, \quad t > 0 \tag{11}$$

$$u(x,0) = 1, \qquad x > 0 \tag{12}$$

$$u(0,t) = \frac{1}{1+t^2}, \qquad t > 0 \tag{13}$$

The PDE again reduces to the total derivative $du/dt = 0$ on the characteristic curves defined by $dx/dt = u$. Thus

$$u = \text{constant} \quad \text{on} \quad \frac{dx}{dt} = u$$

As we determined earlier, the characteristics are straight lines with speed u. Therefore, all the characteristics emanating from the x-axis, where $u = 1$, have speed 1. Figure 2.9 shows the characteristic diagram for (11)–(13). This means that the constant solution $u = 1$ is carried into the region $x > t$ by these characteristics. For $x < t$ the characteristics fan out because the speed u decreases along the t-axis according to the boundary condition (13). Let us now select an arbitrary point (x, t) in the region $x < t$ where we want to determine the solution. Following this characteristic back to the boundary $x = 0$, we denote its intersection point by $(0, \tau)$. The equation of this characteristic is given by

$$x = u(0,\tau)(t - \tau)$$

or

$$x = \frac{1}{1+\tau^2}(t - \tau) \tag{14}$$

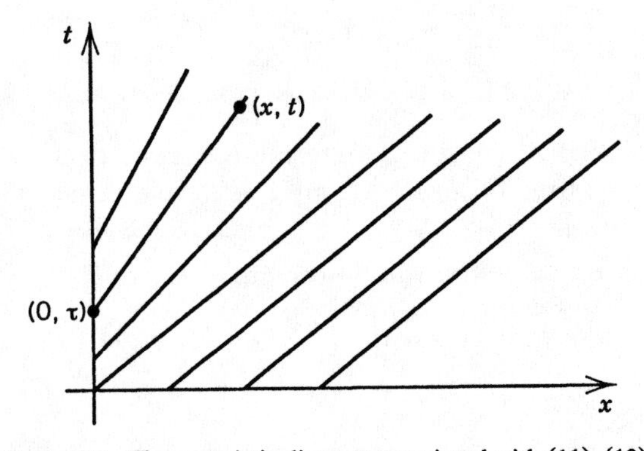

Figure 2.9. Characteristic diagram associated with (11)–(13).

By the constancy of u along the characteristics, we know that

$$u(x,t) = u(0,\tau) = \frac{1}{1+\tau^2}$$

and $\tau = \tau(x,t)$ is given implicitly by formula (14). In the present case we may solve (14) for τ to obtain

$$\tau = \frac{-1 + [1 + 4x(t-x)]^{1/2}}{2x} \tag{15}$$

[note that (14) reduces to a quadratic equation in τ which can be solved by the quadratic formula, taking the plus sign on the square root in order to satisfy the boundary condition]. Consequently, a solution to the problem (11)–(13) is

$$u(x,t) = 1 \qquad \text{for } x \geq t$$

$$u(x,t) = \frac{1}{1+\tau^2} \qquad \text{for } x < t$$

where τ is given by (15).

EXERCISES

1. Obtain the solution to the initial value problem

$$u_t + uu_x = -ku^2, \qquad x \in R, \quad t > 0$$
$$u(x,0) = 1, \qquad x \in R$$

where k is a positive constant. Sketch the characteristics.

2. Consider the initial value problem

$$u_t + uu_x = -ku, \qquad x \in R, \quad t > 0$$
$$u(x,0) = u_0(x), \qquad x \in R$$

where k is a positive constant. Determine a condition on u_0 so that a smooth solution exists for all $t > 0$ and $x \in R$, and determine the solution in parametric form.

3. Find a solution to the initial value problem

$$u_t + cu_x = xu, \qquad x \in R, \quad t > 0$$
$$u(x,0) = u_0(x), \qquad x \in R$$

where c is a positive constant. Sketch the characteristics.

4. Find a solution $u = u(x, t)$ of the initial value problem

$$u_t - xtu_x = x, \qquad x \in R, \quad t > 0$$
$$u(x,0) = u_0(x), \qquad x \in R$$

5. Solve the signaling problem

$$u_t + u_x = u^2, \qquad x > 0, \quad t \in R$$
$$u(0,t) = u_0(t), \qquad t \in R$$

2.4 PROPAGATION OF SINGULARITIES

In all of the preceding discussion and examples we have assumed that the initial and boundary data were given by smooth functions. Now we consider the case where the initial or boundary data are continuous but may have discontinuities in the derivatives. The question we address is how those discontinuities on the boundary of the region are propagated into the region by the PDE. A simple example shows us what to expect.

Example 1. Consider the simple advection equation

$$u_t + cu_x = 0, \qquad x \in R, \quad t > 0 \tag{1}$$

subject to an initial condition given by a piecewise smooth function u_0 defined by

$$u(x,0) = u_0(x) = \begin{cases} 1, & x < 0 \\ 1 - x, & 0 < x < 1 \\ 0, & x > 1 \end{cases}$$

Because the general solution of (1) is $u(x, t) = f(x - ct)$, a right traveling wave, u is constant on the characteristic curves $x - ct =$ constant, which carry the initial data into the region $t > 0$ (see Figure 2.10). Thus the discontinuities in u' at $x = 0$ and $x = 1$ are carried along the characteristics into the region $t > 0$.

In summary, characteristics carry data from the boundary into the region. Therefore, abrupt changes in the derivatives on the boundary produce corresponding abrupt changes in the region. In other words, discontinuities in derivatives propagate along the characteristics. For a general nonlinear equation, as we shall observe in Chapter 4, discontinuities in the boundary functions themselves do not propagate along characteristics; these kinds of singularities are shocks and they propagate along different spacetime curves.

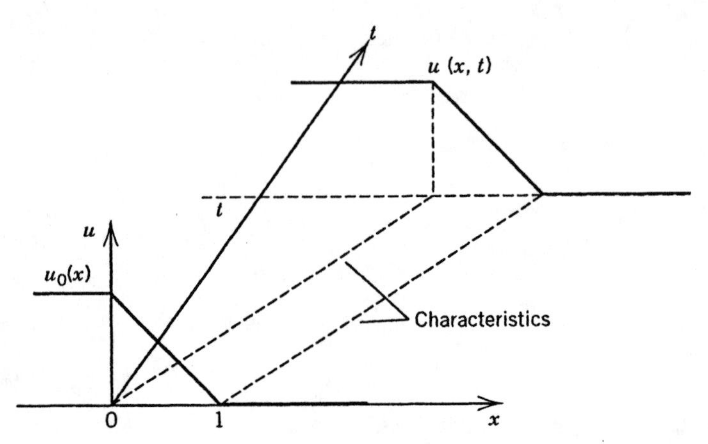

Figure 2.10. Discontinuities in derivatives propagating on characteristics.

For the simple nonlinear kinematic wave equation

$$u_t + c(u)u_x = 0 \tag{2}$$

we now demonstrate that discontinuities in the derivatives propagate along characteristics. Let $\xi = \xi(x, t)$ be a curve in xt-space separating two regions where a solution $u = u(x, t)$ is C^1, and suppose that u has a crease in the surface; that is, u is continuous across the curve, but there is a simple jump discontinuity in the derivatives of u across the curve. Such a curve is sometimes called a *wavefront* (see Figure 2.11). To analyze the behavior of u along the wavefront, we introduce a new set of curvilinear coordinates given

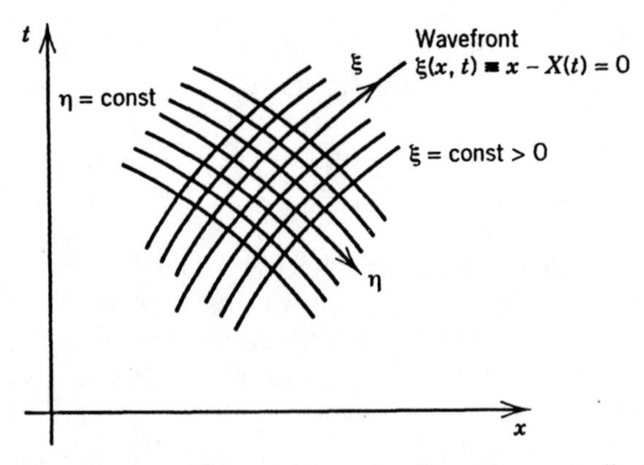

Figure 2.11. Wavefront $x = X(t)$ and the associated curvilinear coordinates ξ and η.

by

$$\xi = \xi(x,t), \qquad \eta = \eta(x,t) \tag{3}$$

For example, $\eta(x,t)$ = constant can be taken as the family of curves orthogonal to $\xi(x,t)$ = constant. Then the chain rule for derivatives implies that

$$u_t = u_\xi \xi_t + u_\eta \eta_t, \qquad u_x = u_\xi \xi_x + u_\eta \eta_x$$

Here there should be no confusion in using the same variable u for the transformed function of ξ and η. The PDE (2) therefore becomes

$$(\xi_t + c(u)\xi_x)u_\xi + (\eta_t + c(u)\eta_x)u_\eta = 0 \tag{4}$$

This equation is valid in the regions $\xi > 0$ and $\xi < 0$. Also, by hypothesis,

$$u(0+,\eta) = u(0-,\eta) \tag{5}$$

and therefore

$$u_\eta(0+,\eta) = u_\eta(0-,\eta) \tag{6}$$

Consequently, the tangential derivatives are continuous across the wavefront $\xi(x,t) = 0$.

Now we introduce some notation for the jump in a quantity across a wavefront. Let Q be some quantity that has a value Q_+ just ahead (to the right) of the wavefront, and a value Q_- just behind (to the left) of the wavefront. Then the *jump in the quantity* Q across the wavefront is defined by

$$[Q] = Q_- - Q_+$$

Continuing with our calculation, we now take the limit of equation (4) as $\xi \to 0+$ and then take the limit of (4) as $\xi \to 0-$. Subtracting these two results gives

$$(\xi_t + c(u)\xi_x)[u_\xi] = 0 \tag{7}$$

where we have used (5) and (6). If we assume that $[u_\xi] \neq 0$, then

$$\xi_t + c(u)\xi_x = 0 \tag{8}$$

In particular, if the wavefront $\xi(x,t) = 0$ is given by $x = X(t)$, then (8) becomes

$$\frac{dX}{dt} = c(u) \tag{9}$$

That is, $\xi(x,t) = 0$ must be a characteristic. We summarize the result in the following theorem.

Theorem 1. Let D be a region of spacetime and let $\xi(x,t) = x - X(t) = 0$ be a smooth curve lying in D that partitions D into two disjoint regions D^+ and D^- (see Figure 2.12). Let u be a smooth solution to (2) in D^+ and D^- that is continuous in D, and assume that the derivatives of u suffer simple jump discontinuities across $\xi(x,t) = 0$. Then $\xi(x,t) = 0$ must be a characteristic curve.

The next question of interest concerns the magnitude of the jump $[u_x]$ as the wave propagates via (2) along a characteristic. Under special assumptions we now derive an ordinary differential equation for the magnitude of the jump that governs its change. Let us assume that the wavefront is at the origin $x = 0$ at time $t = 0$, and that the state ahead of the wave is a constant state $u = u_0$. Then the characteristics ahead of the wavefront have constant speed $c_0 = c(u_0)$, and the wavefront itself has equation $\xi(x,t) = x - c_0 t = 0$ (see Figure 2.13). Thus $u = u_0$ for $\xi > 0$. The coordinate lines $\xi =$ constant are parallel to the wavefront, and we want to examine the behavior of the wave near the wavefront (where ξ is small and negative) as time increases. Therefore, using the assumption that ξ is small, we make the Ansatz

$$u = \begin{cases} u_0, & \xi > 0 \quad \text{(ahead)} \\ u_0 + u_1(t)\xi + \frac{1}{2}u_2(t)\xi^2 + \cdots, & \xi < 0 \quad \text{(behind)} \end{cases}$$

Computing the derivatives yields

$$u_t = u_1(t)(-c_0) + \xi u_1'(t) + u_2(t)\xi(-c_0) + O(\xi^2) \tag{10}$$

$$u_x = u_1(t) + u_2(t)\xi + O(\xi^2) \tag{11}$$

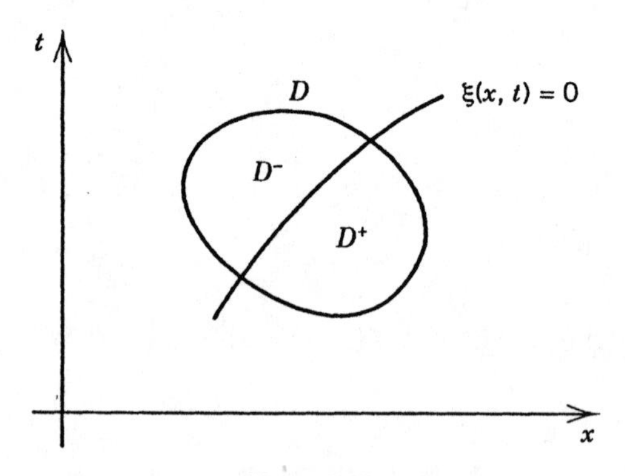

Figure 2.12

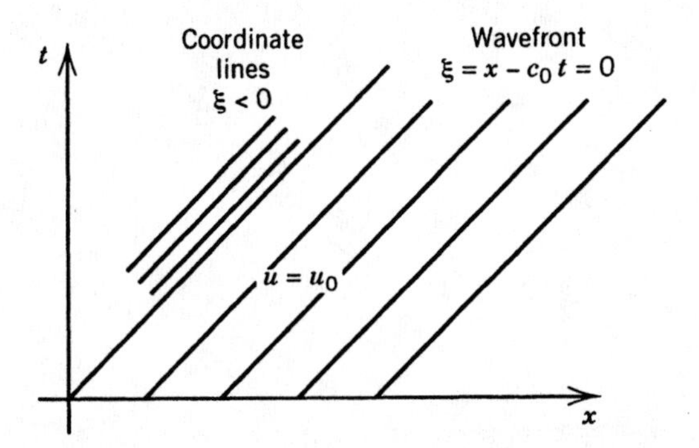

Figure 2.13. Wavefront propagating into a constant state.

and

$$c(u) = c_0 + c'(u_0)(u - u_0) + O\left((u - u_0)^2\right)$$
$$= c_0 + c'(u_0)u_1(t)\xi + O(\xi^2) \tag{12}$$

Note that our approach here is to trade in the independent variables t and x for new variables t and ξ, where ξ is a curvilinear coordinate measuring the distance from the wavefront. Now substitute (10)–(12) into the PDE (2) to obtain

$$-c_0u_1 + \xi(u_1' - c_0u_2) + O(\xi^2)$$
$$+\left((c_0 + c'(u_0)u_1\xi + O(\xi^2)\right)\left(u_1 + u_2\xi + O(\xi^2)\right) = 0$$

Since this equation must hold for all ξ, we may set the coefficients of the powers of ξ equal to zero; the $O(1)$ and $O(\xi)$ coefficients are

$$O(1): \quad -c_0u_1 + c_0u_1 = 0$$
$$O(\xi): \quad u_1' + c'(u_0)u_1^2 = 0$$

The $O(1)$ equation holds identically, and the $O(\xi)$ equation is a differential equation for the first order correction $u_1(t)$ in the expansion for u in region just behind the wavefront ($\xi < 0$). Since u_x is given by (11), the function $u_1(t)$ also will approximate the jump $[u_x]$ to first order. (Recall that $[u_x] = u_x(\xi - , t)$ since $u_x(\xi + , t) = 0$.) Solving the $O(\xi)$ differential equation for u_1 easily gives

$$u_1(t) = \frac{1}{k + c'(u_0)t} \tag{13}$$

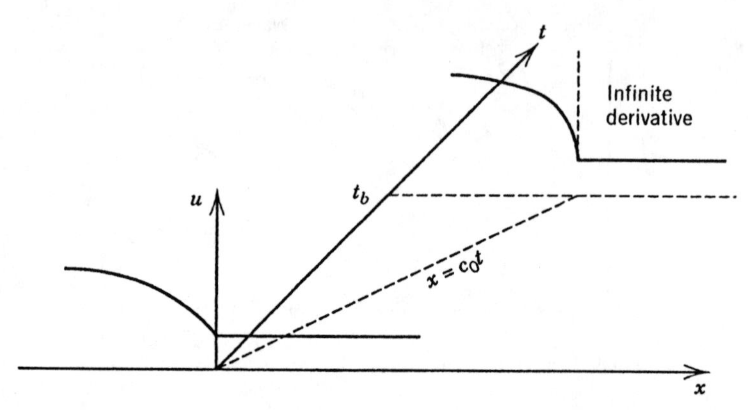

Figure 2.14. Evolution along a characteristic of a jump discontinuity in the derivative until a gradient catastrophe occurs at a breaking time t_b.

where k is a constant of integration that can be determined by initial data, that is, the initial jump in u_x at time $t = 0$. If $c'(u_0) > 0$ and $k < 0$, that is, the wavefront initially has a negative jump at $x = 0$, $t = 0$, as shown in Figure 2.14, then (13) implies that the gradient jump will steepen and the wave and break at a finite time $t_b = -k/c'(u_0)$. If $k > 0$, the jump in the gradient will tend to zero as t goes to infinity.

The wavefront calculation just presented is one of the standard tools used in nonlinear hyperbolic problems to gain information about signal propagation along characteristics. The method lies in the domain of what is often called *weakly nonlinear* theory. When we discuss hyperbolic systems in Chapter 5, we shall find that this technique is a valuable tool in deriving manageable equations which approximate wavefront phenomena in nonlinear problems.

EXERCISE

1. Consider the initial value problem

$$u_t + u^2 u_x = 0, \qquad x \in R, \quad t > 0$$

$$u(x,0) = 1, \quad \text{if } x > 0; \qquad u(x,0) = \frac{2x - 1}{x - 1}, \quad \text{if } x < 0$$

Sketch the initial signal and determine the initial jump in u_x. Sketch the characteristic diagram. Approximate the time t_b that the wave will break along the characteristic $x = t$.

2.5 GENERAL FIRST ORDER EQUATION

Now we consider the general first order nonlinear PDE

$$H(x,t,u,p,q) = 0, \qquad p = u_x, \quad q = u_t \tag{1}$$

on the domain $x \in R$, $t > 0$, subject to the initial condition

$$u(x,0) = u_0(x), \qquad x \in R \tag{2}$$

In the PDE (1) there is not now an obvious directional derivative that defines characteristic directions along which the PDE can be reduced to an ordinary differential equation. However, with some elementary analysis we can discover such directions. Let C be a curve in spacetime given by

$$x = x(s), \qquad t = t(s)$$

where s is a parameter (see Figure 2.15). Then the total derivative of u along C is

$$\frac{du}{ds} = u_x x'(s) + u_t t'(s) = px' + qt'$$

We ask if there is a special direction (x', t') that has special significance for (1). We first calculate the total derivatives of p and q along C. To this end,

$$p' = \frac{d(u_x)}{ds} = u_{xx}x' + u_{xt}t' \tag{3}$$

$$q' = \frac{d(u_t)}{ds} = u_{tx}x' + u_{tt}t' \tag{4}$$

Now take the two partial derivatives (with respect to x and with respect to t)

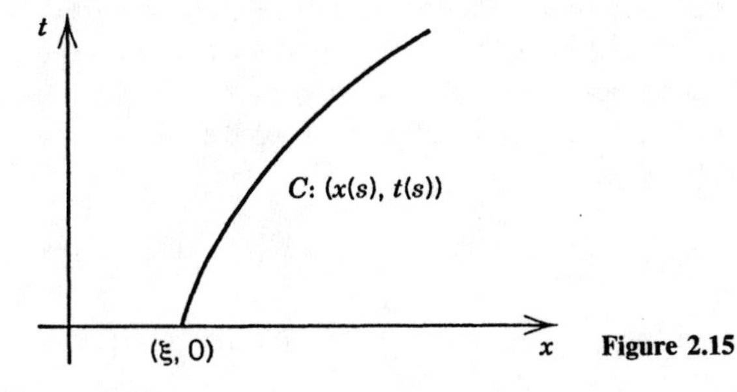

Figure 2.15

of the PDE (1) to obtain

$$H_x + H_u p + H_p u_{xx} + H_q u_{tx} = 0 \tag{5}$$

$$H_t + H_u q + H_p u_{xt} + H_q u_{tt} = 0 \tag{6}$$

Comparing (3) and (4) with the last two terms in (5) and (6) suggests a judicious choice for the direction (x', t') to be

$$x' = H_p, \qquad t' = H_q \tag{7}$$

In this case equations (3) and (4) combine to become

$$p' = -H_x - H_u p, \qquad q' = -H_t - H_u q \tag{8}$$

and the total derivative of u along C becomes

$$u' = pH_p + qH_q \tag{9}$$

Now let us summarize our results. If *characteristic curves* are defined by the system of differential equations (7), then (8) and (9) hold along these curves; the latter are a system of ordinary differential equations that dictate how u, p, and q change along the curves. In other words, (8) and (9) hold along (7). The entire set of equations (7)–(9) is called the *characteristic system* associated with the nonlinear PDE (1). It is a system of five ordinary differential equations for x, t, u, p, and q.

In principle, therefore, we can present an algorithm to solve the initial value problem (1)–(2). We emphasize that the following calculations are made assuming that a smooth solution exists. The initial condition (2) translates into

$$x = \xi, \qquad u = u_0(\xi) \quad \text{at } t = 0 \tag{10}$$

and hence

$$p = u_0'(\xi) \tag{11}$$

where ξ is, as before, a real number that parameterizes the characteristics curves, representing their intersection with the x-axis. To solve the characteristic system we would also need a condition on q at $t = 0$. Such a condition can be obtained from the PDE (1) itself, for we can evaluate the PDE along the $t = 0$ time line to obtain

$$H(\xi, 0, u_0(\xi), u_0'(\xi), q) = 0 \tag{12}$$

Assuming that (12) can be solved for q (the condition $H_q \neq 0$ for $t = 0$ will guarantee this), we obtain

$$q = q(\xi) \qquad \text{at } t = 0 \tag{13}$$

Finally, the characteristic system (7)–(9) can be solved, subject to the initial conditions (10), (11), and (13), to obtain t, x, u, p, and q along a characteristic curve. In some cases the parameters may be eliminated to obtain a formula $u = u(x, t)$ for the solution.

Example 1. Consider the initial value problem

$$u_t + u_x^2 = 0, \qquad x \in R, \quad t > 0 \tag{14}$$
$$u(x, 0) = x, \qquad x \in R$$

Here

$$H(x, t, u, p, q) = q + p^2 = 0 \tag{15}$$

and the initial time line can be parameterized by

$$x = \xi, \qquad t = 0, \qquad u = \xi, \qquad p = 1 \tag{16}$$

The PDE (15) then gives at $t = 0$ a condition on q, namely

$$q = -1 \tag{17}$$

The characteristic system (7)–(9) corresponding to (15) is

$$x' = 2p, \qquad t' = 1, \qquad u' = 2p^2 + q, \qquad p' = 0, \qquad q' = 0 \tag{18}$$

Because (18) is autonomous we may assume that the initial data are given at $s = 0$. Solving this system is straightforward in the present case (for other problems solving the characteristic system may be difficult or impossible). Clearly, $t = s$ and

$$p = c_1 \quad \text{and} \quad q = c_2 \tag{19}$$

where c_1 and c_2 are constants. Therefore, from (18),

$$x' = 2c_1 \quad \text{or} \quad x = 2c_1 t + c_3 \tag{20}$$

where c_3 is another constant. Next, from (18),

$$u' = 2c_1^2 + c_2$$

and consequently,

$$u = (2c_1^2 + c_2)t + c_4 \tag{21}$$

Applying the initial conditions (16) and (17) allows us to determine the constants $c_1, \ldots, c_4$. Easily we obtain

$$c_1 = 1, \quad c_2 = -1, \quad c_3 = \xi, \quad c_4 = \xi$$

and therefore

$$x = 2t + \xi, \quad u = t + \xi$$

In this instance we may eliminate the parameter ξ and obtain the analytic solution

$$u(x, t) = x - t$$

The characteristic curves in this problem are given by $x = 2t + \xi$ and are straight lines in spacetime moving with speed 2.

Example 2. Consider the initial value problem

$$H(x, t, u, p, q) = 2tq + 2xp - p^2 - 2u = 0, \quad x \in R, \quad t > 0 \tag{22}$$

$$u(x, 0) = u_0(x), \quad x \in R \tag{23}$$

At $t = 0$ we have $x = \xi$, $u = u_0(\xi)$, and $p = u_0'(\xi)$. To complete the initial data we need q at $t = 0$. From (22) we have

$$H(\xi, 0, u_0(\xi), u_0'(\xi), q) = 0$$

But q drops out of this equation and we are unable to determine q at $t = 0$. Thus the PDE (22) does not determine the initial derivative in the time direction and therefore not enough information is given to move away from the initial time line. Notice that the characteristic direction $(x', t') = (H_p, H_q) = (2x - 2p, 2t)$ is, at $t = 0$, given by $(2\xi - 2u_0'(\xi), 0)$, which is a vector in the same direction as the initial time line. Thus the characteristic curves, which are to carry the initial data into the region $t > 0$, are tangent to the initial time line and do not have a component in the time direction to carry signals forward. Thus (22)–(23) is not a well-posed problem.

We summarize the comments in the preceding paragraph by making a formal statement.

Remark. A necessary condition that the initial value problem (1)–(2) have a smooth solution is that equation (1) with $t = 0$ uniquely determine q as a

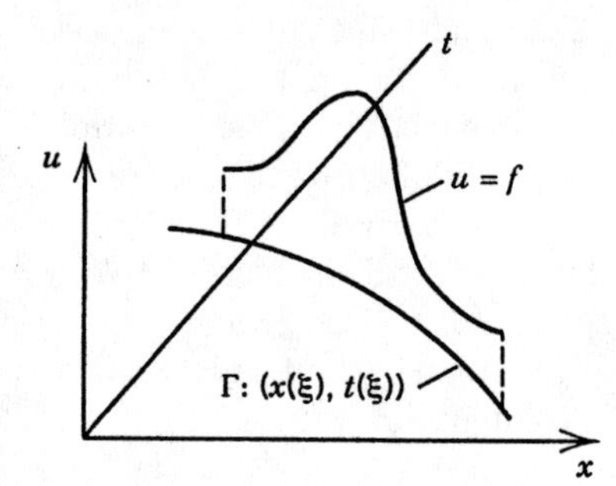

Figure 2.16. Illustration of Cauchy data where u is prescribed along a curve Γ.

function of x, u, and p. A condition that guarantees this solvability is that $H_q \neq 0$ at each point $(x, 0, u, p, q)$. Under this condition it can be shown that a local solution exists about each point $(x, 0)$ on the initial time line.

Finally, we remark that the foregoing procedure can be extended to solve the PDE (1) subject to data given along any curve Γ in spacetime; that is, we consider the PDE (1) with the initial condition (2) replaced by the boundary condition

$$u(x,t) = f(x,t) \qquad \text{on } \Gamma \tag{24}$$

where f is a given function (see Figure 2.16). If we parametrize the curve Γ by $x = x(\xi)$, $t = t(\xi)$, condition (24) becomes

$$u = f(x(\xi), t(\xi)) = F(\xi) \qquad \text{along } \Gamma$$

As in the initial value problem, we must be able to determine p and q along the curve Γ so that we will have enough data to solve the characteristic system. The PDE (1) must hold along Γ, so that

$$H(t(\xi), x(\xi), F(\xi), p(\xi), q(\xi)) = 0 \tag{25}$$

Further, taking $d/d\xi$ of u (i.e., differentiating u along Γ), we have

$$\frac{dF}{d\xi} = p(\xi)\frac{dx}{d\xi} + q(\xi)\frac{dt}{d\xi} \tag{26}$$

Equations (25) and (26) represent two equations for $p(\xi)$ and $q(\xi)$. We may expect to be able to solve these equations when the Jacobian is nonzero, that is,

$$H_p\frac{dt}{d\xi} - H_q\frac{dx}{d\xi} \neq 0 \tag{27}$$

Geometrically, the solvability condition (27) requires that the characteristic direction $(x', t') = (H_p, H_q)$ traverse the boundary curve Γ, which has direction $(dx/d\xi, dt/d\xi)$. In other words, (27) means that the boundary curve Γ must nowhere have characteristic direction. As noted in Example 2, if the curve along which the data are prescribed has characteristic direction at a point on the curve, not enough information is supplied to carry data off that curve into the region where the problem is to be solved.

The initial value problem (1)–(2), or the more general problem (1) with data (24) given on a curve Γ, is called a *Cauchy problem*. The general theory of Cauchy problems (existence, uniqueness, and regularity of solutions) is discussed in many of the references (see, e.g., John [1982]).

EXERCISES

1. Solve the initial value problem

$$u_t + u_x = t, \qquad x \in R, \quad t > 0$$
$$u(x,0) = 0, \qquad x \in R$$

2. Solve the initial value problem

$$u_t + u_x = 0, \qquad x \in R, \quad t > 0$$
$$u(x,0) = -x^2, \qquad x \in R$$

Does the solution exist for all $t > 0$?

3. Find the characteristic system associated with the *Hamilton–Jacobi equation*

$$u_t + h(t, x, u_x) = 0$$

where h is a given function (called the *Hamiltonian*).

4. Determine the characteristic system associated with the nonlinear equation

$$u_t + c(u)u_x = 0$$

and show that it coincides with the results obtained in Section 2.2.

5. Use the method of characteristics to determine two different solutions to the initial value problem

$$u = p^2 - 3q^2, \qquad x \in R, \quad t > 0$$
$$u(x,0) = x^2, \qquad x \in R$$

6. Solve

$$u_t = u_x^3, \qquad x \in R, \quad t > 0$$
$$u(x,0) = 2x^{3/2}, \qquad x \in R$$

2.6 UNIQUENESS RESULT

Up to this point our approach has been algorithmic in nature; that is, most of our efforts have gone into the actual construction of solutions, and we have essentially ignored existence and uniqueness questions. Having constructed a solution in some manner, one may legitimately ask if a different construction could give a different solution. Now, both to answer this question and to give the reader a flavor of a simple theoretical result, we prove a basic uniqueness result due to A. Haar in 1928. The proof given here is an adaptation of the proof given in Courant and Hilbert [1962], and it applies to a specific class of nonlinear equations.

We require one definition, namely what it means for a function to satisfy a Lipschitz condition.

Definition 1. A function $f(x, y)$ satisfies the *Lipschitz condition with respect to y* on a domain D in the plane iff there exists a constant $k > 0$ such that

$$|f(x, y_1) - f(x, y_2)| \leq k|y_1 - y_2|$$

for all (x, y_1) and (x, y_2) in D. The constant k is called the *Lipschitz constant with respect to y.*

The reader may have encountered this concept in elementary ordinary differential equations; it arises as a condition on $f(x, y)$ that guarantees uniqueness of solutions to the initial value problem $dy/dx = f(x, y)$, $y(x_0) = y_0$. It is not difficult to show that the property of being Lipschitz is stronger than continuity, yet weaker than differentiability.

We may extend the concept to require Lipschitz in several variables. For example, a function $f(x, y, z)$ is said to be Lipschitz in y and z iff there exists two positive constants k and m such that

$$|f(x, y_1, z_1) - f(x, y_2, z_2)| \leq k|y_1 - y_2| + m|z_1 - z_2|$$

for all (x, y_1, z_1) and (x, y_2, z_2). The constants k and m are the Lipschitz constants with respect to y and z, respectively.

Now the uniqueness theorem of Haar:

Theorem 1. Consider a PDE of the form

$$u_t = G(x, t, u, p), \qquad p = u_x \tag{1}$$

where G is continuous and satisfies a Lipschitz condition with respect to u and p in R^4, and k is the Lipschitz constant with respect to p. Let u and v

be smooth solutions of (1) such that

$$u(x,0) = v(x,0), \qquad x_1 \le x \le x_2$$

Then $u(x,t) = v(x,t)$ on the triangle $T = \{(x,t)\colon\ t \ge 0,\ t \le k^{-1}(x - x_1),\ t \le k^{-1}(x_2 - x)\}$.

Proof. Let $w = u - v$. We need to show that $w = 0$ on T. First, using the Lipschitz property

$$\begin{aligned} |w_t| = |u_t - v_t| &= |G(t,x,u,u_x) - G(t,x,v,v_x)| \\ &\le a|u - v| + k|u_x - v_x| = a|w| + k|w_x| \end{aligned}$$

At points where $w > 0$ the last inequality can be written

$$w_t < bw + k|w_x| \tag{2}$$

where $b > a$. Now we define the function W by $W = we^{-bt}$. It clearly suffices to show that $W = 0$ on T. We proceed by contradiction and assume without loss of generality that W is positive at some point in T. Let P be a point in T (a closed bounded set) where W assumes a local maximum, where, of course $W > 0$. The point P cannot be on the base of T since $W = 0$ there. Thus P must be in the interior of T or on the lateral sides. In this case (see Figure 2.17) the directions $(-k, -1)$ and $(k, -1)$ both point into T. Therefore, the directional derivatives of W in these two directions must be nonpositive (W has a maximum at P). That is,

$$\begin{aligned} (-k,-1)(W_x, W_t) &= -kW_x - W_t \le 0 \\ (k,-1)(W_x, W_t) &= kW_x - W_t \le 0 \end{aligned}$$

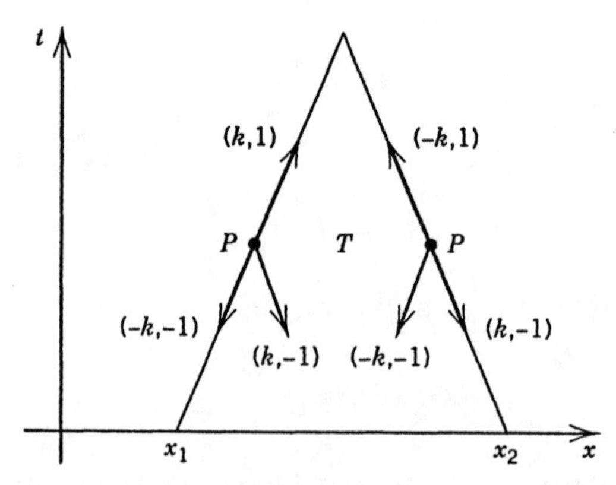

Figure 2.17. Triangle T.

Consequently,

$$W_t \geq k|W_x| \qquad \text{at } P$$

or, in terms of w,

$$w_t \geq bw + k|w_x| \qquad \text{at } P$$

which contradicts (2). So the proof is complete.

REFERENCES

Courant, R., and D. Hilbert, 1962. *Methods of Mathematical Physics*, Vol. 2, Wiley-Interscience, New York.

John, F., 1982. *Partial Differential Equations*, 4th ed., Springer-Verlag, New York.

3

WEAK SOLUTIONS TO HYPERBOLIC EQUATIONS

In Chapter 2 the emphasis was on the role of characteristics in the study of hyperbolic problems and on the development of algorithms for determining the solution to such problems. The underlying assumption was that a continuous, smooth solution exists, at least up to some time when it ceased to be valid. In this chapter we lay out the foundation for the investigation of discontinuous, or weak, solutions, and the propagation of shock waves.

By our very definition, a solution to a first order PDE must be smooth, or have continuous first partial derivatives, so that it makes sense to calculate those derivatives and substitute them into the PDE to check if, indeed, we have a solution. Such smooth solutions are called classical or genuine solutions. Now we want to generalize the notion of a solution and admit discontinuous functions. If a discontinuity in a solution surface exists along some curve in spacetime, there must be some means of checking for a solution along that curve without calculating the partial derivatives, which do not exist along the curve. Our aim in this chapter is to develop such a criterion and formulate the general concept of a weak solution. In Sections 3.1 and 3.2 we take a heuristic approach and derive a jump condition that must hold across a discontinuity; in Section 3.3 we study two applications (traffic flow and chemical reactors) to gain some understanding of characteristics and shock formation in practical settings. In Section 3.4 we come to grips with the underlying mathematical ideas and we state a formal definition of a weak solution to a PDE. Finally, in Section 3.5 we ask about the asymptotic, or long-time, behavior of solutions. The scenario is this: At time $t = 0$ a smooth profile is propagated in time until it evolves into a shock wave, or a discontinuous signal; after this discontinuous wave forms, we ask how the magnitude of the discontinuity decays and along what path in spacetime it is propagated. Understanding this process will give us a total picture of how an initial signal evolves via a first order PDE.

3.1 DISCONTINUOUS SOLUTIONS

In Chapter 2 we observed that characteristics play a fundamental role in understanding how solutions to first order PDEs are propagated in spacetime. There our study presupposed that solutions were smooth, or, at the worst, piecewise smooth and continuous. In this chapter we take up the question of discontinuous solutions. Evidently, as the following example shows, for a *linear* equation it is possible to propagate discontinuous initial or boundary data into the region of interest along the characteristics as well.

Example 1. Consider the advection equation

$$u_t + cu_x = 0, \qquad x \in R, \quad t > 0 \quad (c > 0)$$

subject to the initial condition $u(x,0) = u_0(x)$, where u_0 is defined by $u_0 = 1$ if $x < 0$ and $u_0 = 0$ if $x > 0$. Since the solution to the advection equation is $u = u_0(x - ct)$, the initial condition is propagated along the characteristics $x - ct = \text{const}$, and the discontinuity at $x = 0$ is propagated along the line $x = ct$, as shown in Figure 3.1. Thus the solution to the initial value problem is given by

$$u(x,t) = 0 \quad \text{if } x > t; \qquad u(x,t) = 1 \quad \text{if } x < t$$

Now let us examine a simple nonlinear problem with the same initial data as in Example 1. In general, a hyperbolic system with piecewise constant initial data is called a *Riemann problem*. Such problems receive considerable attention in the PDE literature.

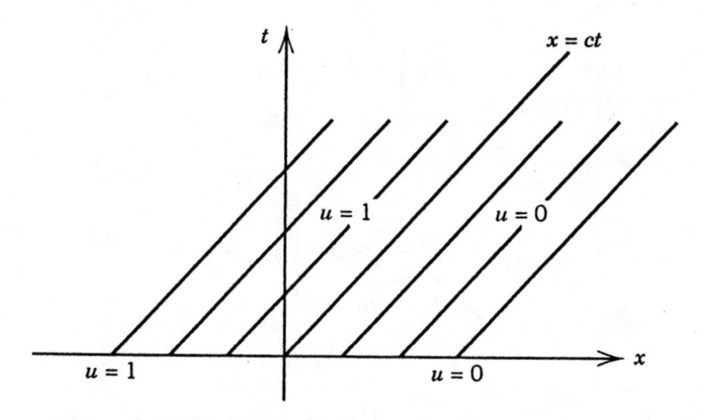

Figure 3.1. Characteristics $x - ct = \text{const}$.

Example 2. Consider the initial value problem

$$u_t + uu_x = 0, \qquad x \in R, \quad t > 0 \tag{1}$$

$$u(x,0) = 1 \quad \text{if } x < 0; \qquad u(x,0) = 0 \quad \text{if } x > 0 \tag{2}$$

We know from Chapter 2 that the PDE (1) implies that $du/dt = 0$ along $dx/dt = u$, or $u =$ constant on the straight-line characteristics having speed u. The characteristics emanating from the x-axis have speed zero (vertical) if $x > 0$, and they have speed unity if $x < 0$. The characteristic diagram is shown in Figure 3.2. Immediately at $t > 0$, one observes that the characteristics collide and a contradiction is implied since u must be constant on characteristics. One way to attempt to avoid this impasse is to insert a straight line $x = mt$ of nonnegative speed along which the initial discontinuity at $x = 0$ is carried. The characteristic diagram is now changed to Figure 3.3. For $x > mt$ we can take $u = 0$, and for $x < mt$ we can take $u = 1$, thus

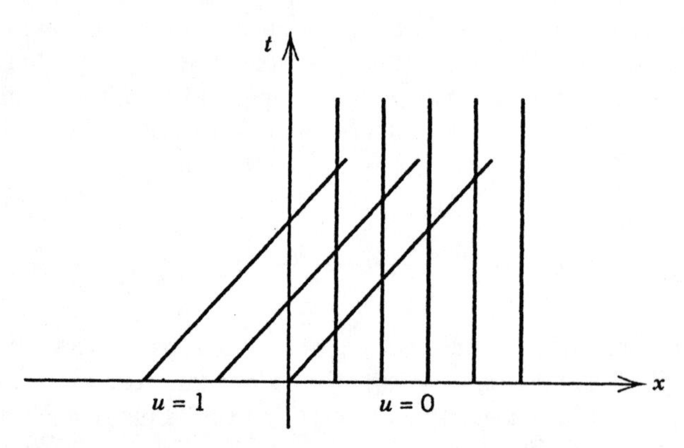

Figure 3.2. Characteristics for the initial value problem (1)–(2).

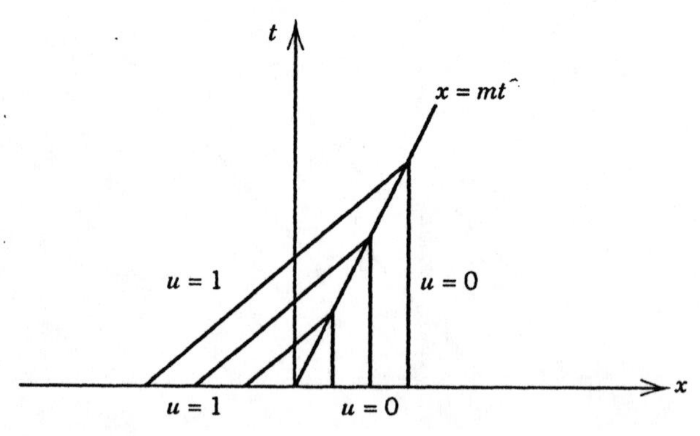

Figure 3.3. Insertion of a line $x = mt$ along which the discontinuity is carried.

giving a solution to the PDE (1) on both sides of the discontinuity. The only question is the choice of m; for any $m \geq 0$ it appears that a solution can be obtained away from the discontinuity and that solution also satisfies the initial condition. Shall we give up uniqueness for this problem? Is there some other solution that we have not discovered? Or, is there some special choice of m?

The answers to these questions lie at the foundation of the notion of a discontinuous solution to a PDE. As it turns out, just as discontinuities in derivatives have to propagate along characteristics, so it is that discontinuities in the solutions themselves must propagate along special loci in spacetime. These curves are called *shock paths*, and they are not, in general, characteristic curves. What dictates these shock paths is the basic conservation law itself. As we have observed, PDEs arise from conservation laws in integral form, and the integral form of these laws hold true even though the functions may not meet the smoothness requirements of a PDE. The integral form of these conservation or balance laws will imply a condition (called a *jump condition*) that will allow a consistent shock path to be fit into the solution that carries the discontinuity. Indeed, conservation must hold even across a discontinuity. So the answer to the questions posed in Example 2 is that there is a special value of m (in fact, $m = \frac{1}{2}$) for which conservation holds along the discontinuity.

Jump Conditions

To obtain a restriction on a solution across a discontinuity we consider the integral conservation law

$$\frac{d}{dt}\int_a^b u(x,t)\,dx = \phi(a,t) - \phi(b,t) \tag{3}$$

where u is the density and ϕ is the flux. We recall from Chapter 1 (Section 1.2) that (3) states that the time rate of change of the total amount of u inside the interval $[a, b]$ must equal the rate that u flows into $[a, b]$ minus the rate that u flows out of $[a, b]$. Under suitable smoothness assumptions (e.g., both u and ϕ continuously differentiable), (3) implies that

$$u_t + \phi_x = 0 \tag{4}$$

which is the differential form of the conservation law. But if u and ϕ have simple jump discontinuities, we still insist on the validity of the integral form (3). We note here that ϕ may depend on x and t through dependence on u [i.e., $\phi = \phi(u)$], and in this case (4) can be written

$$u_t + c(u)u_x = 0, \qquad c(u) = \phi'(u) \tag{5}$$

which is one of the basic PDEs investigated in Chapter 2.

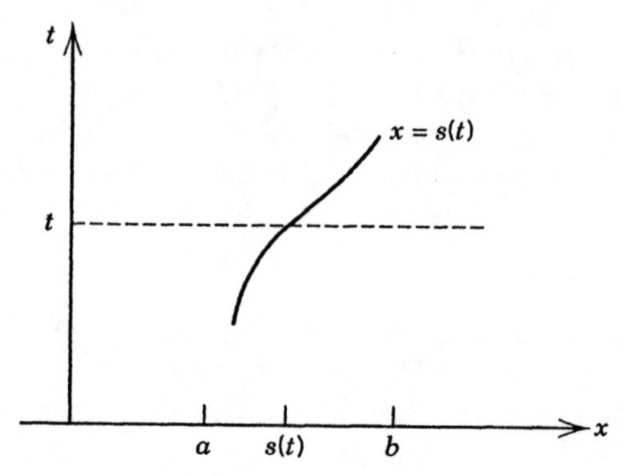

Figure 3.4. Smooth curve in spacetime along which a discontinuity is propagated. Such a curve is called a shock path.

Now assume that $x = s(t)$ is a smooth curve in spacetime along which u suffers a simple discontinuity (see Figure 3.4); that is, assume that u is continuously differentiable for $x > s(t)$ and $x < s(t)$, and that u and its derivatives have finite one-sided limits as $x \to s(t)^-$ and $x \to s(t)^+$. Then, choosing $a < s(t)$ and $b > s(t)$, equation (3) may be written

$$\frac{d}{dt}\int_a^{s(t)} u(x,t)\,dx + \frac{d}{dt}\int_{s(t)}^b u(x,t)\,dx = \phi(a,t) - \phi(b,t) \tag{6}$$

Leibniz' rule for differentiating an integral whose integrand and limits depend on a parameter (here the parameter is time t) can now be applied on the left side of (6), since the integrands are smooth. We therefore obtain

$$\int_a^{s(t)} u_t(x,t)\,dx + \int_{s(t)}^b u_t(x,t)\,dx + u(s^-,t)s' - u(s^+,t)s'$$
$$= \phi(a,t) - \phi(b,t) \tag{7}$$

where $u(s^-,t)$ and $u(s^+,t)$ are the limits of $u(x,t)$ as $x \to s(t)^-$ and $x \to s(t)^+$, respectively, and $s' = ds/dt$ is the speed of the discontinuity $x = s(t)$. In (7) we now take the limit as $a \to s(t)^-$ and $b \to s(t)^+$. The first two terms to go zero because the integrand is bounded and the interval of integration shrinks to zero. Therefore, we obtain

$$-s'[u] + [\phi(u)] = 0 \tag{8}$$

where the brackets denote the jump of the quantity inside across the discontinuity (by our previous notation in Chapter 2, the jump is the value on the left minus the value on the right). Equation (8) is called the *jump condition* (in fluid mechanical problems, conditions across a discontinuity are known as *Rankine–Hugoniot conditions*), and it relates the conditions ahead of the discontinuity and behind the discontinuity to the speed of the discontinuity itself. In this context, the discontinuity in u that propagates along the curve $x = s(t)$ is called a *shock wave*, and the curve $x = s(t)$ is called the *shock path*, or just the *shock*; s' is called the *shock speed*, and the magnitude of the jump in u is called the *shock strength*. We observe that the form of (8) gives the correspondence

$$(\cdot)_t \leftrightarrow -s'[(\cdot)], \qquad (\cdot)_x \leftrightarrow [(\cdot)] \tag{9}$$

between a given PDE (4) and its associated shock condition (8); using (9) makes it easy to remember the jump conditions associated with a given PDE.

Example 3. Consider the PDE

$$u_t + uu_x = 0$$

In conservation form this PDE can be written

$$u_t + \left(\frac{u^2}{2}\right)_x = 0$$

where the flux is $\phi = u^2/2$. According to (8), the jump condition is given by

$$-s'[u] + \left[\frac{u^2}{2}\right] = 0$$

or

$$s' = \frac{u_+ + u_-}{2}$$

So the speed of the shock is the average of the u values ahead and behind the shock. Returning to Example 2, where the initial condition was given by the discontinuous data $u = 1$ for $x < 0$ and $u = 0$ for $x > 0$, we now observe that a shock can be fit with speed

$$s' = \frac{0 + 1}{2} = \frac{1}{2}$$

So a solution, consistent with the jump condition, to the initial value problem (1)–(2) is

$$u(x,t) = 1 \quad \text{if } x < \frac{t}{2}; \qquad u(x,t) = 0 \quad \text{if } x > \frac{t}{2}$$

Rarefaction Waves

We now consider an example that indicates another type of problem with nonlinear equations with discontinuous initial or boundary data.

Example 4. Consider the kinematic wave equation

$$u_t + uu_x = 0, \qquad x \in R, \quad t > 0$$

subject to the initial condition

$$u(x,0) = 0 \quad \text{if } x < 0; \qquad u(x,0) = 1 \quad \text{if } x > 0$$

For this Riemann problem the characteristic diagram looks as shown in Figure 3.5. Since u is constant along characteristics, the data $u = 1$ are carried into the region $x > t$ along characteristics with speed 1, and the data $u = 0$ are carried into the region $x < 0$ along vertical (speed zero) characteristics. There is a region $0 < x < t$ void of the characteristics. In this case there is a continuous solution that can be constructed which connects the solution $u = 1$ ahead to the solution $u = 0$ behind. Simply insert characteristics (straight lines in this case) passing through the origin into the void in such a way that u is constant on the characteristics and u varies continuously from 1 to zero along these inserted characteristics (see Figure 3.6). In other words, along the characteristic $x = ct$, $0 < c < 1$, take $u = c$. Consequently, the solution to the Riemann problem is

$$u(x,t) = 0 \quad \text{if } x < 0;$$

$$u(x,t) = \frac{x}{t} \quad \text{if } 0 < \frac{x}{t} < 1; \qquad u(x,t) = 1 \quad \text{if } x > t$$

A solution of this form is called a *centered expansion wave*, or a *fan*; other

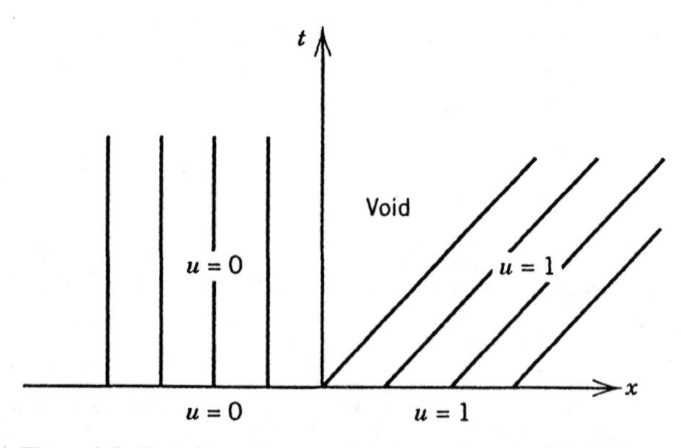

Figure 3.5. Spacetime diagram showing a characteristic void.

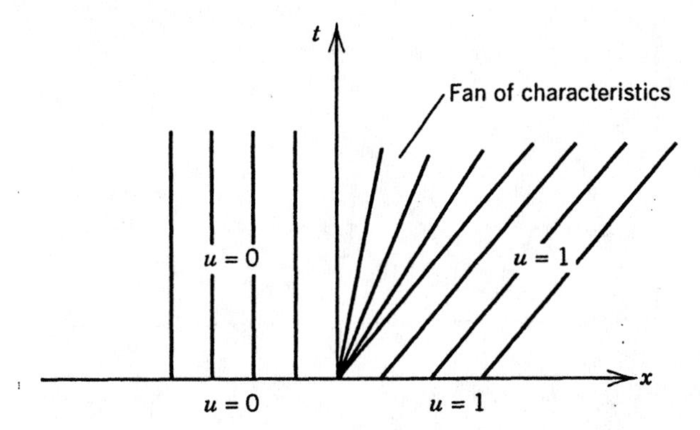

Figure 3.6. Insertion of a characteristic fan in the void in Figure 3.5.

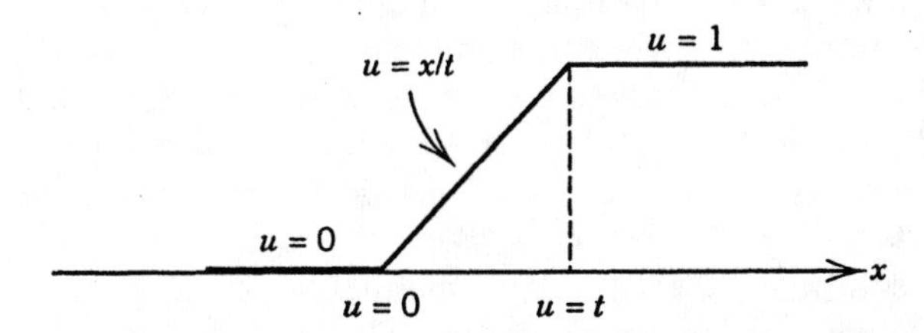

Figure 3.7. Graph of a wave profile at time t corresponding to the characteristic diagram in Figure 3.6.

terms are a *release wave* or a *rarefaction wave*. The idea is that the wave spreads out as time increases; a graph of the solution is shown in Figure 3.7.

Shock Propagation

We end the present discussion with some remarks regarding the formation of shock waves. In air, for example, the speed that finite signals or waves propagate is proportional to the local density of air. (Here we are not referring to small-amplitude signals, as in acoustics, which propagate at a constant sound speed.) Thus, at local points where the density is higher, the signal (say, the value of the density at that point) propagates faster. Therefore, a density wave propagating in air will gradually distort and steepen until it propagates as a discontinuous disturbance, or shock wave. Figure 3.8 depicts various time snapshots of such a density wave. The wave steepens in time because of the tendency of the medium to propagate signals faster at higher density; thus point A moves to the right faster than point B, and finally, a shock forms. This same mechanism causes rarefaction waves to form which release the density and spread out the wave on the back side. This dependence of propagation speed on amplitude is reflected mathematically in the

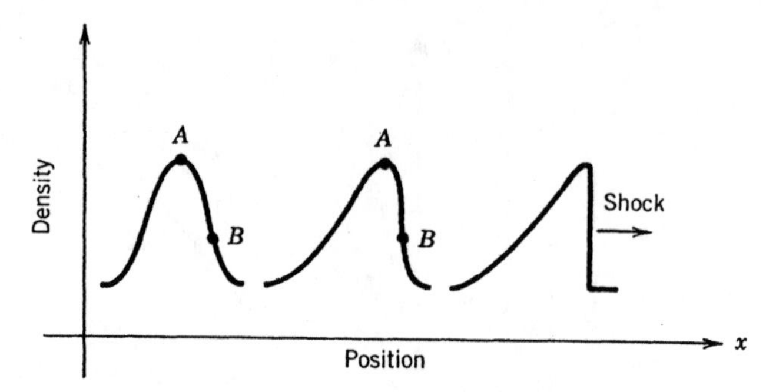

Figure 3.8. Schematic showing how a wave profile steepens into a shock wave. Points where the density is higher are propagated at a faster speed.

conservation law (5) by noting that the characteristic speed $c = c(u)$ depends on u; this phenomenon is typically nonlinear.

Therefore, the discontinuities do not need to be present initially; shocks can form from the distortion of a perfectly smooth solution. The time when the shock forms, usually caused by an infinite spatial derivative or gradient catastrophe, is called the *breaking time*. The determination of conditions under which solutions blow up in this manner is one of the most important problems in nonlinear hyperbolic PDEs. The problem of fitting in a shock (i.e., determining the spacetime location of the shock) after the blowup occurs is a difficult problem, both analytically and numerically. We emphasize again that the distortion of wave profile and the formation of a shock is a distinctly nonlinear phenomenon. In the next section we begin to examine shock formation for certain classes of PDEs and initial data.

We end this section with a more complicated example of shock fitting.

Example 5. Consider the initial value problem

$$u_t + uu_x = 0, \qquad x \in R, \quad t > 0$$

$$u(x,0) = \begin{cases} 1, & x < 0 \\ -1, & 0 < x < 1 \\ 0, & x > 0 \end{cases}$$

Here $c(u) = u$ and the characteristics emanate with speed u from the x-axis as shown in the characteristic diagram in Figure 3.9. The flux is given by $\phi(u) = u^2/2$, and the jump condition is

$$s' = \frac{u_1 + u_2}{2}$$

where u_1 is the value of u ahead of the shock and u_2 is the value of u behind the shock. Clearly, a shock must form at $t = 0$ with speed

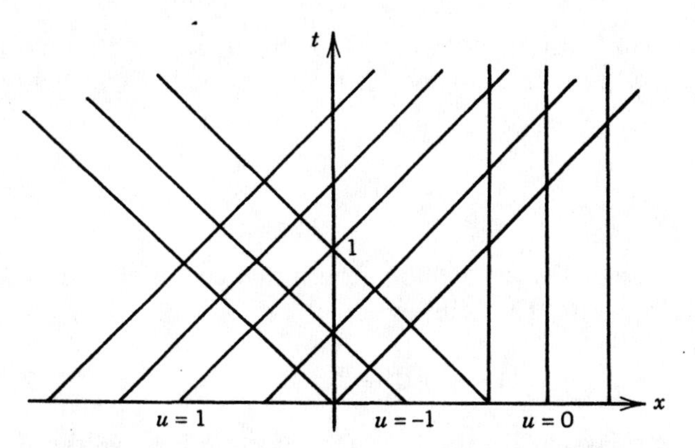

Figure 3.9. Characteristic diagram with intersecting characteristics. Figures 3.10 through 3.12 show how a shock is fit into the diagram.

$s' = (-1 + 1)/2 = 0$ and propagate until time $t = 1$, as shown in Figure 3.10. To continue the shock beyond $t = 1$ we must know the solution ahead of the shock. Therefore, we introduce an expansion wave in the void in Figure 3.10. That is, we take

$$u = \frac{x - 1}{t}$$

in this region; note that the straight-line characteristics issuing from $x = 1$, $t = 0$, have equation $x = -kt + 1$ or $(x - 1)/t = k =$ constant. This expansion fan takes u from the value -1 behind the wave to the value 0 ahead of the wave. The shock beyond $t = 1$ will, according to the jump condition, have speed

$$s' = \frac{(x - 1)/t + 1}{2} \tag{10}$$

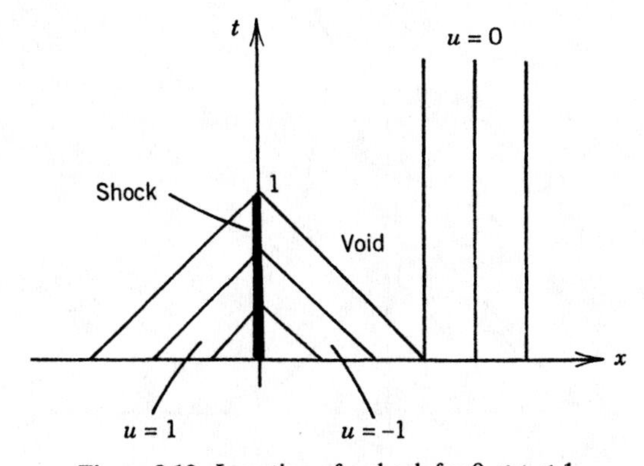

Figure 3.10. Insertion of a shock for $0 < t < 1$.

To find the shock path we note that $s' = dx/dt$, so equation (10) is a first order ordinary differential equation for the shock path. This ODE can be put in the form

$$\frac{dx}{dt} - \frac{x}{2t} = \frac{1 - 1/t}{2}$$

which is a first order linear equation. The initial condition is $x = 0$ at $t = 1$, where the shock starts. The solution is

$$x = s(t) = t + 1 - 2\sqrt{t}, \qquad 1 \leq t \leq 4 \tag{11}$$

and a graph is shown in Figure 3.11. The shock in (11) will propagate until $x = 1$ at $t = 4$. At this instant the shock runs into the vertical characteristics emanating from $x > 1$ and the jump condition (10) is no longer valid. The new jump condition for $t > 4$ is

$$s' = \frac{0 + 1}{2} = \frac{1}{2}$$

Therefore, the shock is a straight line with speed $\frac{1}{2}$ for $t > 4$, and its equation is

$$x - 1 = \frac{t - 4}{2}, \qquad t > 4$$

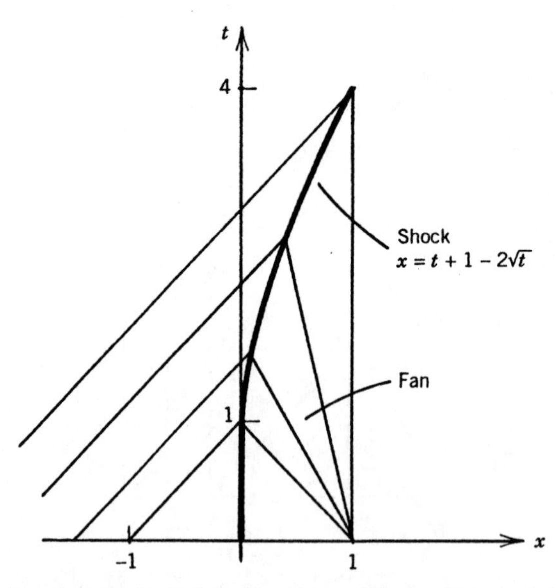

Figure 3.11. Insertion of a fan in the void region and the resulting shock for $1 < t < 4$.

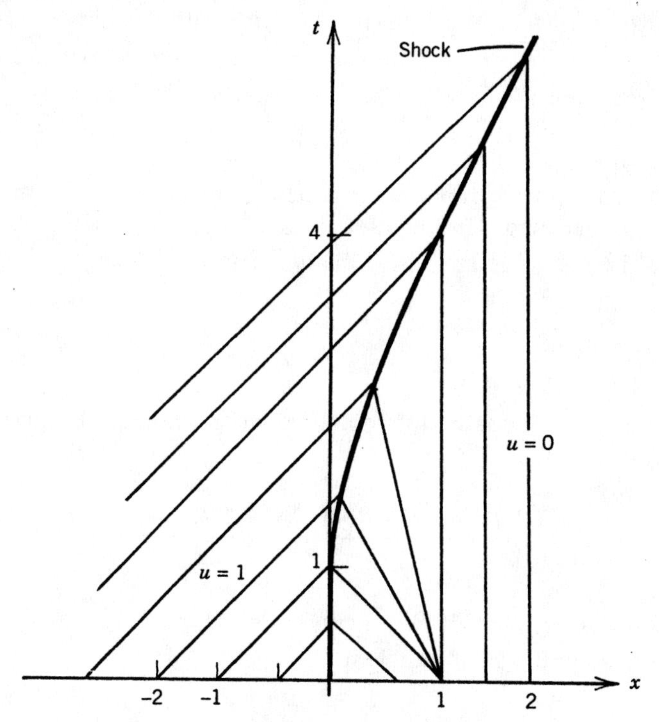

Figure 3.12. Continuation of the shock for $t > 4$.

The entire solution is indicated on the wave diagram in Figure 3.12. We may think of shock propagation in the following way. The speed of the shock in this problem is the average value of the solution ahead and the solution behind; the characteristics ahead and behind carry information to the shock, dictating its path. A similar interpretation may be advanced, in general, for more complicated problems such as aerodynamical flows. The location of the shock path is determined by the flow ahead and the flow behind. Information about the flow is carried by the characteristics to the shock front, fixing its position in spacetime according to the jump conditions, or conservation laws, that hold at the shock front.

EXERCISES

1. Consider the Riemann problem

$$u_t + c(u)u_x = 0, \qquad x \in R, \quad t > 0$$

where $c(u) = \phi'(u) > 0$, and $c'(u) > 0$, with initial condition

(a) $u(x,0) = u_0$, $x > 0$; $u(x,0) = u_1$, $x < 0$; u_0 and u_1 positive constants with $u_1 > u_0$ or

(b) $u(x,0) = u_0$, $x > 0$; $u(x,0) = u_1$, $x < 0$; u_0 and u_1 positive constants with $u_1 < u_0$.

In each case draw a representative characteristic diagram showing the shock path, and find a formula for the solution.

2. How do the results of Exercise 1 change if $c'(u) < 0$?
3. Consider the equation

$$u_t + c(u)u_x = 0$$

with c twice continuously differentiable and $c(u) = \phi'(u)$, where ϕ is the flux. Prove that if

$$\frac{\phi(u_2) - \phi(u_1)}{u_2 - u_1} = \frac{c(u_1) + c(u_2)}{2}$$

for all u_1 and u_2, then $\phi(u)$ is a quadratic function of u.

4. Consider the initial value problem

$$u_t + uu_x = 0, \qquad x \in R, \quad t > 0$$

$$u(x,0) = \begin{cases} 1, & x < 0 \\ 1 - x, & 0 < x < 1 \\ 0, & x > 1 \end{cases}$$

Find a continuous solution for $t < 1$. For $t > 1$ fit in a shock and find the form of the solution. Sketch time snapshots of the wave for $t = 0$, $t = \frac{1}{2}$, $t = 1$, and $t = \frac{3}{2}$.

3.2 SHOCK FORMATION

In Section 3.1 we mainly dealt with the Riemann problem, where the initial data were discontinuous. Now we want to consider the problem of shock formation from smooth data: in particular, the question of when the blowup occurs. We focus on the initial value problem

$$u_t + c(u)u_x = 0, \qquad x \in R, \quad t > 0 \tag{1}$$

$$u(x,0) = u_0(x), \qquad x \in R \tag{2}$$

where $c(u) > 0$, $c'(u) > 0$, and u_0 is class C^1 on R. We have already shown in Chapter 2 that if u_0 is a nondecreasing function on R, a smooth solution

$u = u(x, t)$ exists for all $t > 0$ and is given implicitly by the formulas

$$u(x,t) = u_0(\xi) \qquad \text{where} \quad x - \xi = c(u_0(\xi))t \tag{3}$$

Therefore, for singularities to occur in (1)–(2), necessarily u_0 must be strictly decreasing in some open interval.

Let us consider the case when $u_0(x) > 0$ and $u_0'(x) < 0$ on R. The characteristic equations are

$$u = \text{constant} \quad \text{on} \quad \frac{dx}{dt} = c(u)$$

Therefore, the characteristics, which are straight lines, issuing from two points ξ_1 and ξ_2 on the x-axis with $\xi_1 < \xi_2$ have speeds $c(u_0(\xi_1))$ and $c(u_0(\xi_2))$, respectively. Because u_0 is decreasing and c is increasing, it follows that

$$c(u_0(\xi_1)) > c(u_0(\xi_2))$$

In other words, the characteristic emanating from ξ_1 is faster than the one emanating from ξ_2 (see Figure 3.13). Therefore, the characteristics cross and a contradiction results since u is a different constant on each characteristic. So a smooth solution cannot exist for all $t > 0$.

To determine the breaking time when a gradient catastrophe occurs, we calculate u_x along a characteristic, which has equation

$$x - \xi = c(u_0(\xi))t \tag{4}$$

Let $g(t) = u_x(x(t), t)$ denote the gradient of u along the characteristic

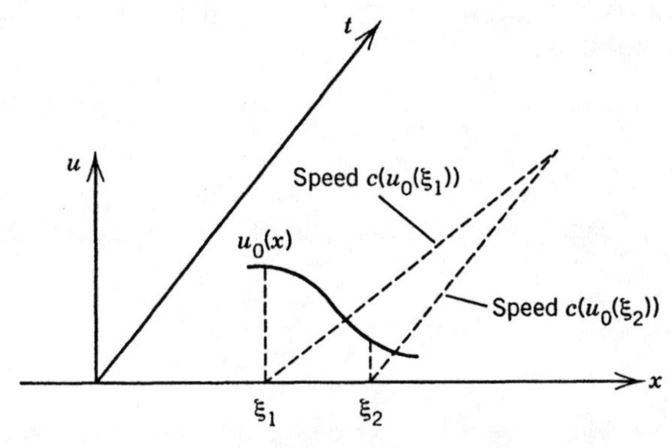

Figure 3.13

$x = x(t)$ given by (4). Then

$$\frac{dg}{dt} = u_{tx} + c(u)u_{xx}$$

But differentiating the PDE (1) with respect to x gives

$$u_{tx} + c(u)u_{xx} + c'(u)u_x^2 = 0$$

Comparing the last two equations gives

$$\frac{dg}{dt} = -c(u)g^2 \tag{5}$$

along the characteristic. The ordinary differential equation (5) can be solved immediately to obtain

$$g = \frac{g(0)}{1 + g(0)c'(u_0(\xi))t} \tag{6}$$

where $g(0)$ is the initial gradient at $t = 0$. Translating (6) into other notation gives

$$u_x = \frac{u_0'(\xi)}{1 + u_0'(\xi)c'(u_0(\xi))t} \tag{7}$$

which is a formula for the gradient u_x along the characteristic (4). Because of our assumptions (u_0 nonincreasing and c increasing), it follows that u_0' and c' in the denominator of (7) have opposite signs, and therefore u_x will blow up at some finite time along the characteristic (4). Consequently, if we examine u_x along *all* the characteristics, the breaking time for the wave will be on the characteristic, parameterized by ξ, where the denominator in (7) first vanishes. Denoting

$$F(\xi) = c(u_0(\xi))$$

we conclude that the wave first breaks along the characteristic $\xi = \xi_b$ for which $|F'(\xi)|$ is a maximum. Then the time of the first breaking is

$$t_b = -\frac{1}{F'(\xi_b)}$$

The positive time t_b is called the *breaking time* of the wave. Figure 3.14 depicts several time snapshots of a wave that breaks at time t_b.

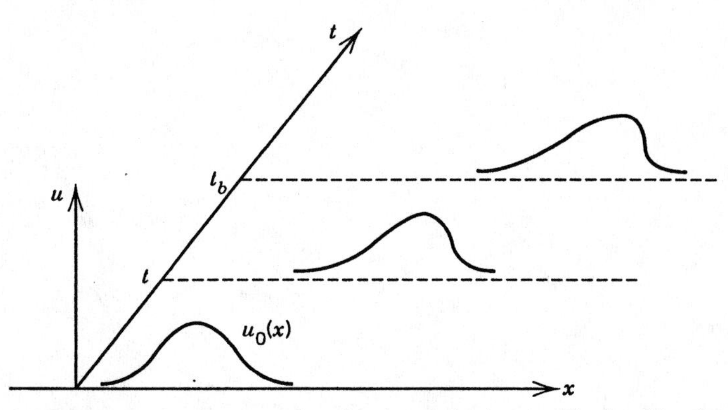

Figure 3.14. Diagram showing how a wave evolves into a shock wave. The breaking time t_b is the time when the gradient catastrophe occurs.

We remark that if the initial function u_0 is not monotone, breaking will first occur on the characteristic $\xi = \xi_b$, for which $F'(\xi) < 0$ and $|F'(\xi)|$ is a maximum.

Example 1. Consider the initial value problem

$$u_t + uu_x = 0, \qquad x \in R, \quad t > 0$$

$$u(x,0) = \exp(-x^2), \qquad x \in R$$

The initial signal is a bell-shaped curve, and characteristics leave the x-axis with speed $c(u) = u$. A characteristic diagram is shown in Figure 3.15. It is clear that the characteristics collide and a shock will form on the front portion of the wave along the characteristics emanating from the positive x-axis. To determine the breaking time of the wave, we compute

$$F(\xi) = c(u_0(\xi)) = \exp(-\xi^2)$$

Then

$$F'(\xi) = -2\xi\exp(-\xi^2) \quad \text{and} \quad F''(\xi) = -(4\xi^2 - 2)\exp(-\xi^2)$$

Thus $F''(\xi) = 0$ when $\xi = \xi_b = \sqrt{1/2}$. This is where F' is a maximum. Therefore, the breaking time is

$$t_b = -\frac{1}{F'(\xi_b)} \cong 1.16$$

So breaking first occurs along the characteristic $\xi = \xi_b$ at time t_b. The

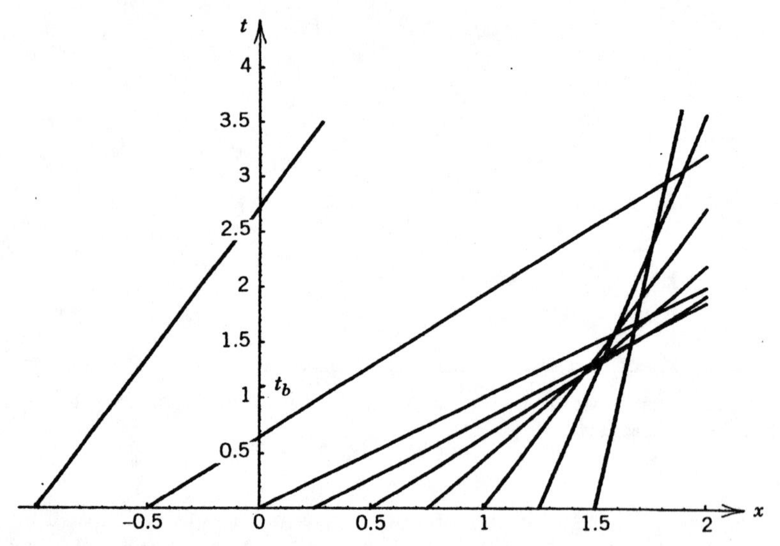

Figure 3.15. MATHEMATICA generated characteristic diagram. The breaking time is approximately $t_b = 1.16$ and breaking first occurs on the characteristic $\xi_b = \sqrt{1/2}$.

solution of the initial value problem up to time t_b is given implicitly by

$$u(x,t) = \exp(-\xi^2)$$

where $\xi = \xi(x, t)$ is the solution of $x - \xi = \exp(-\xi^2)t$.

There is an alternative way to look at the problem of determining the breaking time of a wave for the initial value problem (1)–(2). The solution is given implicitly by the formulas (3). To determine the solution explicitly we must be able to solve the equation

$$x - \xi = c(u_0(\xi))t \tag{8}$$

for ξ and substitute the result into the equation $u = u_0(\xi)$ to find $u = u(x,t)$ (i.e., u as a function of x and t). So we may ask when it is possible to solve (8) for ξ. This is a standard problem involving the implicit function theorem from advanced calculus. If we write (8) as

$$\mathscr{F}(x,t,\xi) = x - \xi - c(u_0(\xi))t = 0 \tag{9}$$

then the condition that allows us to solve (9) for ξ (locally) is that the partial derivative of $\mathscr{F}$ with respect o ξ must be nonzero. Thus we must have

$$-1 - c'(u_0(\xi))u_0'(\xi)t \neq 0 \tag{10}$$

This is the same condition that the denominator of (7) should be nonzero (i.e., u_x is not infinite). Consequently, if condition (10) holds, for each (x,t) we may uniquely determine the characteristic, designated by ξ, that leads backward in time from (x,t) to the point $(\xi,0)$ on the initial time line (x-axis).

EXERCISES

1. Consider the initial value problem

$$u_t + uu_x = 0, \qquad x \in R, \quad t > 0$$

$$u(x,0) = \frac{1}{1+x^2}, \qquad x \in R$$

Find the breaking time t_b of the wave and write the solution for $0 < t < t_b$ in implicit form. Sketch the characteristic diagram.

2. Discuss the existence of a global solution to the initial value problem

$$u_t = -\frac{t}{u}, \qquad x \in R, \quad t > 0$$

$$u(x,0) = u_0(x), \qquad x \in R$$

Sketch a characteristic diagram.

3. Consider the conservation law

$$u_t + \phi(u)_x = 0, \qquad x \in R, \quad t > 0$$

$$u(x,0) = u_0(x) > 0, \qquad x \in R$$

where $\phi', \phi'' > 0$, and where $u_0' > 0$. Prove that there exists a constant E for which $u_x < E/t$ for all $t > 0$.

4. Consider the initial boundary value problem

$$u_t + uu_x = 0, \qquad x > 0, \quad t > 0$$

$$u(x,0) = 1, \quad x > 0; \qquad u(0,t) = t + 1, \quad t > 0$$

Sketch a characteristic diagram and find the breaking time t_b. Determine the solution up to $t = t_b$. After the shock forms, calculate its path in spacetime. *Solution:* $s(t) = (t+3)(3t+1)/16$.

5. Consider the initial value problem

$$u_t + uu_x = -ku^2, \qquad x \in R, \quad t > 0 \quad (k > 0)$$
$$u(x, 0) = u_0(x) > 0, \qquad x \in R$$

(a) Find u implicitly.
(b) Determine conditions on k and u_0 such that a shock will not form.
(c) If a shock forms, what is t_b?

6. Consider the PDE

$$u_t + uu_x = B(x - vt)$$

where v is a positive constant and the function B satisfies the conditions $B(y) = 0$ if $|y| \geq y_0$ and $B(y) > 0$ otherwise. (This PDE is the basic conservation law with a moving source term.)

(a) Prove that a traveling wave solution of the form $u = u(x - vt)$ satisfying $u(+\infty) = u_0 < v$ exists if, and only if,

$$(u_0 - v)^2 > 2\int_R B(y)\, dy$$

(b) Consider the initial condition $u(x, 0) = u_0$, $x \in R$. By transforming to a moving coordinate system (moving with speed v), show that

$$\frac{du}{d\tau} = B(z) \quad \text{on} \quad \frac{dz}{d\tau} = u - v$$

with $u = u_0$ and $z = x$ at $\tau = 0$.

(c) In part (b) take $B(y) = B =$ constant for $|y| \leq a$ and $B(y) = 0$ otherwise. Draw a characteristic diagram in $z\tau$-space in both the case $v < u_0$ and the case $v > u_0$. When do shocks form? Discuss the case $v = u_0$.

7. Consider the initial value problem

$$u_t + uu_x = 0, \qquad x \in R, \quad t > 0$$
$$u(x, 0) = 2 - x^2 \quad \text{for } x < 1; \qquad u(x, 0) = 1 \quad \text{for } x \geq 1$$

Sketch the characteristic diagram and find the breaking time t_b. Write the solution for $t < t_b$.

8. Consider the initial value problem

$$u_t + uu_x = 0, \qquad x \in R, \quad t > 0$$
$$u(x, 0) = 1 \quad \text{for } x > 1; \qquad u(x, 0) = |x| - 1 \quad \text{for } x < 1$$

Find a solution containing a shock and determine the shock path.

9. Consider the initial value problem

$$u_t + uu_x = 0, \qquad x \in R, \quad t > 0$$

$$u(x,0) = 2 \quad \text{if } x < 0; \qquad u(x,0) = 1 \quad \text{if } 0 < x < 2;$$

$$u(x,0) = 0 \quad \text{if } x > 2$$

Find a solution containing a shock and determine the shock path. In this case two shocks merge into a single shock.

3.3 TRAFFIC FLOW; PLUG FLOW REACTORS

There are many applications of first order PDEs in problems of science and engineering. Rather than present some exhaustive list, we focus on two such applications in this section, one involving traffic flow and one involving a plug flow chemical reactor. In a later chapter, where we consider *systems* of first order equations, there is more fertile ground for applications.

Traffic Flow

What was once perhaps a novel application of first order PDEs—namely, the application to the flow of traffic—has become almost commonplace in treatments of first order PDEs. Nevertheless, the application is interesting and it provides the novice with a simple, concrete example where intuition and mathematics can come together in a familiar setting. Therefore, we give a brief discussion of this topic.

We imagine a single-lane roadway occupied by cars whose density $u = u(x,t)$ is given in vehicles per unit length, and the flux of vehicles $\phi(x,t)$ is given in vehicles per unit time. The length and time units are often taken to be miles and hours. Note that we are making a continuum hypothesis here by regarding u and ϕ as continuous functions of the distance x. (A more satisfying treatment might be to regard the vehicles as discrete entities and attempt to write down finite difference equations to model the flow; this approach has been developed, but we shall not follow it here.) The basic conservation law states that the time rate of change of the total number of vehicles in any interval $[a,b]$ must equal the flow rate in at $x = a$ minus the flow rate out at $x = b$, or, in integral form,

$$\frac{d}{dt}\int_a^b u(x,t)\,dx = \phi(a,t) - \phi(b,t) \tag{1}$$

If u and ϕ are smooth functions, then in the standard way (see Section 1.2) we can obtain a partial differential equation relating u and ϕ that models the flow of the cars. This equation is the conservation law

$$u_t + \phi_x = 0 \tag{2}$$

To make further progress we need to assume a constitutive relation for the flux ϕ. The flux represents the rate that vehicles go by a given point; it seems clear that the flux should depend on the traffic density u; that is, $\phi = \phi(u)$. If $u = 0$, the flux should be zero; if u is large, the flux should also be zero since the traffic is jammed. Therefore, we assume that there is a positive value of u, say u_j, such that $\phi(u_j) = 0$. Otherwise, we assume that ϕ is positive and concave down on the interval $(0, u_j)$, and ϕ has a unique maximum at $u_{\max}$. Thus $u = u_{\max}$ is the value of the density where the greatest number of vehicles get through. In summary, our assumptions on the flux are

$$\begin{gathered}\phi(u) > 0 \quad \text{and} \quad \phi''(u) < 0 \text{ on } (0, u_j); \\ \phi(0) = \phi(u_j) = 0; \qquad \phi'(u) > 0 \text{ on } [0, u_{\max}); \\ \phi'(u) < 0 \text{ on } (u_{\max}, u_j]; \qquad \phi'(u_{\max}) = 0\end{gathered} \tag{3}$$

A generic graph is shown in Figure 3.16.

As before, we may write the PDE (2) as

$$u_t + c(u)u_x = 0, \qquad \text{where} \quad c(u) = \phi'(u), \quad c'(u) < 0 \tag{4}$$

Equation (4) has characteristic form

$$u = \text{constant} \quad \text{on} \quad \frac{du}{dt} = c(u) \tag{5}$$

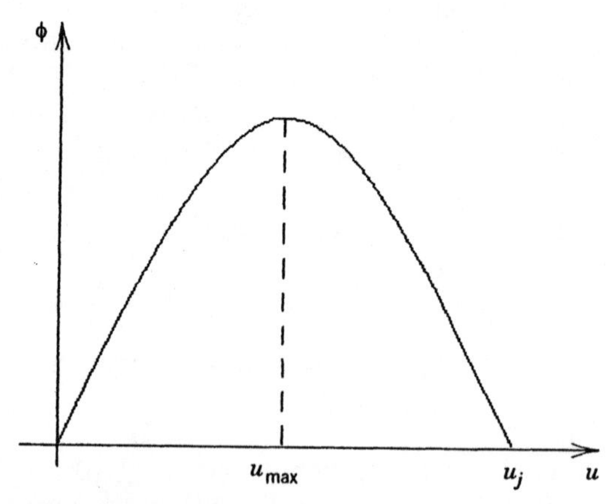

Figure 3.16. Experimentally measured flux ϕ as a function of the vehicle density u.

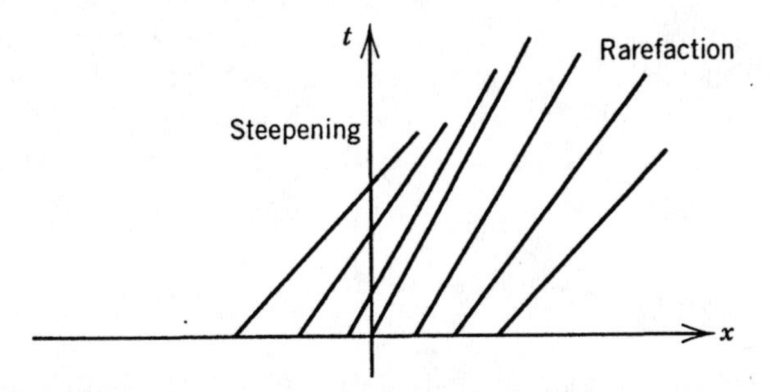

Figure 3.17. Characteristic diagram for an initial density in the form of a bell-shaped curve.

Therefore, signals which travel at characteristic speed $c(u)$ propagate forward into the traffic if $u < u_{max}$, and signals propagate backward if $u > u_{max}$. Because $c'(u) < 0$, waves will show a shocking-up effect on their back side. For example, if there is a density wave in the form of a bell-shaped curve, with the density everywhere less than u_{max}, the wave will propagate as shown in Figures 3.17 and 3.18, which show the characteristic diagram and time snapshots of the wave, respectively.

To observe how a driver might experience and respond to such a density wave, we introduce the *velocity* $V = V(u)$ of the traffic flow defined by the equation

$$\phi(u) = uV(u) \tag{6}$$

Thus like ϕ and u, V is a point function of (x, t), giving a continuum of values at each location x in the flow. With this interpretation, V is the actual

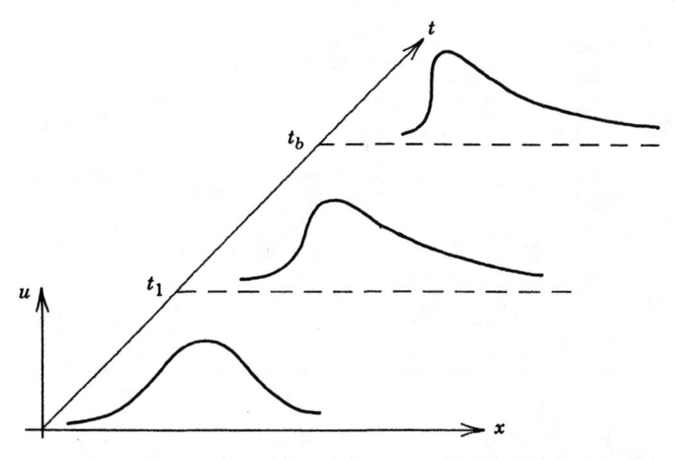

Figure 3.18. Diagram showing how the density wave profile "shocks up" on the back side when $c'(u) < 0$.

velocity of the vehicle at (x, t). Equation (6) just states that density times velocity equals flux, which is a correct statement. We now place one additional restriction on ϕ that will guarantee that V is a decreasing function of u; in particular, we assume that $u\phi'(u) < \phi(u)$ (see Exercise 1). Using (6) we find that

$$c(u) = \phi'(u) = uV'(u) + V(u)$$

Then, since $V' < 0$, it follows that $c(u) < V(u)$. Consequently, the traffic moves at a speed greater than the speed that signals are propagated in the flow. For example, a driver approaching a density wave like the one shown in Figure 3.18 will decelerate rapidly through the steep rear of the wave and then accelerate slowly through the rarefaction on the front side.

Some typical experimental numbers for values of the various constants are quoted in Whitham [1974]. For example, on a highway, $u_j = 225$ vehicles/mile, $u_{max} = 80$ vehicles/mile, and the maximum flux is 1500 vehicles/hour. Interestingly enough, the velocity that gives this maximum flux is a relatively slow 20 miles/hour. In another case study, in the Lincoln Tunnel, linking New York and New Jersey, a flux given by $\phi(u) = au \ln(u_j/u)$, with $a = 17.2$ and $u_j = 228$ was found to fit the experimental data.

A close examination of another common traffic phenomenon will aid in an understanding of the role of characteristics in nonlinear hyperbolic problems. Consider the problem of a stream of traffic of constant density $u_0 < u_{max}$ suddenly encountering a traffic light that turns red (at time $t = 0$). Figure 3.19 shows the resulting shock wave that is propagated back into the flow. Ahead of the shock the cars are jammed at density $u = u_j$ and velocity zero; behind the shock the cars have density u_0 and corresponding velocity $V(u_0) > c(u_0) > 0$. Figure 3.20 illustrates the characteristic diagram and a typical vehicle trajectory. The speed of the shock is given by the jump

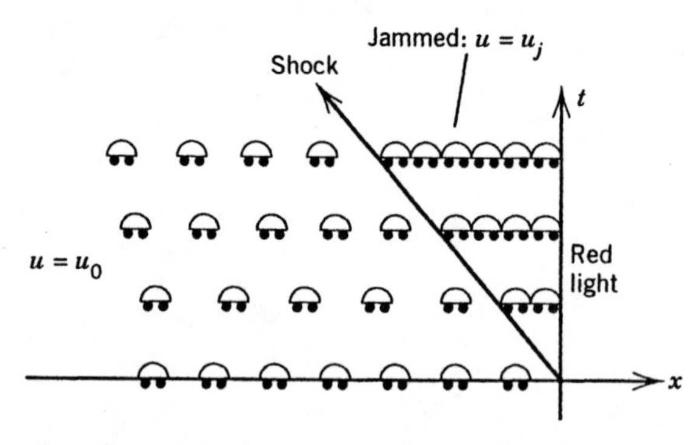

Figure 3.19. Spacetime diagram showing the effect of a red light stopping a uniform stream of traffic. A shock is propagated back into the traffic. This is a shock we experience regularly when we drive a vehicle.

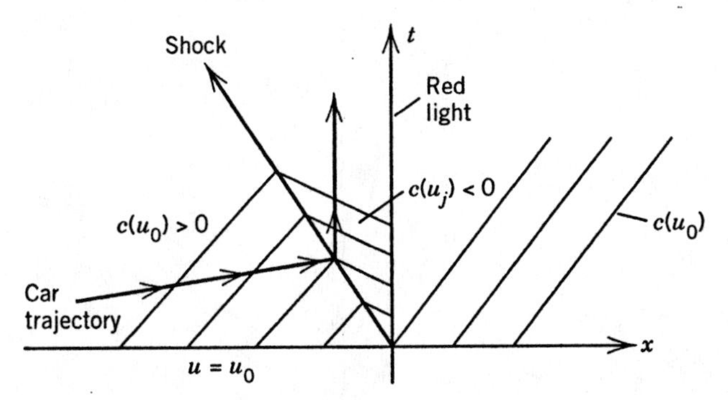

Figure 3.20. Characteristic diagram associated with a red light stopping a uniform stream of traffic.

condition

$$s' = \frac{\phi_{\text{ahead}} - \phi_{\text{behind}}}{u_{\text{ahead}} - u_{\text{behind}}} = \frac{0 - \phi(u_0)}{u_j - u_0}$$

$$= -\frac{\phi(u_0)}{u_j - u_0} < 0$$

If the traffic light changes to green at some point $t = t_0$, a centered rarefaction wave is created that releases the density as the stopped vehicles move forward. The lower value of the density created by the rarefaction ahead of the shock wave will force the shock to change direction and accelerate through the intersection. A characteristic diagram of this situation is shown in Figure 3.21. Notice that the characteristics are straight lines, and the fan must take characteristics of speed $c(u_j)$ behind the rarefaction to characteristics of speed $c(0)$ ahead of the rarefaction. The density u must change from the jammed value u_j behind to the fan to the value zero ahead (since there are no cars ahead). The characteristic speed $c(u)$ in the rarefaction is therefore given by the equation

$$c(u) = \frac{x}{t - t_0}$$

[Note that u is constant on straight lines, so $c(u)$ is constant on straight lines; the straight lines in the fan emanating from $(0, t_0)$ are given by $x/(t - t_0) =$ const.] The curved shock path can be computed by the jump condition

$$s' = \frac{dx}{dt} = \frac{\phi(u) - \phi(u_0)}{u - u_0}$$

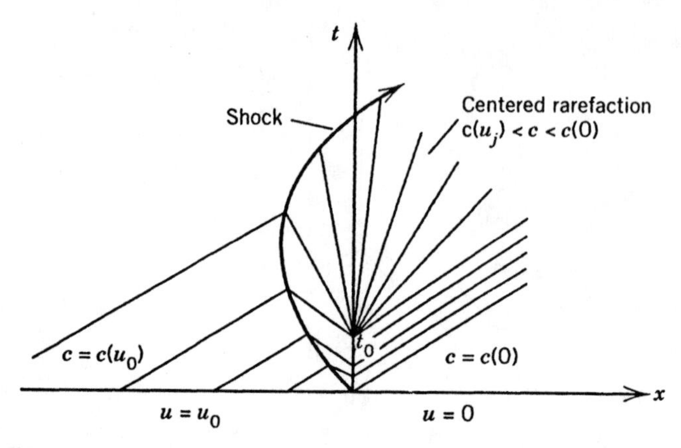

Figure 3.21. When the light turns green a rarefaction wave releases the jammed traffic and the shock turns to go through the intersection itself. At a busy intersection, the shock does not get through the intersection before the light turns red again.

The exercises at the end of the section will give the reader further practice in traffic flow problems.

Plug Flow Chemical Reactors

There are many types of chemical reactors used in industry to produce products from given reactants. One of these is called a *plug flow reactor* (also called a piston flow, slug flow, or ideal tubular reactor). The model of such a reactor is a long tube where a reactant enters at a fixed rate and a chemical reaction takes place as the material moves through the tube, producing the final product as it leaves the other end. In such a reactor the flow is assumed to be one-dimensional; that is, all state variables are assumed to be constant in any cross section; the variation takes place only in the axial direction. A model of a plug flow reactor of cross-sectional area A and length L is shown in Figure 3.22. A reactant $\mathscr{A}$ is fed in the left end at $x = 0$ at the constant volumetric rate Q (volume/second). Upon entry into the reactor, $\mathscr{A}$ reacts chemically and produces a product $\mathscr{B}$; that is, $\mathscr{A} \to \mathscr{B}$. The rate of

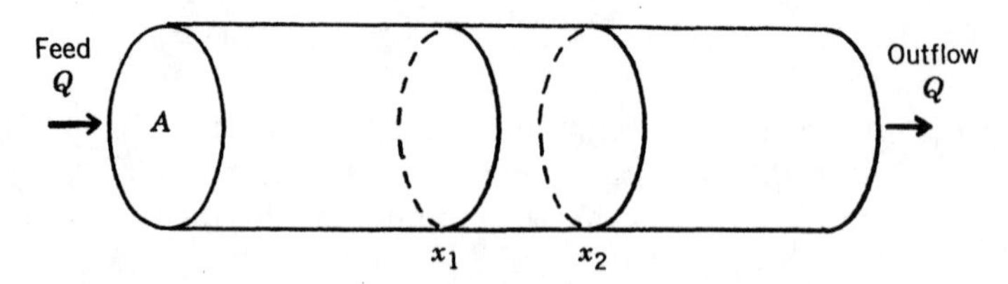

Figure 3.22. Plug flow reactor with material entering and leaving at the same volumetric flow rate Q.

production, or the chemical reaction rate, is given by r, measured in mass/volume/time. The material leaves the reactor at $x = L$ at the same volumetric rate Q. We let $a = a(x,t)$ denote the concentration of the chemical species $\mathscr{A}$, given in mass per unit volume, and we note that the flux is then given by Qa. A model equation that governs the evolution of the concentration $a(x,t)$ can be written down from a basic conservation law, which states that the time rate of change of the total amount of $\mathscr{A}$ in the section $[x_1, x_2]$ must equal the net flux of $\mathscr{A}$ through x_1 and x_2, plus the rate that $\mathscr{A}$ is consumed by the chemical reaction in $[x_1, x_2]$. In symbols,

$$\frac{d}{dt}\int_{x_1}^{x_2} Aa(x,t)\,dx = Qa(x_1,t) - Qa(x_2,t) + \int_{x_1}^{x_2} Ar\,dx \tag{7}$$

Assuming that a is sufficiently differentiable and using the arbitrariness of the interval $[x_1, x_2]$, we can reduce (7) to a PDE

$$a_t + va_x = r, \qquad v = \frac{Q}{A} \tag{8}$$

Equation (8) is the advection equation with a source term. We now make some assumptions regarding the reaction rate r. Generally, the reaction rate depends on the concentration a as well as the temperature at which the reaction occurs. For first order kinetics, the usual assumption is that $r = -ka$, where k is a temperature-dependent rate factor (usually called the rate constant, which is a slight misnomer). Rather than introduce another variable (temperature), however, we shall assume that k is indeed a constant, so we consider the problem

$$a_t + va_x = -ka \tag{9}$$

To equation (9) we append the initial condition

$$a(x,0) = a_0, \qquad 0 < x < L \tag{10}$$

and the boundary condition

$$a(0,t) = f(t), \qquad t > 0 \tag{11}$$

Hence the initial concentration in the reactor is a_0, a constant, and the concentration of the feed at $x = 0$ is $f(t)$; therefore, the flux at $x = 0$ is $Qf(t)$. We can easily solve this problem. The characteristic form of (9) is

$$\frac{da}{dt} = -ka \quad \text{on} \quad x = vt + c_1 \tag{12}$$

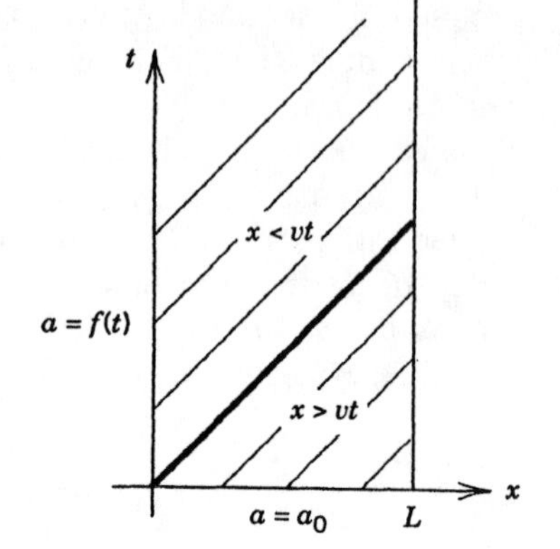

Figure 3.23. Characteristic diagram for the initial–boundary value problem (9)–(11).

where c_1 is a constant. Whence

$$a = c_2 e^{-kt} \quad \text{on} \quad x = vt + c_1 \tag{13}$$

where c_2 is another constant. We note (see Figure 3.23) that the initial and boundary data are carried along the straight-line characteristics at speed v to the boundary at $x = L$, which is the right end of the reactor. Therefore, we divide the problem into two regions, $x > vt$ and $x < vt$. For $x > vt$ we may parameterize the initial data by $a = a_0$, $x = \xi$, at $t = 0$. Then, from (13), the constants are given by

$$c_1 = a_0, \qquad c_2 = \xi \tag{14}$$

Thus

$$a(x,t) = a_0 e^{-kt}, \qquad x > vt \tag{15}$$

For $x < vt$ we parameterize the boundary data (11) by $a = f(\tau)$, $t = \tau$, at $x = 0$. Then the constants c_1 and c_2 in (13) are given by

$$c_1 = -v\tau \qquad c_2 = f(\tau)e^{k\tau}$$

Then

$$a = f(\tau)e^{-k(t-\tau)} \quad \text{on} \quad x = v(t-\tau)$$

or

$$a(x,t) = f\left(t - \frac{x}{v}\right)e^{-kx/v}, \qquad x < vt \tag{16}$$

Equations (15) and (16) represent the solution to the problem (9)–(11).

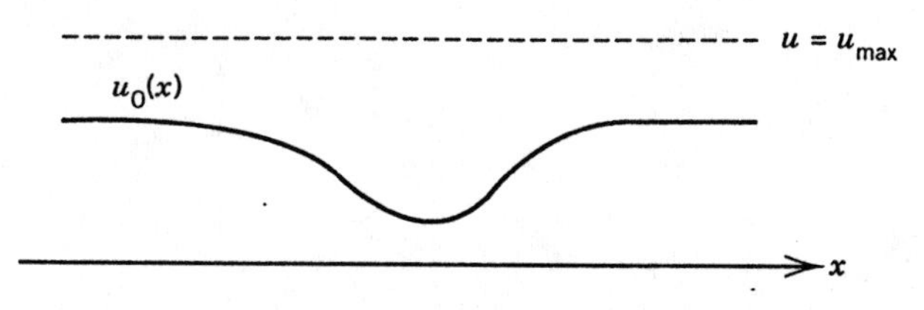

Figure 3.24. Exercise 2.

Physically, the concentration decays from its initial or boundary value as it moves through the reactor at constant speed $v = Q/A$.

In Chapter 5 we consider more complicated reactors that are non-isothermal. In that case an energy balance equation will be required to go with the species equation derived above, and a coupled system for the species concentration and the temperature will be obtained.

EXERCISES

1. In traffic flow the flux is given by $\phi(u) = uV(u)$, where V is the velocity, and conditions (3) hold for the flux. Determine a growth condition on ϕ that is implied by the assumption that $V' < 0$.
2. In the traffic flow problem assume that the flux is given by $\phi(u) = au(u_j - u)$, where a and u_j are positive constants. Find the vehicle velocity $V(u)$ and the characteristic speed $c(u)$. Describe how a density wave like the one shown in Figure 3.24 evolves in time. On a characteristic diagram indicate the path of a vehicle approaching such a wave from behind. How does the situation change if the density wave is like the one in Figure 3.25?
3. It has been determined experimentally that traffic flow through the Lincoln Tunnel is approximately described by the conservation law $u_t + \phi_x = 0$

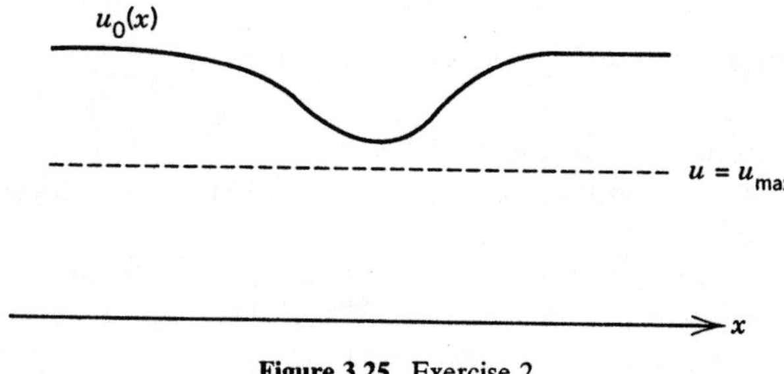

Figure 3.25. Exercise 2.

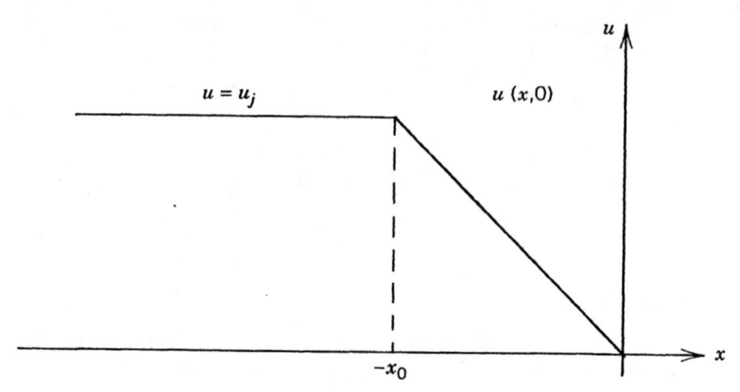

Figure 3.26. Exercise 3.

with flux

$$\phi(u) = au \ln \frac{u_j}{u}$$

where a and u_j are positive known constants. Suppose that the initial density u of the traffic varies linearly from bumper-to-bumper traffic (behind $x = -x_0$) to no traffic (ahead of $x = 0$), as sketched in Figure 3.26. Two hours later, where does $u = u_j/2$?

4. An age-structured population in which older persons are removed at a faster rate than younger persons, and no one survives past age $x = L$ can be modeled by the initial boundary value problem

$$u_t + u_x = -\frac{cu}{L - x}, \qquad 0 < x < L, \quad t > 0$$

$$u(0, t) = b(t), \quad t > 0; \qquad u(x, 0) = f(x), \quad 0 < x < L$$

Find the age-structured population density $u(x, t)$ and give physical interpretations of the positive constant c and the positive functions $b(t)$ and $f(t)$.

3.4 WEAK SOLUTIONS: A FORMAL APPROACH

We were somewhat glib in the first two sections in this chapter about the concept of a discontinuous solution to a PDE. The implication was that a discontinuous solution is a function that is smooth and satisfies the PDE on both sides of a curve, the shock path; along the shock path the function suffers a simple discontinuity and, as well, satisfies a jump condition relating its values on both sides of the discontinuity to the speed of the discontinuity

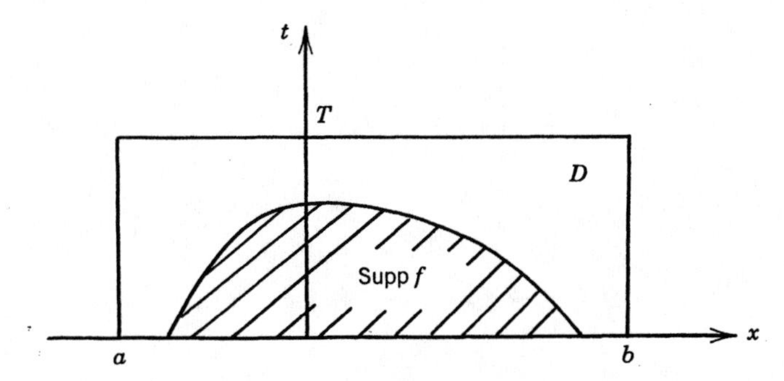

Figure 3.27. Rectangular region containing the support of the function f.

itself. In this section we wish to formulate a more formal, mathematical definition of these concepts. Our result will be an extension of the notion of a solution to a PDE to include nonclassical solutions.

For definiteness we consider the initial value problem

$$u_t + \phi(u)_x = 0, \qquad x \in R, \quad t > 0 \tag{1}$$

$$u(x,0) = u_0(x), \qquad x \in R \tag{2}$$

where ϕ is a continuously differentiable function on R. By a *classical*, or *genuine*, solution to (1)–(2) is meant a smooth function $u = u(x,t)$ that satisfies (1)–(2). Let us assume for the moment that $u = u(x,t)$ is a classical solution. Now, let $f(x,t)$ be any smooth function that vanishes outside some closed, bounded set in the plane. Such sets in the plane are *compact* sets, and the closure of the set of points where a function f is nonzero is called the *support* of f, denoted by supp f. Therefore, our assumption is that f is a function with compact support. The set of all smooth functions with compact support is denoted by C_0^1. Therefore, we may choose a rectangle $D = \{(x,t): a \le x \le b,\ 0 \le t \le T\}$, where $f = 0$ along $x = a$, $x = b$, and $t = T$ (see Figure 3.27). If we now multiply (1) by f and integrate over D, we obtain

$$0 = \int_0^T \int_a^b (fu_t + f\phi_x)\,dx\,dt \tag{3}$$

Integrating both terms in (3) by parts then gives

$$\int_0^T \int_a^b fu_t\,dx\,dt = \int_a^b \int_0^T fu_t\,dt\,dx = \int_a^b \left[fu|_0^T - \int_0^T f_t u\,dt \right] dx$$

$$= -\int_a^b f(x,0)u_0(x)\,dx - \int_a^b \int_0^T f_t u\,dt\,dx$$

and

$$\int_0^T \int_a^b f\phi_x \, dx \, dt = \int_0^T \left[f\phi|_a^b - \int_a^b f_x \phi \, dx \right] dt = -\int_0^T \int_a^b f_x \phi \, dx \, dt$$

Therefore, from (3), we have

$$\iint_{t \geq 0} (uf_t + f_x \phi) \, dx \, dt + \int_{t=0} u_0 f \, dx = 0 \tag{4}$$

To summarize, if u is a genuine solution to (1)–(2) and f is any smooth function with compact support, the integral relation (4) holds. We now observe that the integrands in (4) do not involve any derivatives of either u or ϕ. Thus (4) remains well defined even if u and $\phi(u)$ or their derivatives have discontinuities. We use equation (4), therefore, as a basis of a definition for a generalized solution of the initial value problem.

Definition 1. A bounded piecewise smooth function $u(x, t)$ is called a *weak solution* of the initial value problem (1)–(2), where u_0 is assumed to be bounded and piecewise smooth, iff (4) is valid for all smooth functions f with compact support.

Consequently, we have a more general definition of a solution that does not require smoothness. In a more general analysis context using the Lebesgue integral, we can replace the condition piecewise smooth by *measurable*.

To verify that Definition 1 is indeed an extension of the usual notion of a classical solution, one should actually check that if u is smooth and satisfies (4) for all functions f with compact support, then u is a solution to the initial value problem (1)–(2). We leave this verification as Exercise 1 at the end of the section.

An important question that we can now ask is: Does the definition of a weak solution, involving equation (4), result in a formula for the shock speed that is consistent with the jump condition derived in Section 3.2? The answer is, of course, affirmative, and we now present that argument. This will give an alternative way to derive the jump condition. To carry out the analysis we need to recall Green's theorem in the plane.

Theorem 1 (*Green's Theorem*). Let C be a piecewise smooth, simple closed curve in the xt-plane, and let D denote the domain enclosed by C. If $p = p(x, t)$ and $q = q(x, t)$ are smooth functions in $D \cup C$, then

$$\int_C p \, dx + q \, dt = \iint_D (q_x - p_t) \, dx \, dt$$

where the line integral over C is taken in the counterclockwise direction.

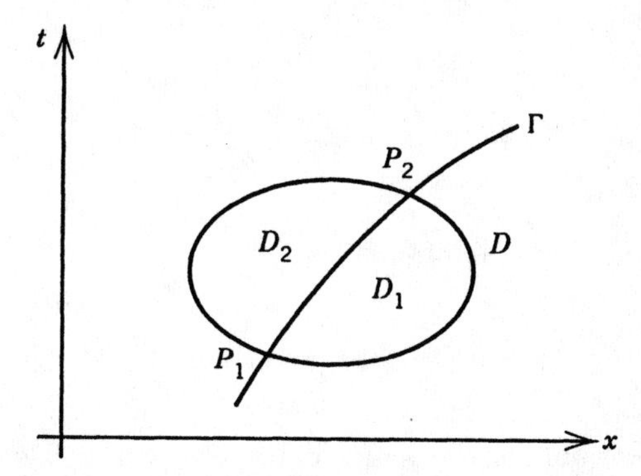

Figure 3.28. Ball D centered on a curve Γ along which a discontinuity in u occurs.

Now, let Γ be a smooth curve in spacetime given by $x = s(t)$ along which u has a simple jump discontinuity; also, let D be a circular region (a ball) centered at some point on Γ and lying in the $t > 0$ plane. (see Figure 3.28). Further, let D_1 and D_2 denote the disjoint subsets of D on each side of Γ. Now, choose $f \in C_0^1$ in D. From condition (4) we have

$$0 = \iint_{D_2} (uf_t + \phi(u)f_x)\,dx\,dt$$

$$= \iint_{D_2} (uf_t + \phi(u)f_x)\,dx\,dt + \iint_{D_2} (uf_t + \phi(u)f_x)\,dx\,dt \tag{5}$$

From the product rule for derivatives and the fact that u is smooth in D_2, the second integral in (5) can be written

$$\iint_{D_2} (uf_t + \phi(u)f_x)\,dx\,dt = \iint_{D_2} ((uf)_t + (f\phi)_x)\,dx\,dt$$

$$= \int_{\mathrm{Bd}\,D_2} -uf\,dx + f\phi\,dt$$

where in the last step we used Green's theorem. But the line integral is only nonzero along Γ because of our choice of f. Therefore, denoting by u_1 and u_2 the limiting values of u on the front and back sides of Γ, respectively [i.e., $u_1 = u(s(t)+, t)$ and $u_2 = u(s(t)-, t)$], we have

$$\iint_{D_2} (uf_t + \phi(u)f_x)\,dx\,dt = \int_{P_1}^{P_2} f(-u_2\,dx + \phi(u_2)\,dt) \tag{6}$$

In a similar manner,

$$\iint_{D_1} (uf_t + \phi(u) f_x)\, dx\, dt = \int_{P_2}^{P_1} f(-u_1\, dx + \phi(u_1)\, dt) \tag{7}$$

Equations (5), (6), and (7) imply that

$$\int_\Gamma f(-[u]\, dx + [\phi(u)]\, dt) = 0 \tag{8}$$

where the brackets denote the jump in the quantity inside (e.g., $[u] = u_2 - u_1$). Finally, the fact that f may be chosen arbitrarily leads to the conclusion that

$$-[u]\, dx + [\phi(u)]\, dt = 0$$

or

$$-s'(t)[u] + [\phi(u)] = 0 \tag{9}$$

which is the jump condition obtained in Section 3.2. Therefore, we have shown that *the definition of a weak solution leads to the jump condition across a shock*.

Consequently, the weak form of solutions given in integral form by (4) seems to tie down the notion of a solution; where the solution is smooth, the PDE holds, and where the solution is discontinuous, the shock condition is implied. The following example shows that care must be taken when going from a PDE to the integrated form of the weak solution.

Example 1. Consider the PDE

$$u_t + uu_x = 0 \tag{10}$$

Any positive smooth solution of this equation is a smooth solution of both the conservation laws

$$u_t + \left(\frac{u^2}{2}\right)_x = 0 \quad \text{and} \quad \left(\frac{u^2}{2}\right)_t + \left(\frac{u^3}{3}\right)_x = 0 \tag{11}$$

each of which satisfies different jump conditions, because the flux is different. Thus to one PDE there are associated different weak solutions, depending on the form in which the equation is written.

Thus we cannot tell what the correct shock condition is if the only thing given is the PDE. This means that the most basic, relevant piece of information is the integral form of the conservation law or the PDE in conservation form, where the flux is identified. Fortunately, in real applications in engineering or the physical or biological sciences, it is generally the case that the integral form of the conservation law comes out initially from the modelling process. Without knowledge of the underlying physical process, a PDE by itself does not uniquely determine the conservation law or shock conditions.

Exercise 6 shows that the uniqueness problem is in fact even worse than expected. Weak solutions are not unique, and another condition is required if we want otherwise. Such a condition is called an entropy condition, and it dictates how the speed of the characteristics ahead of and behind a shock wave must relate to the shock speed itself. For the PDE (1) with $\phi''(u) > 0$, it can be proved that there is a unique solution to the initial value problem (1)–(2) which also satisfies the condition

$$u(x+a,t) - u(x,a) \leq \frac{aE}{t}, \qquad x \in R, \quad t > 0 \tag{12}$$

where $a > 0$ and E is a positive constant independent of x, t, and a. Equation (12) is called the *entropy condition*, and it requires that a solution can only jump down as we traverse a discontinuity from left to right. In terms of characteristic speeds, (12) implies (see Exercise 7) the *entropy inequality*

$$\phi'(u_2) > s' > \phi'(u_1) \tag{13}$$

where s' is the shock speed and u_1 and u_2 are the states ahead and behind the shock, respectively. Equation (13) requires that the shock speed be intermediate between the speeds of the characteristics ahead and behind. The entropy condition originates from a condition in gas dynamics which requires that the entropy must increase (as required in the second law of thermodynamics) across a shock wave, the latter of which is basically an irreversible process. This condition precludes the existence of rarefaction shocks, where there is a positive jump from left to right across the shock wave. The proof of existence and uniqueness of a weak solution satisfying the entropy condition is fairly difficult and is beyond the scope of this book. We refer the reader to Lax [1973] or Smoller [1983] for highly readable treatments.

EXERCISES

1. Let u and u_0 be smooth functions that satisfy (4) for all $f \in C_0^1$. Prove that u and u_0 satisfy the initial value problem (1)–(2). *Hint:* Choose f judiciously, use integration by parts, and use the fact that if $\int hf\,dx = 0$ for all $f \in C_0^1$, where h is continuous, then $h = 0$.

2. Find a weak solution to each of the following initial value problems:

(a) $u_t + (e^u)_x = 0, \quad x \in R, \quad t > 0$
$u(x,0) = 2 \quad \text{if} \quad x < 0; \quad u(x,0) = 1 \quad \text{if} \quad x > 0.$

(b) $u_t + 2uu_x = 0, \quad x \in R, \quad t > 0$
$u(x,0) = (-x)^{1/2} \quad \text{if} \quad x < 0; \quad u(x,0) = 0 \quad \text{if} \quad x > 0.$

3. *Porous Media*. We consider a problem involving flow of an incompressible fluid through a porous medium (e.g., water flowing through soil). The soil is assumed to contain a large number of pores through which the water can infiltrate downward, driven by gravity. With the positive x-axis oriented downward, we let $u = u(x,t)$ denote the ratio of the water-filled pore space to the total pore space at the depth x, at time t. Thus $u = 0$ is dry soil and $u = 1$ is *saturated* soil where all the pore space is filled with water. The basic conservation law holds for volumetric water content u, namely

$$u_t + \phi_x = 0$$

where ϕ is the vertical flux. In one model of partially saturated, gravity-dominated flow, the flux is assumed to have the form

$$\phi = ku^n$$

where k and n are positive constants depending on the soil characteristics; k is called the hydraulic conductivity. In the following, take $n = 2$ and $k = 1$. If the moisture distribution in the soil is $u = \frac{1}{2}$ initially, and if at the surface $x = 0$ the water content is $u = 1$ (saturated) for $0 < t < 1$ and $u = \frac{1}{2}$ for $t > 1$, what is the water content $u(x,t)$ for $x, t > 0$? Plot the jump in u across the shock as a function of time. (Compare this exercise to the development of the porous media equation in Section 1.3.)

4. *Traffic Flow*. In the model of traffic flow presented in Section 3.3, the velocity V is found experimentally to be related to the density u by the constitutive relation $V = v_0(1 - u/U)$, where v_0 is the maximum speed and U is the maximum density. At time $t = 0$ cars are traveling along the roadway with constant velocity $2v_0/3$, and a truck enters the roadway at $x = 0$ traveling at speed $v_0/3$. It continues at this constant speed until it exits at $x = L$.

(a) On a spacetime diagram sketch the path of the truck and the path of the shock while the truck is on the road.

(b) After the truck leaves the road, find the time that the car following the truck takes to catch the traffic ahead of the truck.

(c) Describe the qualitative behavior of the weak solution to this problem for all $t > 0$.

5. Determine the jump conditions associated with the two equations (11).

6. Consider the initial value problem

$$u_t + uu_x = 0, \qquad x \in R, \quad t > 0$$

$$u(x,0) = 0 \quad \text{if } x < 0; \qquad u(x,0) = 1 \quad \text{if } x > 0$$

and let u and v be defined by

$$u(x,t) = 0 \quad \text{if } x < \frac{t}{2}; \qquad u(x,0) = 1 \quad \text{if } x > \frac{t}{2}$$

$$v(x,t) = 0 \quad \text{if } x < 0; \qquad v(x,t) = \frac{x}{t} \quad \text{if } 0 < x < t;$$

$$v(x,t) = 1 \quad \text{if } x > t$$

Sketch u and v and the characteristics. Are both u and v weak solutions to the initial value problem? Discuss. If the entropy condition (12) is assumed, which is the *correct* solution?

7. Prove that the entropy condition (12) implies the entropy inequality (13).

8. Consider the initial value problem

$$u_t + \left(\frac{u^2}{2}\right)_x = 0 \qquad x \in R, \quad t > 0$$

$$u(x,0) = 0 \quad \text{if } x > 0 \quad \text{and} \quad u(x,0) = U(x) \quad \text{if } x < 0$$

where U is unknown. Determine conditions on a shock $x = s(t)$, $t > 0$, in the upper half plane that is consistent with definition of a weak solution, and then, under these conditions, calculate U in terms of $s(t)$. What is $U(x)$ if $s(t) = (t+1)^{1/2} - 1$?

9. Consider the conservation law

$$\frac{d}{dt}\int_a^b u(x,t)\,dx = \frac{1}{2}\left[u(a,t)^2 - u(b,t)^2\right] + \int_a^b g(x)\,dx$$

where g is a discontinuous source term given by $g(x) = 1$ if $x > \frac{1}{2}$ and $g(x) = C = \text{constant}$ if $x < \frac{1}{2}$. Calculate the weak solution when the initial condition is given by $u(x,0) = U = \text{constant}$ for all $x \in R$.

10. Consider the initial boundary value problem

$$u_t + u_x = -u + u(1,t), \qquad 0 < x < 1, \quad t > 0$$

$$u(x,0) = u_0(x), \quad 0 < x < 1; \qquad u(0,t) = u_b(t), \quad t > 0$$

This problem has an unknown source term evaluated on the boundary of the domain, and it arises in a problem in detonation theory [see

W. Fickett, *Quart. Appl. Math.* **XLVI** (3), 459–471 (1988)]. It can be handled by the method of characteristics.

(a) Find an analytic expression for $u(1,t)$, $0 < t < 1$, and determine the solution in $t < x < 1$.

(b) Find a differential-difference equation for $u(1,t)$, $t > 1$, and a formula for the solution in terms of $u(1,t)$ in the region $0 < x < t$, $x < 1$. *Answer:* $u'(1,t) + au(1,t-1) = a[u_b(t-1) + u_b'(t-1)]$, $a = 1/e$.

(c) If $u_b(0) = u_0(0)$, show that the jump condition across the line $x = t$ is given by

$$u_+(t) - u(t) = \exp(-t)[u_b(0) - u_0(0)]$$

(d) Find the solution to the given problem in the case $u_b(t) = 0$ and $u_0(x) = x$.

11. Show that $u(x,t) = x/t$ if $-\sqrt{t} < x < \sqrt{t}$, and $u(x,t) = 0$ otherwise, is a weak solution to the equation $u_t + (u^2/2)_x = 0$ on the domain $x \in R$, $t \geq t_0 > 0$. Where are the shocks, and what are their speeds? Sketch a characteristic diagram and a typical wave profile for $t > t_0$, and show that the entropy condition holds.

12. Solve the Riemann problem

$$u_t + \phi_x = 0, \qquad u(x,0) = 1 \quad \text{if } x < 0, \quad \text{and} \quad u(x,0) = 0 \quad \text{if } x > 0$$

where the flux ϕ is given by

$$\phi(u) = \frac{u^2}{u^2 + (1-u)^2/2}$$

3.5 ASYMPTOTIC BEHAVIOR OF SHOCKS

Equal-Area Principle

In this section we examine, in a simple context, the evolution of an initial waveform over its entire history, as propagated by the nonlinear equation $u_t + c(u)u_x = 0$. The scenario is this: An initial signal $u_0(x)$ at $t = 0$ begins to distort and "shocks up" at breaking time $t = t_b$; after the shock forms, the shock discontinuity follows a path $x = s(t)$ in spacetime. One of the fundamental problems in nonlinear analysis is to determine the shock path and how the strength of the shock varies along that path for large times t.

Central to the calculation of the asymptotic form is an interesting geometric result called the equal-area principle. To formulate this principle in a simple manner we limit the types of initial data that we consider. We say that an initial profile $u = u_0(x)$ *satisfies property* (P) if $u_0(x)$ is a smooth, nonnegative pulse; that is, u_0 belongs to $C^1(R)$ and has a single maximum (say at $x = x_0$); u_0 is nondecreasing for $x < x_0$ and nonincreasing for $x > x_0$, and u_0 approaches a nonnegative constant value as $|x| \to \infty$. Therefore, we consider the initial value problem

$$u_t + c(u)u_x = 0, \qquad x \in R, \quad t > 0 \tag{1}$$

$$u(x,0) = u_0(x), \qquad x \in R \tag{2}$$

where u_0 satisfies property (P) and $c(u) = \phi'(u) > 0$, $c'(u) \geq 0$. We already know from Section 3.2 that a classical solution exists only up to some time $t = t_b$ when a gradient castastrophe occurs. This classical solution is given implicitly by the formulas [see (3) of Section 3.2]

$$u = u_0(\xi) \tag{3}$$

$$x = \xi + F(\xi)t, \qquad F(\xi) = c(u_0(\xi)) \tag{4}$$

Equation (4) defines the straight-line characteristic curves propagating from $x = \xi$ at $t = 0$ into the spacetime domain with speed $F(\xi)$.

Accordingly, as the initial signal propagates, the wave breaks at the first time t_b where the characteristics collide. Now, for $t > t_b$, instead of forming the weak solution with a shock, let us assume that the wave actually breaks (like an ocean wave), forming a multivalued wavelet where the intersecting characteristics are carrying their constant values of u. We may understand the formation of this wavelet as shown in Figure 3.29. A point $P:(\xi, u)$ on the initial wave at $t = 0$ is propagated to a point $P':(x, u)$ on the multivalued wavelet at time t, where the coordinate x at time t is given by (4); points having height u are propagated at speed $c(u)$. We may now state:

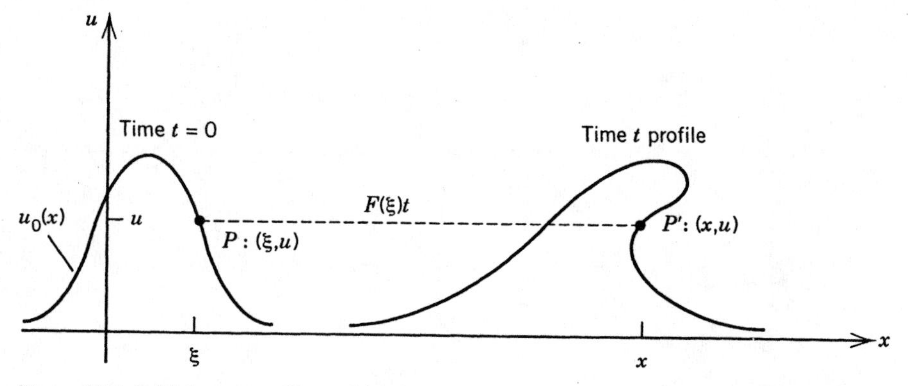

Figure 3.29. Initial wave profile evolving into a multivalued wavelet. Points at height u on the wave are propagated at speed $F(\xi) = c(u_0(\xi))$.

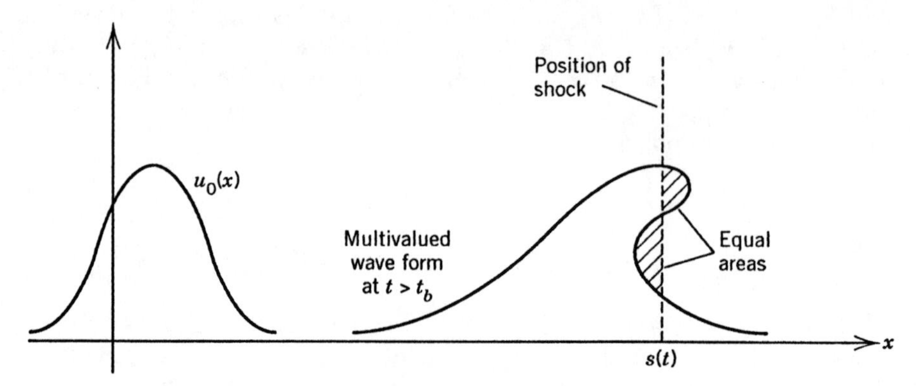

Figure 3.30. Equal-area rule.

Equal-Area Principle. The location of the shock $x = s(t)$ at time t is the position where a vertical line cuts off equal area lobes of the multivalued wavelet.

Figure 3.30 shows a geometrical interpretation of this rule.

Two simple facts are required to demonstrate the equal-area principle, and these are recorded formally in the following two lemmas. The first states that the area under a section of the wavelet remains unchanged as that section is propagated in time (see Figure 3.31), and the second states that the area under the weak solution (i.e., the discontinuous solution with a shock) remains unchanged as it is propagated in time (see Figure 3.32).

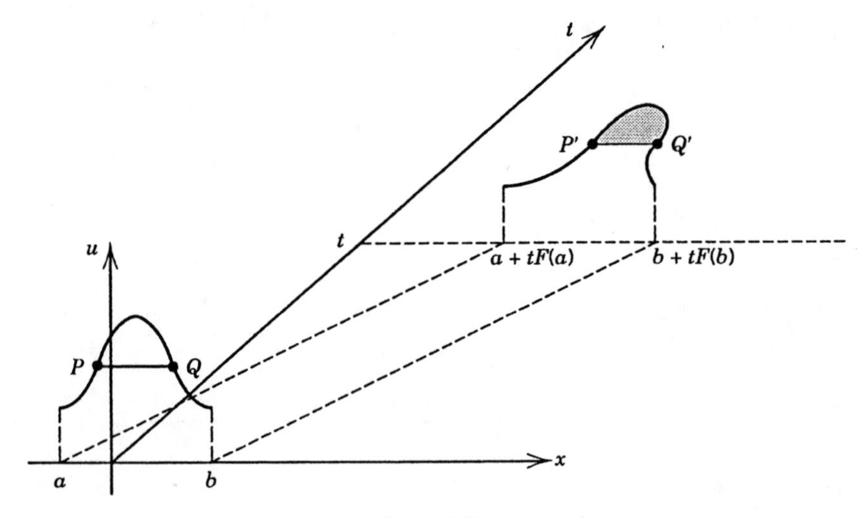

Figure 3.31

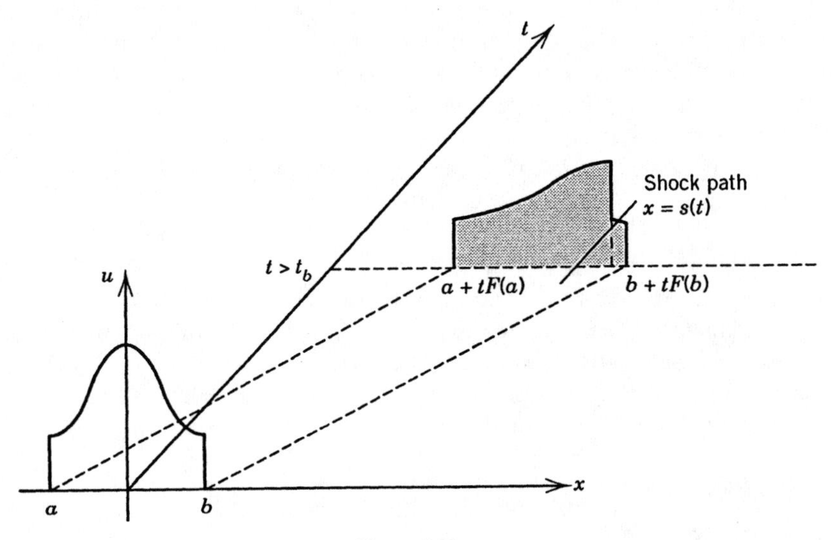

Figure 3.32

Lemma 1. Consider a section of the initial waveform between $x = a$ and $x = b$, with $u_0(a) = u_0(b)$, with u_0 satisfying property (P). Then for any $t > 0$ the area under this section of the wave remains constant as it propagates in time.

The proof of this fact is geometrically obvious (see Figure 3.31). From (4), the x coordinates a and b are moved in time t to $a + F(a)t$ and $b + F(b)t$, respectively. Any horizontal line segment PQ at $t = 0$ has the same length as $P'Q'$ at time t because points on the wave at the same height u move at the same speed $c(u)$.

Lemma 2. Let a and b be chosen such that $u_0(a) = u_0(b)$, where u_0 satisfies property (P), and assume that the shock locus is given by $x = s(t)$ with $a + tF(a) < s(t) < b + tF(b)$ for t in some open interval I. Then

$$\int_{a+tF(a)}^{b+tF(b)} u(x,t)\, dx = \text{constant}, \qquad t \in I \tag{5}$$

where $u = u(x, t)$ is the weak solution.

The proof of (5) can be carried out in a straightforward manner by showing the derivative of the left side of (5) is zero. We leave this calculation as an exercise.

The equal-area principle now follows easily from Lemmas 1 and 2 since both the multivalued waveform shown in Figure 5.31 and the weak solution shown in Figure 5.32 both encompass the same area, the area under the initial waveform from $x = a$ to $x = b$. We remark that these lemmas and the equal-area rule hold for more general initial data than that satisfying property (P). Furthermore, in the argument above we assumed that $c(u) > 0$; equally valid is the case $c(u) < 0$, with the appropriate adjustments.

Shock Fitting

The general problem of determining the position of the shock locus is one of the fundamental problems for hyperbolic systems. The equal-area principle leads to a quantitative condition whereby the shock path $x = s(t)$ can be calculated directly for certain simple equations. In this subsection we consider the initial value problem for the inviscid Burgers' equation:

$$u_t + uu_x = 0, \qquad x \in R, \quad t > 0 \tag{6}$$

$$u(x,0) = u_0(x), \qquad x \in R \tag{7}$$

where, as before, u_0 satisfies property (P). Thus we have taken $c(u) = u$ in (1); the general case for arbitrary $c(u)$ is discussed in detail in Whitham [1974].

Let us suppose that the shock is at some yet undetermined location $x = s(t)$ at some time $t > t_b$, and assume that the two characteristics carrying data to the shock from ahead and behind are specified by $\xi_1 = \xi_1(t)$ and $\xi_2 = \xi_2(t)$, respectively (see Figure 3.33). By construction, we know that the equations of these two characteristics are

$$x = \xi_1 + u_0(\xi_1)t, \qquad x = \xi_2 + u_0(\xi_2)t$$

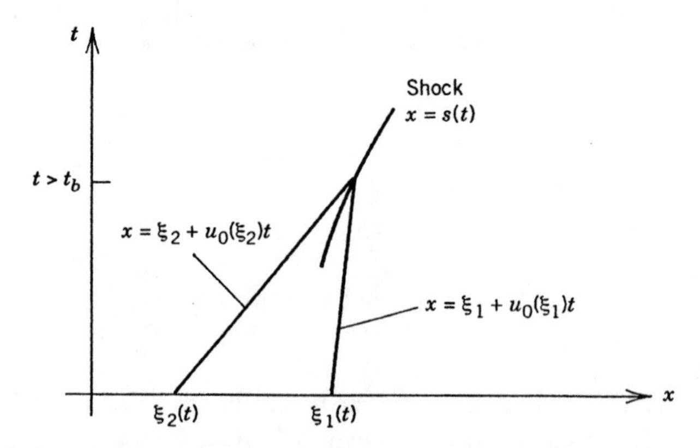

Figure 3.33. Two characteristics from behind and ahead that intersect the shock at time t.

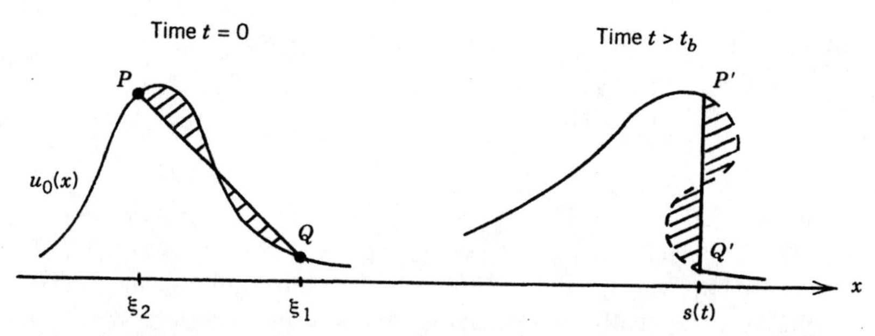

Figure 3.34. Whitham's rule. The chord PQ on the initial curve that cuts off equal areas has slope $-1/t$, where t is the time at the shock.

and therefore we have

$$s(t) = \xi_1 + u_0(\xi_1)t \qquad s(t) = \xi_2 + u_0(\xi_2)t \tag{8}$$

It follows immediately from (8) that

$$t = -\frac{\xi_1 - \xi_2}{u_0(\xi_1) - u_0(\xi_2)} > 0 \tag{9}$$

which is an equation for the time at the shock in terms of the known initial signal and the still unknown values of ξ_1 and ξ_2. The ratio on the right side of (9) is the reciprocal of the slope of the chord PQ on the initial wave profile u_0 connecting the points $(\xi_2, u_0(\xi_2))$ and $(\xi_1, u_0(\xi_1))$, as shown in Figure 3.34. Because the movement of the wavelet in xu-space is governed by the linear transformation $(x, u) \to (x + ut, u)$, straight lines are mapped to straight lines, and therefore the chord PQ is mapped to a vertical segment $P'Q'$ at time t. [This argument cannot be made in the case for arbitrary $c(u)$.] This vertical segment is the shock at time t, and the shock strength is clearly given by the formula

$$\text{shock strength} = u_0(\xi_2) - u_0(\xi_1) \tag{10}$$

Remember, ξ_1 and ξ_2 are still undetermined. Equation (8) gives two equations for ξ_1, ξ_2, and $s(t)$, and therefore we need one additional relation to determine these quantities and thus to determine the shock path. This additional relation comes from the equal-area principle. Since the shock cuts off equal-area lobes in the multivalued wavelet, the chord PQ must cut off equal-area lobes on the initial wave profile, as shown in Figure 3.34. This means that the area under the initial wave profile u_0 from ξ_2 to ξ_1 must equal the area of the trapezoid whose upper boundary is the chord PQ. Or, quantitatively,

$$\int_{\xi_2}^{\xi_1} u_0(\xi)\, d\xi = \tfrac{1}{2}[u_0(\xi_1) + u_0(\xi_2)](\xi_1 - \xi_2) \tag{11}$$

Therefore, we have, in principle, an *algorithm* to determine the shock position $s(t)$: For each fixed $t > t_b$, solve the three equations in (8) and (11) to determine $s(t)$, $\xi_1(t)$, and $\xi_2(t)$. It is clear that one cannot always carry out this calculation analytically, and consequently, a numerical procedure may be advisable.

However, one can proceed geometrically by *Whitham's rule*, a simple construction on the initial waveform to determine the location and time evolution of the shock. This rule is based on equation (9).

Whitham's Rule. For a succession of values of time t, draw a chord on the initial profile u_0 of slope $-1/t$ cutting off equal-area lobes. Then shift the endpoints of the chord to the right by the amounts $tu_0(\xi_2)$ and $tu_0(\xi_1)$, respectively, obtaining the location of the shock at the time t.

Some practice applying this rule will be given in the exercises.

Asymptotic Behavior

Again limiting the discussion to the initial value problem (6)–(7) for the inviscid Burgers' equation, we now investigate the long-time behavior of the shock path and the shock strength. To fix the notion we consider an initial wave profile u_0 satisfying property (P) with the additional stipulation that $u_0(x) = u^*$ for $x \le 0$ and for $x \ge a > 0$. Here u^* is a fixed positive constant. Thus $u_0(x)$ is a single hump, and we assume that the area under the hump, that is, the area bounded by $u_0(x)$, $u = u^*$, $x = 0$, and $x = a$, is A. From equation (9) we observe that the shock strength must vanish as t gets large; the question is the rate of decay and the path of the shock. We shall demonstrate that for large t,

$$\text{shock strength} \sim \left(\frac{2A}{t}\right)^{1/2} \tag{12}$$

$$s(t) \sim u^*t + (2At)^{1/2} \tag{13}$$

Thus the shock strength decays like $t^{-1/2}$ and the shock path is eventually parabolic. We shall also show that the wave profile behind the shock is given by, for large t,

$$u(x,t) \sim \frac{x}{t} \tag{14}$$

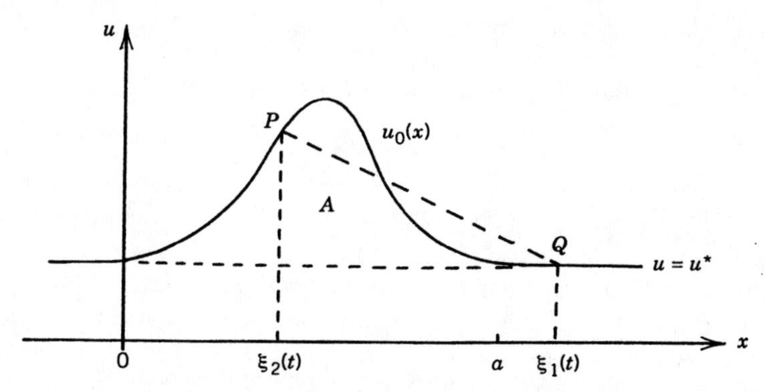

Figure 3.35. Initial wave profile $u_0(x)$ having area A.

Here we are using the symbol $\sim$ to mean in a limiting or asymptotic manner; we shall use the notation $t \gg 1$ to indicate that t is large.

The initial curve is shown in Figure 3.35. Using the equal-area rule applied to the initial profile, along with equation (9), we observe that $\xi_1(t)$ will eventually be larger than a since the slope of the chord must go to zero for large t. Therefore,

$$u_0(\xi_1(t)) = u^* \qquad \text{for } t \gg 1 \tag{15}$$

Now, subtraction of $u^*(\xi_1 - \xi_2)$ from both sides of equation (11) yields

$$\int_{\xi_2}^{\xi_1} [u_0(\xi) - u^*]\, d\xi = \tfrac{1}{2}(\xi_1 - \xi_2)[u_0(\xi_2) - u^*] \qquad \text{for } \xi_1 > a$$

Now using (9) to replace $\xi_1 - \xi_2$ gives

$$\int_{\xi_2}^{\xi_1} [u_0(\xi) - u^*]\, d\xi = \frac{t}{2}[u_0(\xi_2) - u^*]^2 \qquad \text{for } \xi_1 > a \tag{16}$$

But the left side of (16) can be written as

$$\int_{\xi_2}^{a} [u_0(\xi) - u^*]\, d\xi + \int_{a}^{\xi_1} [u_0(\xi) - u^*]\, dx = \int_{\xi_2}^{a} [u_0(\xi) - u^*]\, d\xi$$

Hence (16) becomes

$$\int_{\xi_2}^{a} [u_0(\xi) - u^*]\, d\xi = \frac{t}{2}[u_0(\xi_2) - u^*]^2 \qquad \text{for } \xi_1 > a \tag{17}$$

Now, for $t \gg 1$ the left side of (17) approaches the area A since ξ_2

approaches 0 for $t \gg 1$. Further, the quantity $u_0(\xi_2) - u^*$ is the shock strength. Thus for $t \gg 1$ the estimate (12) holds.

To reduce (13) we use equation (8) in the form

$$s(t) = \xi_2(t) + tu_0(\xi_2(t))$$

But from (12) we have $u_0(\xi_2(t)) \sim u^* + (2A/t)^{1/2}$ for $t \gg 1$, and we also have $\xi_2(t) \sim 0$ for $t \gg 1$. Relation (13) follows.

To prove that the wave has the form (14) we note that the point on the wavelet at $x = 0$ at $t = 0$ moves at speed u^* and hence is located at $x = u^*t$ at time t. Clearly, $u = u^*$ for $x < u^*t$. At the same time the shock is at $s(t) = u^*t + (2At)^{1/2}$, and $u = u^*$ for $x > s(t)$. So we need the solution in the interval $u^*t < x < u^*t + (2At)^{1/2}$. But the solution in this interval is, from (3) and (4),

$$u = u_0(\xi), \qquad \text{where} \quad x = \xi + tu_0(\xi)$$

Thus

$$u = \frac{x - \xi}{t}$$

But for $t \gg 1$ we have $\xi = \xi_2 \sim 0$, and therefore

$$u(x,t) \sim \frac{x}{t}, \qquad u^*t < x < u^*t + (2At)^{1/2}, \quad t \gg 1 \tag{18}$$

Figure 3.36 shows the triangular wave that develops for large times.

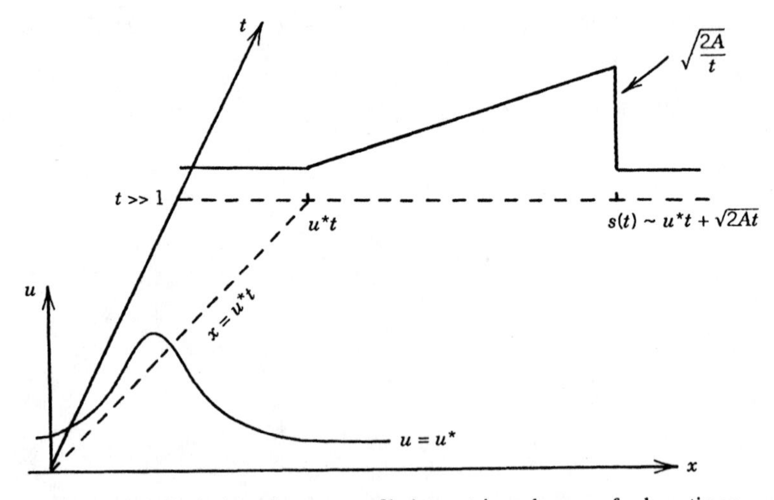

Figure 3.36. Evolution of a wave profile into a triangular wave for long times.

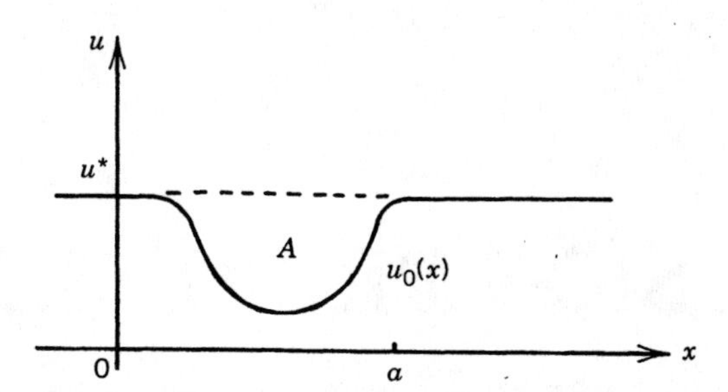

Figure 3.37. Initial wave profile $u_0(x)$ with a dip of area A.

EXERCISES

1. Prove Lemma 2. *Suggestion*: Break up the integral into two parts, each having a smooth integrand. Differentiate using Leibniz's rule and then use the jump conditions that hold across the shock.
2. Use Whitham's rule to determine the shock position geometrically and the wave profile for the initial value problem

$$u_t + uu_x = 0, \qquad u(x,0) = \frac{2}{1+x^2}$$

at the times $t = 1, 3, 5$, and 7. When does this wave break?
3. Consider the initial value problem (6)–(7) where the initial wavelet has the form shown in Figure 3.37. Determine the form of the wavelet, the strength of the shock, and the shock locus for $t \gg 1$.

REFERENCES

Practically all of the general texts referenced at the end of Chapter 1 discuss first order partial differential equations and characteristics. Particularly relevant is Whitham [1974], especially with regard to the asymptotic behavior of shocks and the equal-area rule. Also recommended are Lax [1973] and Lighthill [1978]. The latter contains a detailed discussion of the equal-area rule. The classic treatise by Landau and Lifshitz [1987] discusses these topics from an intuitive physical viewpoint.

Landau, L. D., and E. M. Lifshitz, 1987. *Fluid Mechanics*, 2nd ed., Pergamon Press, Oxford.

Lax, P., 1973, *Hyperbolic Systems of Conservation Laws and the Mathematical Theory of Shocks Waves*, (Society for Industrial and Applied Mathematics, Philadelphia).

Lighthill, J., 1978. *Waves in Fluids*, Cambridge University Press, Cambridge.

Smoller, J., 1983. *Shock Waves and Reaction–Diffusion Equations*, Springer-Verlag, New York.

Whitham, G. B., 1974. *Linear and Nonlinear Waves*, Wiley-Interscience, New York.

4

DIFFUSION PROCESSES

In this chapter we examine some equations that model basic diffusion processes. We also take the opportunity to introduce some mathematical methods that are useful in studying the structure of the solutions to diffusion problems, and, as a matter of fact, other problems as well. These methods include similarity methods, the asymptotic expansion of integrals, and phase-plane methods. The latter is included in an appendix to the chapter.

Until now our experience with diffusion processes has been centered around linear equations. In the first section we review some of the fundamentals and relate diffusion to the basic probability concept of a random walk. In Section 4.2 the concept of invariance of a PDE under a one-parameter family of stretching transformations is introduced, and it is seen how this type of invariance leads to a reduction of the PDE to an ordinary differential equation. Solutions obtained in this way are called similarity solutions; the mathematical method of determining solutions through invariance properties is one of the basic techniques of applied mathematics and is applicable to many classes of PDEs. In Section 4.3 we apply this method to nonlinear diffusion models. One of the most significant results here is that unlike their linear counterparts, nonlinear diffusion equations can propagate signals that resemble a wavefront traveling at finite speed. In Sections 4.4 and 4.5 we extend the study to reaction–diffusion processes and convection–diffusion processes, respectively. The Fisher equation is studied as a prototype of reaction–diffusion processes, and we investigate the existence of traveling wave solutions and their stability. In the same way, Burgers' equation, examined in Section 4.5, is the basic model equation for nonlinear convection–diffusion processes. In Section 4.6 the asymptotic expansion of integrals containing a large parameter is discussed briefly and Laplace's theorem is demonstrated; this theorem is then applied to understand the behavior of the integral representation of the solution to the initial value problem for Burgers' equation. In particular, we investigate the solution in the limit of small values of the diffusion constant, and the initial value problem with a point source is solved.

Finally, in an appendix we develop some concepts associated with two-dimensional systems of ordinary differential equations and their related phase-plane analysis. Many problems in PDEs, especially those associated with the determination of traveling wave solutions, can best be studied in the context of a two-dimensional dynamical system. The ideas presented in the appendix are elementary and should be familiar to most readers.

4.1 REVIEW OF BASIC RESULTS

Linear Diffusion

In this section, to set the stage for a study of nonlinear diffusion processes, we review briefly some of the basic results from linear diffusion theory, some of which has been discussed in Chapter 1. The linear diffusion equation is the parabolic PDE

$$u_t - Du_{xx} = 0$$

which follows from the basic conservation law

$$u_t + \phi_x = 0$$

and Fick's law,

$$\phi = -Du_x$$

The diffusion constant D is a measure of how fast the quantity measured by u (particles, chemicals, animals, energy, etc.) diffuses from high concentrations to low concentrations.

Example 1 (*Classical Heat Equation*). In the case that u is an energy density (i.e., a quantity with dimensions of energy per unit volume), the diffusion equation describes the diffusion of energy in a one-dimensional medium. For example, in a homogeneous medium whose *density* is ρ and whose *specific heat* (at constant volume) is C, the energy density is given by

$$u(x,t) = C\rho T(x,t)$$

where T is the *temperature*. Recall that $[C]$ = energy/(mass · degree) and $[\rho]$ = mass/volume. The conservation law may therefore be written

$$C\rho T_t + \phi_x = 0$$

In heat conduction, Fick's law has the form

$$\phi = -KT_x(x,t)$$

where K is the *thermal conductivity* [measured in energy/(mass · time · degree)]. In the context of heat conduction, Fick's law is called *Fourier's law of heat conduction*. Therefore, it follows that the temperature T satisfies the PDE

$$C\rho T_t - KT_{xx} = 0$$

or

$$T_t - kT_{xx} = 0, \qquad k = \frac{K}{C\rho} \tag{1}$$

Equation (1) is called the *heat equation*, and the constant k, which plays the role of the diffusion constant, is called the *diffusivity*. Thus the diffusivity k in heat flow problems is the analog of the diffusion constant, and the heat equation is just the diffusion equation. In this discussion we have assumed that the physical parameters C, K, and ρ of the medium are constant; however, these quantities could depend on the temperature T, which is an origin of nonlinearity in the problem. We discuss this possibility in Section 4.4.

In Section 1.5 we showed that the initial value problem for the diffusion equation

$$u_t - Du_{xx} = 0, \qquad x \in R, \quad t > 0 \tag{2}$$

$$u(x,0) = u_0(x), \qquad x \in R \tag{3}$$

is

$$u(x,t) = \int_R u_0(\xi)K(x-\xi,t)\,d\xi \tag{4}$$

where the function $K(y,t)$ is defined by

$$K(y,t) = (4\pi Dt)^{-1/2}\exp\left(-\frac{y^2}{4Dt}\right)$$

and is called the *diffusion kernel*. It is interesting to note that for any ξ and any $t > 0$ the kernel $K(x-\xi,t)$ is itself a solution of (2); it is called the *fundamental solution* of the diffusion equation. Moreover, one can show without much difficulty that K satisfies the following properties:

(i) $\lim_{t\to 0^+} K(x-\xi,t) = 0$ for each fixed $x \neq \xi$

(ii) $\lim_{t\to 0^+} K(x-\xi,t) = +\infty$ for $x = \xi$

(iii) $\lim_{|x|\to\infty} K(x-\xi,t) = 0$ for each fixed $t > 0$

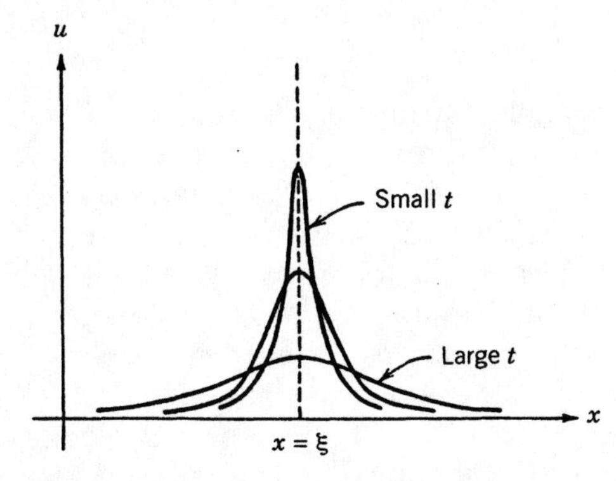

Figure 4.1. Time snapshots of the diffusion kernel $K(x - \xi, t)$ for various times t.

Time snapshots of the kernel function are shown in Figure 4.1. It is also easy to check that the area under each profile is unity, that is,

$$\text{(iv)} \quad \int_R K(x - \xi, t)\, dx = 1, \qquad t > 0$$

These properties imply that the diffusion kernel $K(x - \xi, t)$ is a solution to the initial value problem

$$u_t - Du_{xx} = 0, \qquad x \in R, \quad t > 0$$
$$u(x,0) = \delta(x - \xi), \qquad x \in R$$

where $\delta(x - \xi)$ is the delta function, that is, a point source of unit intensity at $x = \xi$. That is, $\delta(x - \xi) = 0$ for $x \neq \xi$, and $\delta(x - \xi) = +\infty$ for $x = \xi$. Figure 4.1, therefore, depicts how a unit amount (of energy, say) applied at $x = \xi$ at $t = 0$ diffuses through a medium with diffusion constant D. We also remark that the solution (4) of the initial value problem (2)–(3) can therefore be regarded as the superposition of a continuum distribution of sources $u_0(\xi)$, $\xi \in R$. Furthermore, the results also imply that linear diffusion is a process where initial signals are propagated with infinite speed; for example, the solution $K(x - \xi, t)$ is nonzero for any x, regardless of how large, and for any time $t > 0$, regardless of how small. Therefore, we must always be aware of the limitations of the diffusion equation as a model for physical processes (e.g., heat flow).

Another important result associated with the diffusion equation is the maximum principle. We shall take up this principle in Chapter 6, where reaction–diffusion equations are studied. For the present we remark that a solution to the diffusion equation must take on its maximum value on the boundary of the domain over which the problem is defined. Physically, this means that the density function u cannot clump in the interior of the domain; this interpretation seems plausible since the diffusion equation includes Fick's law, a condition that forces movement from high to low concentrations.

Diffusion and Probability

The time snapshots of the fundamental solution of the diffusion equation shown in Figure 4.1 strongly suggest that the diffusion equation does indeed describe diffusion-like processes. However, there is stronger evidence, based on a probability model, that we now discuss. In a diffusion process we invent a mental image of an assemblage of particles, or whatever, moving about in a random manner and spreading throughout the medium. On a microscopic level this irregular motion may be caused by collisions or other interactions among the particles. From this microscopic description we can determine a phenomenological model, based on the notion of a random walk, that describes the observed gross behavior of the process. The subsequent discussion follows the treatment in Murray [1989, pp 232–236].

Suppose that we label the x-axis with discrete points $x_m = m\,\Delta x$, where Δx is a small, fixed length, and $m = 0, \pm 1, \pm 2, \ldots$ (see Figure 4.2). Assume that a particle starts at $x = 0$ and moves randomly at each interval Δt of time either to the left or right with equal probability. Let $p(m, n)$ denote the probability of being at the location $m\,\Delta x$ at time $n\,\Delta t$. To calculate $p(m, n)$ we use a simple counting argument. Let

$$a = \text{the number of steps taken to the right}$$

$$b = \text{the number of steps taken to the left}$$

Then

$$m = a - b \qquad n = a + b$$

Now, by the definition of probability, $p(m, n)$ must be given by

$$p(m, n) = \frac{\text{number of } n\text{-step paths leading to the location } m\,\Delta x}{\text{total number of } n\text{-step paths}}$$

Easily there are 2^n possible n-step paths. Furthermore, the total number of n-step paths, choosing a to the right, is

$$C_a^n = \frac{n!}{a!(n-a)!} = \frac{n!}{a!b!}$$

Therefore,

$$p(m, n) = \frac{n!}{2^n a! b!}, \qquad a = \frac{m+n}{2}$$

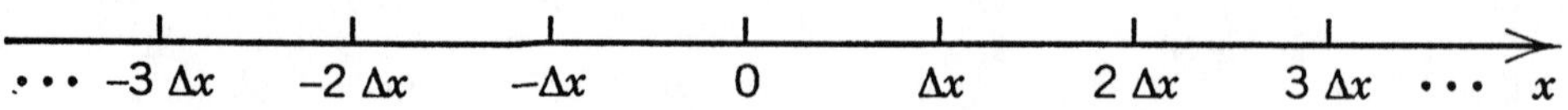

Figure 4.2

Those familiar with probability theory will notice that $p(m, n)$ is the *binomial distribution*. It is easy to compare this experiment with coin tossing. If a fair coin is tossed n times, there are 2^n possible outcomes; if a represents the number of heads and b represents the number of tails, C_a^n is the number of ways that we can get a heads out of the n tosses. The analogy is that the number of heads corresponds to the number of steps to the right in the random walk. An estimate for $p(m, n)$ for large m and n can be obtained from *Stirling's formula*,

$$n! \sim (2\pi n)^{1/2} n^n e^{-n}, \qquad n \text{ large}$$

which is an asymptotic formula for n factorial. The estimate for $p(m, n)$ is

$$p(m,n) \sim \left(\frac{2}{\pi n}\right)^{1/2} e^{-m^2/2n}, \qquad m, n \text{ large} \tag{5}$$

which is the Gaussian probability distribution (the *normal* distribution). The proof of (5) is left as an exercise.

In summary, we have determined an asymptotic expression for the probability of the particle being located at $m\,\Delta x$ after n time steps. Now set

$$x = m\,\Delta x, \qquad t = n\,\Delta t$$

and keep x and t fixed while letting $m, n \to \infty$ and Δx, $\Delta t \to 0$. Then the *probability density* (probability per length) of the particle being between $x - \Delta x$ and $x + \Delta x$ is given by

$$\frac{p(x/\Delta x, t/\Delta t)}{2\Delta x} = \left[\frac{1}{4\pi t(\Delta x^2/2\Delta t)}\right]^{1/2} \exp\left[\frac{-x^2}{4t(\Delta x^2/2\Delta t)}\right] \tag{6}$$

Now assume that the limit as $\Delta x \to 0$ and $\Delta t \to 0$ can be taken in such a way that

$$\lim_{\Delta x, \Delta t \to 0} \frac{\Delta x^2}{2\Delta t} = D \tag{7}$$

where D is a constant. Then, upon defining

$$u(x,t) = \lim_{\Delta x, \Delta t \to 0} \frac{p(x/\Delta x, t/\Delta t)}{2\Delta x}$$

we have

$$u(x,t) = \frac{1}{\sqrt{4\pi D t}} e^{-x^2/4Dt}$$

We note that $u(x,t)$ is the probability density associated with a particle released at the origin at time $t = 0$, moving randomly at each instant to the left or right with equal probability; it coincides with the fundamental solution of the diffusion equation $K(x,t)$. The discrete process described above is called a one-dimensional random walk, and its continuous counterpart is modeled by a solution to the diffusion equation, thereby providing a strong connection between diffusion and random processes.

Another view of this connection can be obtained directly from the expression

$$p(x,t) = p\left(\frac{x}{\Delta x}, \frac{t}{\Delta t}\right)$$

which is the expression for the probability of the particle being at x at time t, given that it started at $x = 0$ at time $t = 0$. If the particle is at x at time t, then at time $t - \Delta t$ it must have been at $x - \Delta x$ or $x + \Delta x$, with equal probability. Consequently, we must have

$$p(x,t) = \frac{p(x - \Delta x, t - \Delta t) + p(x + \Delta x, t - \Delta t)}{2} \tag{8}$$

The right side of (8) may now be expanded in a Taylor series to obtain

$$p_t = \frac{\Delta x^2}{2\Delta t} p_{xx} + \frac{1}{2} p_{tt} \Delta t + \cdots \tag{9}$$

or

$$\left(\frac{p}{2\Delta x}\right)_t = \frac{\Delta x^2}{2\Delta t}\left(\frac{p}{2\Delta x}\right)_{xx} + \frac{1}{2}\Delta t\left(\frac{p}{2\Delta x}\right)_{tt} + \cdots$$

or

$$u_t = \frac{\Delta x^2}{2\Delta t} u_{xx} + \frac{1}{2}\Delta t u_{tt} + \cdots$$

Now, as before, take the limit as $\Delta x \to 0$ and $\Delta t \to 0$ while maintaining the condition (7); we obtain

$$u_t - D u_{xx} = 0$$

which is the diffusion equation. We remark that the approach above depends decisively on the limit (7) being taken in a specific way. Alternative approaches based on Markov processes not depending on this limit may also be developed.

EXERCISES

1. If u_0 is a bounded, continuous function on R, show that the solution $u(x, t)$ of the initial value problem (2)–(3) satisfies the condition

$$\inf u_0(z) \le u(x,t) \le \sup u_0(z)$$

where the sup and inf are taken over all of R.

2. Find all solutions of the diffusion equation $u_t - u_{xx} = 0$ having the form

$$u = \frac{1}{\sqrt{t}} f(z), \qquad \text{where} \quad z = \frac{x}{2\sqrt{t}}$$

3. Find the solution of the initial value problem (2)–(3) when $D = 1$ and the initial condition is given by

$$u_0(x) = 0 \quad \text{if } x < 0, \qquad u_0(x) = 1 \quad \text{if } x > 0$$

The solution is

$$u = \frac{1 + \operatorname{erf}(x/\sqrt{4t})}{2}$$

4. Verify that $\sum_{m=-n}^{n} p(m,n) = 1$ and interpret the result.
5. Verify formulas (5), (6), and (9).
6. Consider the initial value problem (2)–(3) for the diffusion equation, and assume $u_0 > 0$, $D = 1$, and $\int_R u_0(x)\,dx = 1$. Let

$$\bar{x}(t) = \int_R x u(x,t)\,dx \qquad \sigma^2(t) = \int_R (x - \bar{x})^2 u(x,t)\,dx$$

denote the mean and variance of the distribution $u(x, t)$, respectively. Prove that $\bar{x}(t) = \bar{x}(0)$ and $\sigma^2(t) = 2t + \sigma^2(0)$.

7. A model of nonlocal diffusion is the equation

$$u_t = (k * u)(x,t), \qquad x \in R, \quad t > 0$$

where $k * u$ is the *convolution* of $k = k(x)$ and $u = u(x, t)$ defined by $(k * u)(x, t) = \int_R k(x - y)u(y, t)\,dy$. The convolution averages u against the translates of the given continuous function k. Assume that k is an even function with the property that $k(x)x^n$ is absolutely integrable on R for each $n \ge 0$, and assume that u is sufficiently smooth. Show that the nonlocal equation may be approximated by the local equation

$$u_t = m_0 u + m_2 u_{xx} + m_4 u_{xxxx} + \cdots$$

where m_n is the nth moment of k defined by

$$m_n = \frac{1}{n!}\int_{-\infty}^{\infty} k(y)y^n\,dy, \qquad n = 0,1,2,\ldots$$

Take $k(x) = \exp(-x^2)$ and numerically compute the even moments m_0 through m_6.

8. Consider the Barenblatt equation

$$u_t = \begin{cases} ku_{xx} & \text{if } u_t \geq 0 \\ mu_{xx} & \text{if } u_t < 0 \end{cases}$$

where k and m are distinct positive constants. Is this equation linear? Show that the Barenblatt equation can be written

$$au_t + b|u_t| = u_{xx}$$

for some constants a and b.

4.2 SIMILARITY METHODS

In this section we introduce a powerful method for determining transformations that reduce PDEs to ordinary differential equations. This method, called the similarity method, takes advantage of the natural symmetries in a PDE and allows us to define special variables that give rise to the reduction. Equations that model physical problems often inherit symmetries from the underlying physical system; for example, a physical system that is translational invariant often produces governing equations that are unchanged under a translation of coordinates. Equations with symmetries, or equivalently, equations that are invariant under a given transformation, usually have a simple structure that can be used to advantage to simplify the problem. The similarity method is applicable to all types of PDEs, but it is particularly useful in obtaining solutions to diffusion problems, which is the reason that we place this topic in this chapter. In the next section we apply this technique to some nonlinear models of diffusion.

We shall not present the most general similarity method, but rather, refer the reader to one of the books listed in the references at the end of the section. In this section we focus on a method developed by G. D. Birkhoff in the 1930s and consider the special case of scale transformations, or transformations of the independent and dependent variables that are represented by simple multiples of those variables. These are often called stretching transformations, and Birkhoff's method is sometimes called the *method of stretchings*.

At the outset we consider a first order PDE of the form

$$G(x, t, u, p, q) = 0, \qquad p = u_x, \quad q = u_t \tag{1}$$

Definition 1. A *one-parameter family of stretching transformations*, denoted by T_ε, is a transformation on *xtu*-space of the form

$$\bar{x} = \varepsilon^a x, \qquad \bar{t} = \varepsilon^b t, \qquad \bar{u} = \varepsilon^c u \tag{2}$$

where a, b, and c are constants and ε is a real parameter restricted to some open interval I containing $\varepsilon = 1$.

We remark that the stretching transformation (2) automatically induces a transformation on the derivatives p and q via the formulas

$$\bar{p} = \varepsilon^{c-a} p, \qquad \bar{q} = \varepsilon^{c-b} q \tag{3}$$

Now we address the question of what it means for the PDE (1) to be invariant under T_ε. An example will motivate the definition.

Example 1. Consider the nonlinear transport equation

$$u_t + uu_x = 0$$

or, equivalently,

$$q + up = 0 \tag{4}$$

In the transformed coordinate system the operator defining (4) is $\bar{q} + \bar{u}\bar{p}$. According to (2) and (3),

$$\bar{q} + \bar{u}\bar{p} = \varepsilon^{c-b} q + \varepsilon^{2c-a} up \tag{5}$$

If we require that $-b = c + a$, then (5) can be written

$$\bar{q} + \bar{u}\bar{p} = \varepsilon^{2c-a}(q + up) \tag{6}$$

Therefore, under the transformation

$$\bar{x} = \varepsilon^a x, \qquad \bar{t} = \varepsilon^{a-c} t, \qquad \bar{u} = \varepsilon^c u \tag{7}$$

for any constants a and c, the expression defining the PDE in the transformed coordinate system is a multiple of the original expression defining the PDE. This is what is meant by invariance of (4) under (7).

Definition 2. The PDE (1) is *invariant* under the one-parameter family T_ε of stretching transformations defined by (2) iff there exists a smooth function $f(\varepsilon)$ such that

$$G(\bar{x}, \bar{t}, \bar{u}, \bar{p}, \bar{q}) = f(\varepsilon)G(x, t, u, p, q) \tag{8}$$

for all ε in I, with $f(1) = 1$. If $f(\varepsilon) = 1$ for all ε in I the PDE is said to be *absolutely invariant*.

Now we can state and prove the basic reduction theorem.

Theorem 1. If the PDE (1) is invariant under T_ε given by (2), then the transformation

$$u = t^{c/b}y(z), \qquad z = \frac{x}{t^{a/b}} \tag{9}$$

reduces the PDE (1) to a first order ordinary differential equation in $y(z)$ of the form

$$g(z, y, y') = 0 \tag{10}$$

Terminology. The new independent variable z defined in (9) under which (1) reduces to an ordinary differential equation is called a *similarity variable.* Equation (9) is called a *similarity transformation*. After solving (10) for the unknown function y, substitution into (9) yields the *self-similar* form of the solution u.

Before proving the theorem we consider an example.

Example 2. Referring to Example 1, we observed that the transport equation (4) was invariant under the stretching transformation (7). Then the similarity transformations is given by

$$u = t^{c/(a-c)}y(z), \qquad z = \frac{x}{t^{a/(a-c)}} \tag{11}$$

The constants a and c are arbitrary at this point, and they are usually chosen so that (11) can satisfy certain initial or boundary conditions that may be given with the problem. Substituting (11) into the PDE (4) requires the partial derivatives u_t and u_x. The chain rule gives

$$u_x = \frac{y'}{t}, \qquad u_t = \frac{c}{a-c}t^{c/(c-a)-1}y + \frac{ax}{c-a}t^{c/(a-c)}t^{a/(c-a)-1}y'$$

Consequently, (4) becomes

$$yy' + \frac{a}{a-c}zy' + \frac{c}{c-a}y = 0 \tag{12}$$

which is an ordinary differential equation for $y = y(z)$.

Proof of Theorem 1. The proof of the basic reduction theorem is straightforward. By invariance of (2) under T_ε we infer that (8) holds; and since (8) holds for all ε in some interval I containing $\varepsilon = 1$, we may differentiate (8) with respect to ε and afterward set $\varepsilon = 1$ to obtain

$$axG_x + btG_t + cuG_u + (c-a)pG_p + (c-b)qG_q = f'(1)G \tag{13}$$

This first order linear PDE for G is a consequence of the invariance assumption. As expected, not every PDE (2) will be invariant under T_ε, and (13) imposes a condition on the form of G. The characteristic system associated with (13) is

$$\frac{dx}{ax} = \frac{dt}{bt} = \frac{du}{cu} = \frac{dp}{c-a} = \frac{dq}{c-b} = \frac{dG}{f'(1)G}$$

and there are five independent first integrals given by

$$\frac{x}{t^{a/b}}, \quad ut^{-c/b}, \quad pt^{(a-c)/b}, \quad qt^{1-b/c}, \quad \text{and} \quad Gt^{-f'(1)/b}$$

Therefore, the general solution of (13) is

$$G = t^{f'(1)/b}\Psi\left(z, u^{t-c/b}, pt^{(a-c)/b}, qt^{1-b/c}\right) \tag{14}$$

where Ψ is an arbitrary function and $z = x/t^{a/b}$. This equation fixes the form of G. Now u is given by (9) and the partial derivatives p and q may be calculated to give

$$p = u_x = t^{(c-a)/b}y', \qquad q = u_t = \left(\frac{c}{b}\right)t^{-1+c/b}y - \frac{a}{b}t^{c/b-a/b-1}xy'$$

Substituting these expressions into (14) yields the equation

$$\Psi\left(z, y, y', \frac{cy}{b} - \frac{azy'}{b}\right) = 0$$

which is an ordinary differential equation of the form (10), completing the proof.

The method is easily extended to second order PDEs. We state the basic reduction theorem in this case and leave the proof to the reader.

Theorem 2. If the second order PDE

$$G(x,t,u,u_x,u_t,u_{xx},u_{xt},u_{tt}) = 0 \tag{15}$$

is invariant under the one-parameter family T_ε of stretching transformations (2), the transformation (9) reduces the PDE (15) to a second order ordinary differential equation of the form

$$g(z,y,y',y'') = 0 \tag{16}$$

We remark that the one-parameter family of transformations (2) on *xtu*-space induces a transformation on the second derivatives, just as it did on the first derivatives [see (3)]. If we denote the second derivatives by

$$r = u_{xx}, \qquad s = u_{xt}, \qquad v = u_{tt}$$

it follows immediately that

$$\bar{r} = \varepsilon^{c-2a}r, \qquad \bar{s} = \varepsilon^{c-a-b}s, \qquad \bar{v} = \varepsilon^{c-2b}v$$

Example 3. Consider the diffusion equation

$$u_t - Du_{xx} = 0 \tag{17}$$

which in terms of our notation for derivatives, can be written

$$q - Dr = 0 \tag{18}$$

We now determine a stretching transformation (2) under which (18) is invariant. We have

$$\bar{q} - D\bar{r} = \varepsilon^{c-b}q - D\varepsilon^{c-2a}r = \varepsilon^{c-b}(q - Dr)$$

provided that

$$b = 2a \tag{19}$$

Therefore (18) is invariant under the stretching transformation

$$\bar{x} = \varepsilon^a x, \qquad \bar{t} = \varepsilon^{2a}t, \qquad \bar{u} = \varepsilon^c u \tag{20}$$

for any choice of the constants a and c. The similarity transformation is then given by

$$u = t^{c/2a} y(z), \qquad z = \frac{x}{\sqrt{t}} \tag{21}$$

Substitution into (17) gives

$$Dy'' + \frac{z}{2} y' - \frac{c}{2a} y = 0 \tag{22}$$

Now let us impose boundary conditions on (17) and consider the problem on the domain $x > 0$, $t > 0$ subject to the initial condition

$$u(x,0) = 0, \qquad x > 0 \tag{23}$$

and the boundary conditions

$$u(0,t) = 1, \qquad u(\infty,t) = 0, \qquad t > 0 \tag{24}$$

Physically, this problem models diffusion into the region $x > 0$ where the concentration u is zero initially and a constant concentration $u = 1$ is imposed at $x = 0$. From (21) it follows that (23) and (24b) both translate into the condition

$$y(\infty) = 0 \tag{25}$$

On the other hand, (24a) forces

$$u(0,t) = t^{c/2a} y(0) = 1, \qquad t > 0$$

The only possibility that the left side can be independent of t is to force $c = 0$. Therefore, the ordinary differential equation (22) becomes

$$y'' + \frac{z}{2D} y' = 0, \qquad z > 0 \tag{26}$$

subject to the boundary conditions

$$y(0) = 1, \qquad y(\infty) = 0 \tag{27}$$

It is straightforward to solve (26) to obtain

$$y(z) = c_1 + c_2 \int_0^z \exp\left(-\frac{\eta^2}{4D}\right) d\eta \tag{28}$$

where c_1 and c_2 are constants of integration. Equation (28) may be written in terms of the *error function* erf(s), which is defined by

$$\operatorname{erf}(s) = \frac{2}{\sqrt{\pi}} \int_0^s \exp(-\eta^2)\, d\eta$$

as

$$y(z) = c_1 + c_2\sqrt{\pi D}\, \operatorname{erf}\left(\frac{z}{\sqrt{4D}}\right)$$

The condition $y(0) = 1$ forces $c_1 = 1$, and subsequently the condition $y(\infty) = 0$ implies that

$$1 + c_2\sqrt{\pi D}\, \operatorname{erf}(\infty) = 0$$

or $c_2 = -1/\sqrt{\pi D}$. The solution to (26)–(27) is therefore

$$y(z) = 1 - \operatorname{erf}\left(\frac{z}{\sqrt{4D}}\right)$$

Consequently, the solution to the initial boundary value problem (17), (23), (24)

$$u(x,t) = 1 - \operatorname{erf}\left(\frac{x}{\sqrt{4Dt}}\right)$$

The exercises will give the reader more practice with the similarity method for linear diffusion problems. In the next section we examine some nonlinear differential equations.

The preceding discussion centered on an algorithm for determining a group of stretching transformations under which a given PDE is invariant, and then using the similarity transformation to reduce the PDE to an ordinary differential equation. No mention was made of the underlying reasons for this reduction other than to say that invariance or symmetry implies simplicity in the equations. The basic theory goes back to the late nineteenth century with the seminal work of S. Lie on the invariance of ordinary differential equations under one-parameter groups of transformations. Quite generally, invariance leads to simplifications in terms of new variables which are *invariants* of the family of transformations. Notice, for example, that the quantities defining the similarity transformation (9), namely $ut^{-c/b}$ and $x/t^{a/b}$, are both invariants of (2); that is,

$$\bar{u}\bar{t}^{-c/b} = ut^{-c/b}, \qquad \frac{\bar{x}}{\bar{t}^{a/b}} = \frac{x}{t^{a/b}}$$

The method of stretchings presented above can be extended to much more general transformations, and we refer the interested reader to the references (see, e.g., Logan [1987]).

One should not conclude that the similarity method is a universal method that is applicable in all cases. Even though a given PDE admits a symmetry of the form (2), it may be impossible to find constants a, b, and c that force u, given by (9), to satisfy given initial or boundary conditions; that is, auxiliary conditions, especially on finite domains, sometimes break the symmetry. However, the method is applicable on many problems, and it is one of the basic techniques for solving nonlinear PDEs on infinite domains.

EXERCISES

1. Verify equation (22).
2. Use the similarity method to find the solution of the problem

$$\begin{aligned} u_t - u_{xx} &= 0, \qquad x > 0, \quad t > 0 \\ u(x,0) &= 0, \qquad x > 0 \\ u_x(0,t) &= -1, \qquad u(\infty,t) = 0, \quad t > 0 \end{aligned}$$

   Give a physical interpretation of this problem.
3. Using the similarity method, derive the fundamental solution

$$u(x,t) = \left(\frac{1}{4\pi Dt}\right)^{1/2} \exp\left(-\frac{x^2}{4Dt}\right)$$

   of the diffusion equation $u_t - Du_{xx} = 0$. *Hint:* Consider the condition $\int_R u(x,t)\,dx = 1$, $t > 0$.
4. A first order ordinary differential equation of the form $p - f(x,y) = 0$, where $p = y'(x)$, is said to be absolutely invariant under the stretching transformation $\bar{x} = x\varepsilon$, $\bar{y} = \varepsilon^a y$ iff $\bar{p} - f(\bar{x},\bar{y}) = p - f(x,y)$ for all ε in I, where I is an open interval containing $\varepsilon = 1$. In this case prove that the ordinary differential equation can be reduced to a separable equation of the form $ds/s = dr/[F(r) - ar]$ for appropriately chosen r and s.

4.3 NONLINEAR DIFFUSION MODELS

So far we have only considered linear diffusion where the process is governed by the basic conservation law

$$u_t + \phi_x = 0 \tag{1}$$

and Fick's law

$$\phi = -Du_x \qquad (D \text{ constant}) \tag{2}$$

which lead to the linear diffusion equation

$$u_t - Du_{xx} = 0 \tag{3}$$

We have already observed how nonlinearities arise in reaction–diffusion equations by the inclusion of a nonlinear source term. In this section we examine other sources of nonlinear diffusion by studying two models, one in biology and one in heat conduction.

Example 1 (*Insect Dispersal*). The study of insect and animal dispersal forms the basis of a natural extension of the linear diffusion model to a nonlinear problem. If there is an increase in diffusion due to population pressure, it seems reasonable to assume that the diffusion coefficient D is a function of the density u and take, in lieu of Fick's law (2), the constitutive equation

$$\phi = -D(u)u_x \tag{4}$$

When the flux (4) is substituted in (1) we obtain the nonlinear diffusion equation

$$u_t - (D(u)u_x)_x = 0 \tag{5}$$

Writing out the derivatives in (5) gives

$$u_t - D(u)_x u_x - D(u)u_{xx} = 0 \tag{6}$$

so the variable diffusion constant gives rise to a nonlinear convective term $-D(u)_x u_x$ or $-(D'(u)u_x)u_x$, which would propagate signals at speed $-D'(u)u_x$.

Example 2 (*Heat Conduction*). For linear heat flow the governing equations are (see Section 4.1) the conservation law for energy density

$$(\rho CT)_t + \phi_x = 0 \tag{7}$$

and Fourier's law for the heat flux

$$\phi = -KT_x \tag{8}$$

where $T = T(x, t)$ is the temperature, and the constants ρ, C, and K are the density, specific heat, and thermal conductivity of the medium, respectively. In many applications where the temperature range is limited, the specific heat may be regarded essentially as a constant. However, over wide temperature ranges the specific heat is not constant, but rather, is a function of the temperature. That is, $C = C(T)$. In this case (7) and (8) combine to give

$$(C(T)T)_t - \frac{K}{\rho}T_{xx} = 0 \tag{9}$$

which is a nonlinear diffusion equation for the temperature T.

Equations (5) and (9) represent two nonlinear diffusion models that have attracted much attention in the literature on diffusion equations. The simplest assumptions to make are that the diffusion coefficient D in (5) and the specific heat C in (9) are power functions, that is,

$$D(u) = D_0\left(\frac{u}{u_0}\right)^n, \qquad D_0, u_0 \text{ constants}, \quad n > 0 \tag{10}$$

and

$$C(T) = C_0\left(\frac{T}{T_0}\right)^n, \qquad C_0, T_0 \text{ constants}, \quad n > 0 \tag{11}$$

In another context, (5) along with the constitutive assumption (10) is called the porous media equation, and it governs the motion of a fluid through a porous medium (see Section 1.3).

Fourier's law (8) may be generalized in yet another direction, of which we now give an example.

Example 3. In a low-temperature liquid phase of helium, heat transport is not governed by Fourier's law, but rather, by the nonlinear Gorter–Mellink law,

$$\phi = -KT_x^{1/3}, \qquad K \text{ constant} \tag{12}$$

Combined with (7), this gives the nonlinear heat conduction equation

$$T_t - k(T_x^{1/3})_x = 0$$

where $k = K/C\rho$ and C and ρ are constants. In the sequel, and in the exercises, we examine the properties of some of these nonlinear equations using the similarity method of Section 4.2.

Example 4. We consider the nonlinear diffusion equation

$$u_t - (uu_x)_x = 0 \tag{13}$$

which is a special case of (5) and (10) above. We assume that the constants have been scaled out of the problem. Equation (13) is sometimes called Boltzmann's problem. To compare the nonlinear equation (13) with the linear diffusion equation, we consider (13) over the domain $x \in R$, $t > 0$, subject to the initial condition of a unit point source applied at $x = 0$ at time $t = 0$. That is,

$$u(x,0) = \delta(x)$$

with

$$\int_R u(x,t)\,dx = 1 \qquad \text{for all } t > 0 \tag{14}$$

and

$$u(\pm\infty, t) = 0, \qquad t > 0$$

We know from Section 4.1 that the solution to the *linear* diffusion equation in this case is the fundamental solution. To determine a solution to the nonlinear problem we proceed by the similarity method. It is easy to see that (13) is invariant under the stretching transformation

$$\bar{x} = \varepsilon^a x \qquad \bar{t} = \varepsilon^b t \qquad \bar{u} = \varepsilon^{2b-a} u$$

Consequently, the similarity transformation is given by

$$u = t^{(2a-b)/b} y(z), \qquad z = \frac{x}{t^{a/b}} \tag{15}$$

Rather than immediately determine the ordinary differential equation for y, let us first specialize a and b by imposing the condition (14). We have

$$t^{(2a-b)/b} \int_R y\left(\frac{x}{t^{a/b}}\right) dx = t^{3(a/b)-1} \int_R y(z)\,dz = 1 \tag{16}$$

This condition can only be independent of t provided that $a/b = \frac{1}{3}$. Then the similarity transformation (15) becomes

$$u = t^{-1/3} y(z), \qquad z = x t^{-1/3} \tag{17}$$

Substituting these quantities into the PDE (13) yields an ordinary differential equation for $y = y(z)$:

$$3(yy')' + y + zy' = 0$$

This equation may be integrated at once to give

$$3yy' + zy = \text{constant} \tag{18}$$

Because the solution must be symmetric about $z = 0$ [$u_x(0, t) = 0$ and thus $y'(0) = 0$] we infer that the constant must be zero. Further, we must satisfy the boundary condition $u = 0$ at infinity. Hence from (18) it follows that we may take

$$y(z) = \begin{cases} \dfrac{A^2 - z^2}{6}, & \text{if } |z| < A \\ 0, & \text{if } |z| > A \end{cases} \tag{19}$$

where A is a constant of integration. We can determine A from condition (16). To this end

$$1 = \int_R y(z)\,dz = \int_{-A}^{A} y(z)\,dz = \tfrac{2}{9}A^3$$

which gives $A = (\frac{9}{2})^{1/3}$. Therefore, we have constructed a piecewise smooth solution

$$u(x,t) = \begin{cases} \frac{1}{6}t^{-2/3}(A^2t^{2/3} - x^2), & \text{if } |x| < At^{1/3} \\ 0, & \text{if } |x| > At^{1/3} \end{cases} \tag{20}$$

Time snapshots of this solution are shown in Figure 4.3.

The solution (20) to the nonlinear diffusion equation (13) is fundamentally different from the smooth solution (the fundamental solution) $u(x, t) =$

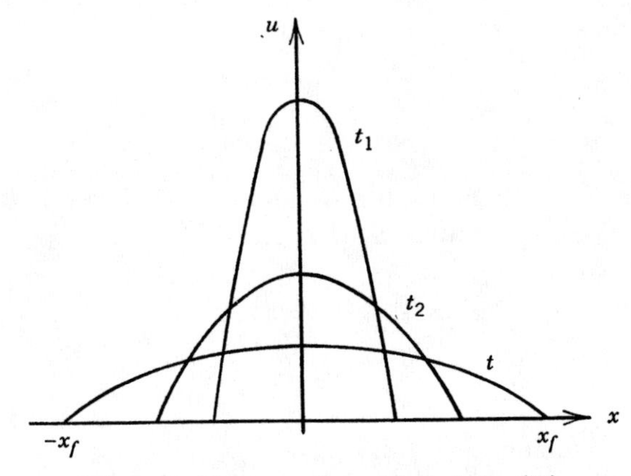

Figure 4.3. Time snapshots of the solution (20).

$(1/4\pi Dt)^{1/2}\exp(-x^2/4Dt)$ to the linear diffusion equation (3) when a point source is present at $x = 0$, $t = 0$. The solution (20) represents a type of sharp wavefront $x_f = At^{1/3}$ propagating into the medium with speed

$$\frac{dx_f}{dt} = \frac{1}{3}At^{-2/3}$$

Ahead of the wave the concentration u is zero, and at the front there is a jump discontinuity in the derivative of u. The wave slows down as t increases. Murray [1989] remarks, for example, that grasshoppers exhibit this type of dispersal behavior. In summary, nonlinear diffusion problems can behave quite differently from linear problems, even showing wavelike structure with propagating wavefronts.

Example 5. Consider the nonlinear diffusion model

$$uu_t - u_{xx} = 0, \qquad x > 0, \quad t > 0 \tag{21}$$

subject to the auxiliary conditions

$$u(x,0) = 0, \qquad x > 0 \tag{22}$$

$$u(\infty, t) = 0, \qquad t > 0 \tag{23}$$

$$u_x(0,t) = -1, \quad t > 0 \tag{24}$$

This problem is a simplified version of the nonlinear heat conduction problem (9) and (11) subject to a heat flux condition (24) imposed at $x = 0$, and the condition that $u = 0$ initially and at infinity. It is straightforward to observe that (22) is invariant under the stretching transformation

$$\bar{x} = \varepsilon^a x, \qquad \bar{t} = \varepsilon^b t, \qquad \bar{u} = \varepsilon^{b-2a}u$$

This invariance leads to the similarity transformation

$$u = t^{1-2a/b}y(z), \qquad z = \frac{x}{t^{a/b}} \tag{25}$$

Restrictions on the constants a and b can be determined by the initial and boundary conditions. Computing $u_x(0,t)$ gives

$$u_x = t^{1-3a/b}y'\frac{x}{t^{a/b}} \tag{26}$$

and therefore

$$u_x(0,t) = t^{1-3a/b}y'(0) = -1 \tag{27}$$

The left side of (27) cannot depend on t, so $a/b = \frac{1}{3}$. Consequently, the similarity transformation (25) becomes

$$u = t^{1/3}y(z), \qquad z = \frac{xt^{-1/3}}{\sqrt{3}} \tag{28}$$

(The inclusion of the factor $\sqrt{3}$ in the denominator of z makes the subsequent calculations more manageable.) Condition (23) implies that

$$y(\infty) = 0 \tag{29}$$

and the initial condition (22) can be written

$$\lim_{t\to 0} u(x,t) = \lim_{z\to\infty} t^{1/3}y(z) = \lim_{z\to\infty}\left(\frac{x}{\sqrt{3}\,z}\right)y(z) = 0$$

since $z \to \infty$ as $t \to 0+$, for each fixed $x > 0$. Therefore, (22) yields the same condition as (23), namely (29). The flux condition (24) becomes

$$y'(0) = -\sqrt{3} \tag{30}$$

Finally, substituting (28) into the PDE (21) gives the ordinary differential equation

$$y'' - y(y - zy') = 0, \qquad z > 0 \tag{31}$$

Therefore, the reduction is complete. The PDE (21) and three auxiliary conditions (22)–(24) have been transformed into a second order ordinary differential equation (31) subject to the two boundary conditions (29) and (30). Once the latter boundary value problem is solved for $y = y(z)$, the solution to the original problem will be given by (28). We also point out that the three original boundary conditions coalesced into two conditions on y; this coalescence had to occur since only two side conditions are required for the second order differential equation (31). Had it not occurred, the boundary value problem for y would have been overdetermined and the similarity method would have failed.

The solution to the nonautonomous equation (31) is by no means a simple matter, and we must resort to numerical methods to determine $y(z)$. This retreat from analytic calculations is common in applied mathematics; it is more typical than not that analytic, closed-form solutions cannot be found, and hence numerical calculations are required in most problems. In the present case, (31) together with the conditions (29) and (30) form a boundary

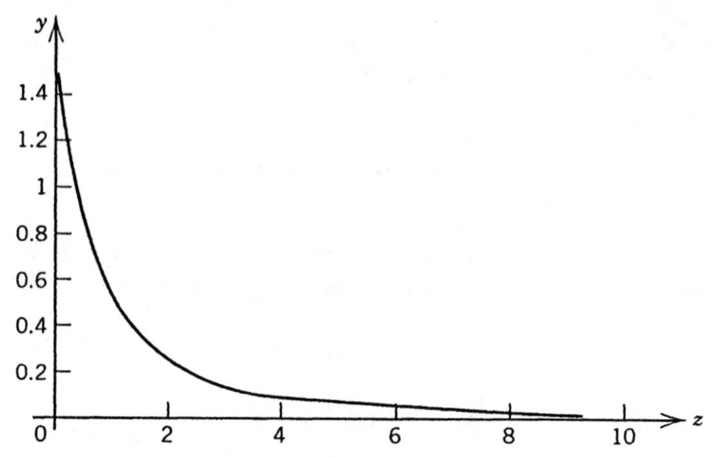

Figure 4.4. MATHEMATICA calculated plot of the solution of the nonlinear equation (31) subject to the boundary conditions (29) and (30).

value problem on the semi-infinite interval $0 < z < \infty$, and we used a *shooting method* to determine a numerical solution. The shooting method can be described briefly as follows. Most software packages (MATHEMATICA was used here) for second order differential equations require initial conditions, that is, conditions on both $y(0)$ and $y'(0)$. Here the initial condition $y(0)$ is not known. Therefore, we compute the numerical solution for several values of $y(0)$, along with the given value of $y'(0)$, until the numerical solution matches the right boundary condition $y(\infty) = 0$. Of course, the shooting method is more delicate on an infinite interval, and one must have confidence that a unique solution to the boundary value problem exists in the first place. See the exercises for verification of this fact. Following the procedure just described, we found that $y(0) = 1.5111$, and the solution to the boundary value problem for y is graphed in Figure 4.4.

The solution to the original PDE (21) subject to (22)–(24) is $u = u(x, t)$ given by (28). For each fixed time t_0, the solution profile is

$$u(x, t_0) = t_0^{1/3} y\left(\frac{x}{\sqrt{3}\, t_0^{1/3}}\right)$$

which is just the profile y in Figure 4.4 scaled horizontally by the factor $1/\sqrt{3}\, t_0^{1/3}$ and then amplified by the factor $t_0^{1/3}$. Hence the time snapshots of the solution are geometrically similar, from whence the terminology *similarity method* can be justified. We remark that the class of self-similar solutions to a given problem is invariant itself; that is, solutions are mapped to solutions under the given stretching transformation.

Finally, the similarity method discussed in this and the preceding sections can be extended to much more general transformations. For example, traveling wave solutions of the form

$$u = y(z), \qquad z = x - ct \tag{32}$$

are actually self-similar solutions and can be shown to originate from an invariance property of the defining PDE under certain one-parameter families of transformations. We have already observed (see, e.g., the Korteweg–deVries equation in Section 1.5) that a transformation of the form (32) reduces some PDEs to an ordinary differential equation for $y(z)$. The reader is referred to the references for more general concepts regarding the invariance of PDEs under one-parameter families of transformations.

EXERCISES

1. Consider the differential equation (31) subject to the boundary conditions (29) and (30). Show that the transformation

$$w = z^2 y, \qquad v = z^2(y - zy')$$

reduces (31) to a first-order autonomous equation

$$\frac{dv}{dw} = \frac{v(2 - w)}{3w - v}$$

and show that the solution trajectory in the wv-plane is along a separatrix connecting the origin to a saddle point at $w = 2$, $v = 6$. Thus show that y behaves like $y \sim 2/z^2$ as $z \to \infty$.

2. Use the similarity method to analyze the nonlinear diffusion problem

$$u_t - \left(u_x^{1/3}\right)_x = 0, \qquad x > 0, \quad t > 0$$

$$u_x(0,t) = -1, \qquad t > 0$$

$$u(x,0) = 0, \quad x > 0, \qquad u(\infty,t) = 0, \quad t > 0$$

Show that a similarity transformation is given by $u = \sqrt{t}\, y(z)$, $z = x/\sqrt{t}$ and that the problem reduces to

$$\tfrac{2}{3}(y')^{-2/3} y'' + zy' - y = 0, \qquad z > 0$$

$$y'(0) = 1, \qquad y(\infty) = 0$$

Use the transformation $w = zy^{1/2}$, $v = z(y')^{1/3}$ to reduce the problem to a wv phase plane, and argue that the unique trajectory is along a separatrix connecting two critical points. Compute $y(0)$ and show that

$y(z) \sim Cz^{-2}$ as $z \to \infty$, for some constant C, and hence for fixed t,

$$u(x,t) \sim \frac{At^{3/2}}{x^2} \qquad \text{as } x \to \infty$$

Sketch a graph of $y = y(z)$.

3. Use the similarity method to find a closed-form analytic solution to the problem

$$\begin{aligned} xu_t - u_{xx} &= 0, & x > 0, \quad t > 0 \\ u(x,0) &= 0, & x > 0 \\ u(\infty, t) &= 0, & t > 0 \\ u_x(0,t) &= -1, & t > 0 \end{aligned}$$

4. The second order ordinary differential equation

$$y'' - G(x, y, y') = 0$$

is *invariant* under the one-parameter family of stretching transformations

$$\bar{x} = \varepsilon x, \qquad \bar{y} = \varepsilon^b y$$

if there exists a smooth function $f(\varepsilon)$, $f(1) = 1$, for which

$$\bar{r} - G(\bar{x}, \bar{y}, \bar{p}) = f(\varepsilon)(r - G(x, y, p))$$

where $r = y''$ and $p = y'$. In this case, prove that the second order equation for y can be reduced to a first order equation of the form

$$\frac{dv}{dw} = \frac{g(v,w) - (b-1)v}{v - bw}$$

where $w = y/x^b$ and $v = y'/x^{b-1}$, and g is a fixed function of v and w. (The variables w and v are called *Lie variables*, and the wv-plane is called the *Lie plane*, after S. Lie.)

5. Consider the nonlinear PDE

$$uu_t + u_x^2 = 0$$

Find the simplest, nontrivial stretching transformation under which the PDE is invariant, determine the similarity transformation, reduce the PDE to an ordinary differential equation, and solve to find the self-similar solutions.

6. Consider the porous media equation

$$u_t = (u^m)_{xx}, \qquad m > 2$$

Discuss the behavior of similarity solutions of the form $u = t^{1/(m-1)}y(z)$, where $z = x/t$.

4.4 REACTION–DIFFUSION; FISHER'S EQUATION

Many natural processes inherently involve the mechanisms of both diffusion and reaction, and such problems are often modeled by so-called *reaction–diffusion* (*R-D*) *equations* of the form

$$u_t - Du_{xx} = f(u) \tag{1}$$

where f is a given, usually nonlinear, function of u. In Section 1.3 we introduced, for example, the *Fisher equation*,

$$u_t - Du_{xx} = ru\left(1 - \frac{u}{K}\right) \tag{2}$$

to model the diffusion of a species (e.g., an insect population u) when the reaction or growth term is given by the logistics law. Here D is the diffusion constant, and r and K are the growth rate and carrying capacity, respectively. R-D equations have become one of the most important classes of nonlinear equations because of their occurrence in many biological and chemical (e.g., combustion) processes.

In this section we carefully examine the Fisher equation and treat it as a prototype of R-D equations in one dimension. In particular, we use the Fisher equation as a vehicle to introduce three different mathematical techniques that are often employed to study R-D equations: The first is the question of existence of wavefront-type solutions through an examination of phase-plane phenomena associated with such equations, the second is a singular perturbation technique to determine the form of the wavefront, and the third is the stability of wavefront-type solutions under small perturbations of the waveform. In the latter case, we superimpose a small change upon the wavefront and then ask how that change evolves in time. Our approach is similar to that of Murray [1989].

Let us rewrite the Fisher equation as

$$\bar{u}_{\bar{t}} - D\bar{u}_{\bar{x}\bar{x}} = r\bar{u}\left(1 - \frac{\bar{u}}{K}\right) \tag{3}$$

using a barred letter $\bar{u}$ to denote the population density, given in number of individuals per unit length, and using barred dimensioned variables $\bar{x}$ and $\bar{t}$ to denote length and time. We have in mind later using unbarred quantities to represent dimensionless variables in the formulation of the main, scaled problem with which we shall work. Although we have sometimes violated this basic principle in the formulation of problems up until now, it is a fundamental fact in modeling that all problems should be reduced to dimensionless form by introducing scaled, dimensionless independent and dependent variables and then recasting the equations in terms of these new scaled variables. In the present case, equation (3) contains three dimensioned constants D, r, and K with the dimensions length-squared per time, time^{-1}, and individuals, respectively. Thus we may define dimensionless quantities t, x, and u by

$$t = \frac{\bar{t}}{r^{-1}}, \qquad x = \frac{\bar{x}}{\sqrt{D/r}}, \qquad u = \frac{\bar{u}}{K} \tag{4}$$

Here r^{-1}, $\sqrt{D/r}$, and K are called the time scale, length scale, and population scale, respectively. Using the transformation (4), the dimensioned equation (3) can be written in dimensionless form as

$$u_t - u_{xx} = u(1 - u) \tag{5}$$

Now, all the variables in (5) are dimensionless; u measures the population relative to the carrying capacity K, t measures time relative to the growth rate r, and x measures distances relative to $\sqrt{D/r}$, which is a diffusion length scale. We notice that all of the constants disappeared in (5); generally, if constants appear, they will be dimensionless combinations of the dimensioned constants in the original problem.

In general, if $\bar{v}$ is a dimensioned variable (dependent or independent) in a given problem, the associated dimensionless variable v is defined by $v = \bar{v}/v_c$, where v_c is a constant having the same dimensions as $\bar{v}$ and is composed of the dimensioned constants in the problem; v_c is called the *characteristic scale* for $\bar{v}$, and for proper scaling v_c should be the same order as $\bar{v}$, so that the new dimensionless v will be of order unity. When a problem is scaled correctly, the dimensionless variables will be $O(1)$, and the magnitude of each term in the differential equation will be represented by the magnitude of its coefficient. Only when a problem has been scaled properly can one be certain of the magnitudes of the terms in an equation; this observation is central in perturbation methods where approximate solutions are sought by neglecting some of the terms in an equation. Some problems require multiple scales (i.e., they must be scaled differently in different regimes). For a thorough discussion of scaling we refer the reader to Lin and Segel [1974] or Logan [1987].

Traveling Wave Solutions

Now we investigate the question of whether equation (5) can admit wave-front-type traveling wave solutions (hereafter, abbreviated TWS). These are solutions of the form

$$u(x,t) = U(z), \qquad \text{where} \quad z = x - ct \tag{6}$$

where c is a positive constant and $U(z)$ has the property that it approaches constant values at $z = \pm\infty$. The function U representing the waveform, and to be determined, should be twice continuously differentiable on R. A priori, the wave speed c is unknown, and it must be determined as part of the solution to the problem. Substituting (6) into (5) yields a second order ordinary differential equation for $U(z)$:

$$-cU' - U'' = U(1 - U), \qquad -\infty < z < \infty \tag{7}$$

where prime denotes the derivative d/dz. This equation cannot be solved in closed form, and the best approach to analyze the problem is in a two-dimensional phase plane. In the standard way, this nonlinear differential equation can be reduced to a pair of first order equations by introducing a new dependent variable V defined by $V = U'$. Then we have the autonomous system

$$\begin{aligned} U' &= V \\ V' &= -cV - U(1 - U) \end{aligned} \tag{8}$$

The critical points of this system in a UV phase plane are $P:(0,0)$ and $Q:(1,0)$. The Jacobi matrix of the linearized system is

$$J(U,V) = \begin{pmatrix} 0 & 1 \\ 2U - 1 & -c \end{pmatrix}$$

It is easy to check that the eigenvalues of $J(1,0)$ are

$$\lambda_{\pm} = \frac{-c \pm \sqrt{c^2 + 4}}{2}$$

which are real and of opposite sign; therefore, $(1,0)$ is a saddle point. The eigenvalues of $J(0,0)$ are

$$\lambda_{\pm} = \frac{-c \pm \sqrt{c^2 - 4}}{2}$$

and therefore $(0,0)$ is a stable node if $c^2 \geq 4$ (the eigenvalues are real and both negative), and $(0,0)$ is a stable spiral if $c^2 < 4$ (the eigenvalues are

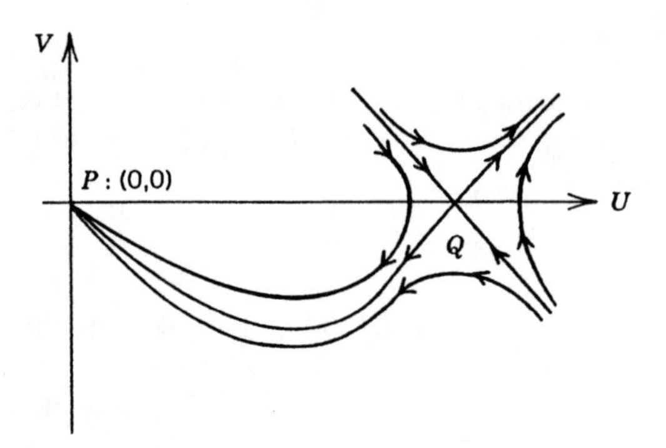

Figure 4.5. Phase portrait for (8) when $c > 2$.

complex with negative real part). The phase portrait for the system (8) is shown in Figure 4.5 in the case $c \geq 2$. It is clear that there is a unique separatrix connecting the saddle point Q to the stable node P. (Actually, this fact can be proved rigorously by a geometric argument using the properties of the vector field defining the system (8), along with two fundamental theorems from the theory of differential equations, the Poincaré–Bendixson theorem and the stable and unstable manifold theorem. We refer the reader to, for example, Hartman [1964].)

Now, having an understanding of the behavior of the system (8), we return to the problem of TWS. We note that along a path in phase space, the parameter z must tend to $+\infty$ or $-\infty$ as the path enters or exits a critical point. Therefore, the separatrix connecting Q to P is described by functions $U = U(z)$, $V = V(z)$, with $U \to 1$ as $z \to -\infty$, and $U \to 0$ as $z \to +\infty$. Moreover, because P and Q are critical points, the right side of (8) vanishes at P and Q; therefore, $U' \to 0$ as $|z| \to \infty$. Consequently, the separatrix connecting Q to P represents a monotone, TWS whose general profile is shown in Figure 4.6 for the case $c \geq 2$.

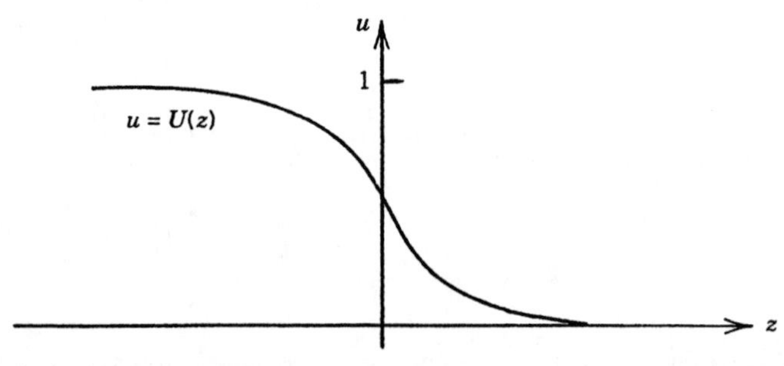

Figure 4.6. Traveling wave solution to Fisher's equation when $c > 2$. This corresponds to the separatrix connecting Q to P in Figure 4.5.

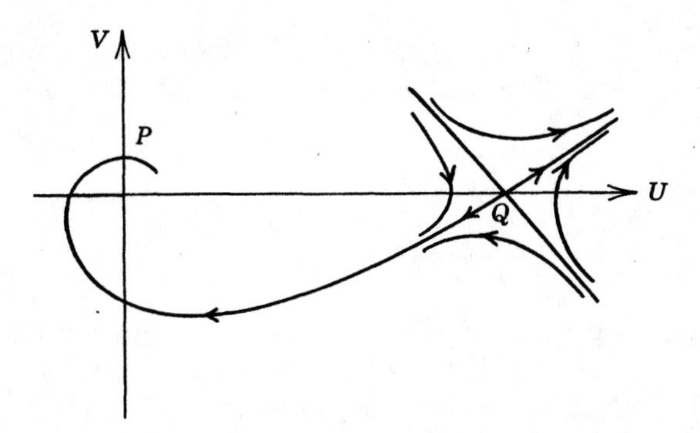

Figure 4.7. Phase portrait for (8) when $0 < c < 2$.

If $c < 2$, then $(0, 0)$ is a stable spiral; now the wave oscillates about $U = 0$ (see Figures 4.7 and 4.8), and if we are only interested in positive values of U in order to have a physically realistic solution (e.g., if U is a population), we would have to reject this case. In summary, we have the following result.

Theorem 1. For each $c \geq 2$ there exists a unique TWS $u(x, t) = U(x - ct)$ to equation (5) with the property that U is monotonically decreasing on R with $U(-\infty) = 1$, $U(\infty) = 0$, and $U'(\pm\infty) = 0$.

Perturbation Solution

Although we have shown the existence of a TWS by appealing to a geometric argument in a phase plane, we do not have a formula for the solution. Using a perturbation method we can obtain an approximation of the solution in the case $c \geq 2$. Basically, a perturbation method is an attempt to find an expansion for the solution of a problem in terms of a power series in some small parameter in the problem. The reader without any formal instruction in perturbation methods should be able to follow the discussion and use the

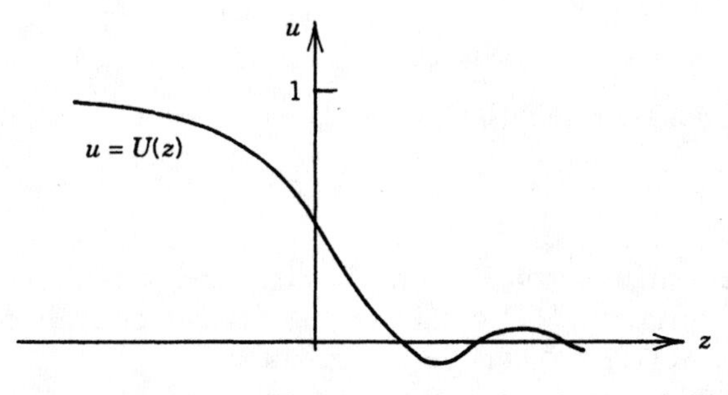

Figure 4.8. Traveling wave solution to Fisher's equation when $0 < c < 2$. There is a decaying oscillating tail as $z \to +\infty$ because P is a spiral point.

technique as a basis for gaining additional experience in this important method in the sequel. We seek an approximate solution to the boundary value problem

$$U'' + cU' + U(1 - U) = 0, \qquad \infty < z < +\infty \tag{9}$$

$$U(-\infty) = 1, \qquad U(+\infty) = 0 \tag{10}$$

where $c \geq 2$. Because equation (9) is autonomous (independent of explicit dependence on z), then if $U(z)$ is a solution, so is $U(z + z_0)$ for any fixed constant z_0. That is, the solution curve may be translated to the left or right, and we still obtain a solution. Consequently, we may take

$$U(0) = \tfrac{1}{2} \tag{11}$$

since the value at $z = 0$ can be chosen to be any number in the range of U. Thus we append condition (11), along with (10), to the second order equation (9). Now we identify a small parameter in the problem by taking

$$\varepsilon = \frac{1}{c^2} \leq 0.25$$

Then the differential equation (9) becomes

$$\sqrt{\varepsilon}\, U'' + U' + \sqrt{\varepsilon}\, U(1 - U) = 0 \tag{12}$$

In (12) the dominant term is U' when ε is small, and the equation is, approximately, $U' = 0$, which has constant solutions; thus this approximation holds for large z since $U = 0$ for z large and positive, and $U = 1$ for z large and negative. In the large interval where U is changing from 1 to zero, a different dominant balance must occur. Therefore, we shrink this large interval to an order one interval by introducing the change of variables

$$s = \sqrt{\varepsilon}\, z = \frac{z}{c}, \qquad g(s) = U\left(\frac{s}{\sqrt{\varepsilon}}\right) \tag{13}$$

The differential equation (12) then becomes

$$\varepsilon g'' + g' + g(1 - g) = 0 \tag{14}$$

where prime, of course, denotes d/ds. Thus in terms of the order 1 variable s, the last two terms in (14) dominate the second derivative term. The auxiliary conditions (10) and (11) transform into

$$g(-\infty) = 1, \qquad g(0) = \tfrac{1}{2}, \qquad g(+\infty) = 0 \tag{15}$$

To find an approximate solution of (14)–(15) in this case we assume a perturbation series

$$g(s) = g_0(s) + \varepsilon g_1(s) + \varepsilon^2 g_2(s) + \cdots \tag{16}$$

which is an expansion in powers of the small parameter ε, and where the coefficients g_0, g_1, g_2, and so on, are to be determined. Substituting the Ansatz (16) into the differential equation (14) and the auxiliary conditions (15) and setting the coefficients of the powers of ε equal to zero gives the following sequence of problems for g_0, g_1, and so on:

$$g_0' = -g_0(1 - g_0), \qquad g_0(0) = \tfrac{1}{2}, \quad g_0(-\infty) = 1, \quad g_0(+\infty) = 0 \tag{17}$$

$$g_1' = -g_1(1 - g_0) + g_0'', \qquad g_1(-\infty) = g_1(0) = g_1(+\infty) = 0, \qquad \text{etc.} \tag{18}$$

Equation (17) can be solved to obtain

$$g_0(s) = (1 + e^s)^{-1} \tag{19}$$

which meets all three boundary conditions in (17). Then (18) can be solved to get

$$g_1(s) = e^s(1 + e^s)^{-2} \ln \frac{4e^s}{(1 + e^s)^2} \tag{20}$$

Consequently, the expansion (16) becomes, in terms of the original variables U and z, and the wave speed c,

$$U(z) = \frac{1}{1 + e^{z/c}} + \frac{1}{c^2} e^{z/c}(1 + e^{z/c})^{-2} \ln \frac{4e^{z/c}}{(1 + e^{z/c})^2} + O\left(\frac{1}{c^4}\right) \tag{21}$$

Therefore, we have obtained an approximate, asymptotic form of the traveling wave $U(z)$ when $c \geq 2$. The approximation is most accurate for large c and least accurate for $c = 2$. It can be shown that the $O(1/c^4)$ term is uniform for all $z \in R$ (i.e., the first two terms in the expansion give a uniform approximation that is valid for all z). Actually, the first term, $(1 + e^{z/c})^{-1}$, is remarkably close (within a few percent) to a numerically computed solution, even in the case $c = 2$. The reader is invited to use a software package or graphing calculator to sketch graphs of (21).

Those familiar with perturbation methods will recognize equation (12) as a standard singular perturbation problem, because the small parameter ε multiplies the highest derivative. However, since the boundary conditions at infinity are automatically satisfied by the scaled problem (14), equation (12) is really just a regular perturbation problem, and the leading order term in the expansion gives a uniformly valid approximation on the entire domain.

Stability of Traveling Waves

One of the most important and interesting questions in applied mathematical problems is that of the stability of a given state of a system. For example, the traveling wave solution that we derived for Fisher's equation represents a state, realized by a surface $u = U(x - ct)$ defined over spacetime, of the physical reactive–diffusive system modeled by the equation. We say that a state is *asymptotically stable* if a small perturbation or change imposed on the state at some time (say $t = 0$) eventually decays away and the system returns to its original state. We now show that the TWS to Fisher's equation is asymptotically stable to small perturbations imposed in the moving coordinate frame of the wave, subject to the condition that the perturbation vanish outside some closed interval. Thus we are not showing stability with respect to all perturbations, just those we described.

We first write Fisher's equation (5) in a moving coordinate frame by changing variables according to

$$t = t, \qquad z = x - ct$$

where $c \geq 2$. Then (5) becomes

$$u_t - u_{zz} - cu_z = u(1 - u) \tag{22}$$

Here we are abusing the notation and reusing u to denote the dependent variable as a function of t and z; technically, of course, a new letter should be selected, but it is common practice to adhere to this notational abuse unless possible confusion dictates more care in defining our symbols. Now, we have obtained a wavefront solution $U(z)$ to (23), so we consider solutions of (22) of the form

$$u = U(z) + V(z, t) \tag{23}$$

where V is a small perturbation, or deviation, from the known state $U(z)$. We assume that

$$V(z, t) = 0 \qquad \text{for } |z| \geq L, \quad \text{for some } L > 0$$

which is the assumption that the perturbation vanishes outside some finite interval in the moving frame. Figure 4.9 shows a typical perturbation V on the wave $U(z)$. It is of course implied by this argument that $V(z, 0)$, the perturbation at time $t = 0$, is given. Substituting (23) into (22) gives a PDE for the perturbation $V(z, t)$, namely

$$V_t - V_{zz} - cV_z = (1 - 2U)V - V^2 \tag{24}$$

Equation (24) is called the *nonlinear perturbation equation*, and it governs the small deviations V. However, because V was assumed to be small, we may

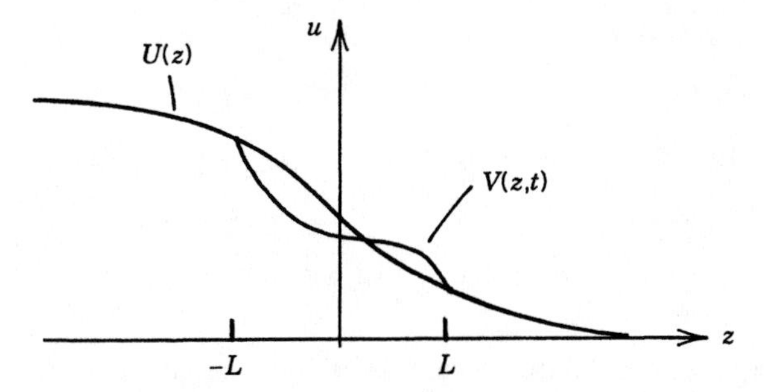

Figure 4.9. Traveling wave solution to Fisher's equation with an imposed small perturbation $V(z,t)$ in the moving coordinate frame.

neglect the V-squared term, which is a much smaller term than the remaining first degree terms in (24). Thus we obtain the *linearized perturbation equation*

$$V_t - V_{zz} - cV_z = (1 - 2U)V \tag{25}$$

We look for solutions of the linear equation (25) of the form

$$V(z,t) = v(z)e^{-\lambda t} \tag{26}$$

Substituting (26) into (25) yields a second order equation for the spatial part $v(z)$,

$$v'' + cv' + [\lambda + 1 - 2U(z)] = 0 \tag{27}$$

and the boundary condition on V implies that

$$v(-L) = v(L) = 0 \tag{28}$$

Equations (27) and (28) represent an eigenvalue problem for $v(z)$, where the growth factor λ is interpreted as an eigenvalue. If there exist negative eigenvalues, (26) implies that the perturbation V will grow in time and the TWS $U(z)$ will be unstable. However, if the only eigenvalues to (27)–(28) are positive, V will decay (exponentially) to zero as $t \to \infty$, and $U(z)$ will be asymptotically stable.

To solve the eigenvalue problem (27)–(28) we first use the Liouville–Green transformation to eliminate the first derivative term and put the problem in normal form. In the present case the transformation is

$$v(z) = w(z)e^{-cz/2} \tag{29}$$

Then (27) and (28) become

$$w'' + \left[\lambda - \left(2U(z) + \frac{c^2}{4} - 1\right)\right]w = 0 \tag{30}$$

$$w(-L) = w(L) = 0 \tag{31}$$

A lot of facts are known about boundary value problems in this form. For example, *if $q(z) > 0$ and q is continuous, the eigenvalues λ of the boundary value problem $w'' + [\lambda - q(z)]w = 0$, $w(-L) = w(L) = 0$, on the interval $-L < z < L$, are all positive* (see, e.g., Hochstadt [1975, p. 148]). In our case

$$2U(z) + \frac{c^2}{4} - 1 \geq 2U(z) > 0 \qquad \text{for } c \geq 2$$

and consequently, the eigenvalue problem (30)–(31), and hence (27)–(28), has positive eigenvalues, which gives our asymptotic stability result.

In summary, we have shown that a small perturbation of finite extent, imposed in the traveling wave frame on the waveform $U(z)$, decays as $t \to \infty$, and thus the system returns, asymptotically, to its original state. This type of small perturbation argument is common in applied mathematics, and the stability calculation for Fisher's equation is representative of many of them.

EXERCISES

1. Show that the traveling wave solution $U(z)$ obtained for Fisher's equation in the case $c \geq 2$ has the property that the slower the wave moves (i.e., the smaller c is), the steeper the wavefront.
2. Find an exact solution to Fisher's equation (5) of the form

$$U(z) = \frac{1}{(1 + ae^{bz})^d}, \qquad z = x - ct$$

   To what wave speed c does this solution correspond?
3. Consider the nonlinear R-D equation

$$u_t - (uu_x)_x = u(1 - u)$$

   Investigate the existence of wavefront-type TWS of speed $c = 1/\sqrt{2}$.

Show that solutions exist of the form

$$u = U(z) = 1 - \exp\left(-\frac{z}{\sqrt{2}}\right), \quad z < 0; \qquad u = 0, \quad z > 0$$

$$(z = x - ct)$$

4. *Project*. Develop a numerical procedure to compute the TWS $u = U(z)$ to Fisher's equation in the case $c = 2$, and sketch the graph.
5. Steady solutions $u = u(x)$ to Fisher's equation on the domain $0 < x < a$ with homogeneous end conditions satisfy the nonlinear boundary value problem

$$u'' + u(1-u) = 0, \qquad 0 < x < a; \quad u(0) = u(a) = 0$$

(a) Show that nontrivial, positive, steady solutions exist provided that $a > \pi$. Suggestion: go to the phase plane and show that

$$a = 2\int_0^{u_m} \frac{du}{\left(u_m^2 - u^2 + \frac{2}{3}(u^3 - u_m^3)\right)^{1/2}}$$

where $u_m = u(a/2) = \max u$.

(b) For $0 < a - \pi \ll 1$ show that the nontrivial solution takes the form $u = \frac{3}{4}(a-\pi)\sin(\pi x/a) + O((a-\pi)^2)$. Suggestion: first rescale the problem to obtain

$$\frac{d^2u}{dy^2} + a^2 u(1-u) = 0, \qquad 0 < y < 1; \quad u(0) = u(1) = 0$$

and then assume $u(y) = u_0(y) + \varepsilon u_1(y) + \varepsilon^2 u_2(y) + \cdots$, where $\varepsilon = a - \pi$.

(c) Using the suggestion in part a, show that $u_m \sim 1 - e^{-a/2}$ as $a \to \infty$.

4.5 CONVECTION–DIFFUSION; BURGERS' EQUATION

The prototype equation for nonlinear convection–diffusion processes is *Burgers' equation*,

$$u_t + uu_x - Du_{xx} = 0 \tag{1}$$

The term uu_x represents a nonlinear convection or transport term, and $-Du_{xx}$ represents a Fickian diffusion term. On one hand, the nonlinear convection term has a *shocking-up* effect on an initial waveform, while the diffusion term attempts to smear out the solution. Thus (1) is a balance between these two effects. Equation (1) is often taken as the analog equation of compressible, viscous fluid flow; in that case the diffusion term is interpreted as a model viscosity term, which is also a dissipative term that tends to smear out signals. In Chapter 5 we derive Burgers' equation in a weakly nonlinear asymptotic limit of governing equations of viscous flow.

Traveling Wave Solution

The competition between the convective term and the dissipative term is best observed by deriving traveling wave solutions (TWS). Therefore, we look for twice continuously differentiable solutions of (1) of the form

$$u(x,t) = U(z) \qquad z = x - ct \tag{2}$$

where c is to be determined. As was the case with other equations, we want wavefront-type solutions that approach positive constant values u_1 at $z = +\infty$, and u_2 at $z = -\infty$, respectively, and assume that $u_1 < u_2$. Substituting (2) into (1) gives an ordinary differential equation for $u = U(z)$,

$$-cU' + UU' - DU'' = 0$$

where prime denotes d/dz. This equation may be integrated immediately to obtain

$$-cU + \tfrac{1}{2}U^2 - DU' = A \tag{3}$$

where A is a constant of integration. Writing (3) in standard form gives

$$U' = D^{-1}\left(\tfrac{1}{2}U^2 - cU - A\right) \tag{4}$$

To evaluate A and c we take the limit as $z \to -\infty$ and $z \to \infty$ in (4) to get

$$A = \tfrac{1}{2}u_1^2 - cu_1 = \tfrac{1}{2}u_2^2 - cu_2$$

whence

$$c = \frac{u_1 + u_2}{2} \tag{5}$$

which gives the wave speed as the average of the two known states at infinity. Therefore, the constant of integration A is given by

$$A = -\frac{u_1 u_2}{2}$$

The differential equation (4) therefore becomes

$$-2DU' = (U - u_1)(u_2 - U) \tag{6}$$

Separating variables and integrating gives

$$\frac{z}{D} = \frac{2}{u_2 - u_1} \ln \frac{u_2 - U}{U - u_1}$$

where the constant of integration was chosen so that $U(0) = c$. Note that c is the average value of u_1 and u_2, and because (6) is autonomous, U may be chosen to be any value between u_1 and u_2 (solutions are translational invariant). Solving for U then gives the waveform

$$U(z) = u_1 + \frac{u_2 - u_1}{1 + \exp[(u_2 - u_1)z/2D]} \tag{7}$$

where $z = x - ct$ is a traveling wave variable. A graph of (7) is shown in Figure 5.1.

Consequently, we have constructed a positive monotonically decreasing TWS that approaches constant states at infinity. The speed c of the wave is the average value of the states u_1 and u_2 at infinity [see (5)]. A waveform of this shape would break and form a shock if the diffusion term were absent; its presence prevents the deformation, and the transport term and diffusion term are exactly balanced in (7). We remark that the diffusion coefficient D affects the shape of the waveform (7); if D is large, there is a greater diffusion effect and the wave in Figure 5.1 has a shallow gradient. On the other hand, if D is small, the gradient is steep. Analytically, we can see this as follows. The thickness of the wave (7) is defined to be the quantity $(u_2 - u_1)/\max|u'(z)|$, which one can easily compute to be $8D/(u_2 - u_1)$, since the maximum derivative occurs at $z = 0$. Because the profile in Figure 4.10 resembles the actual profile in a real shock wave when D is small, the TWS (7) is frequently called the *shock structure* solution; in an

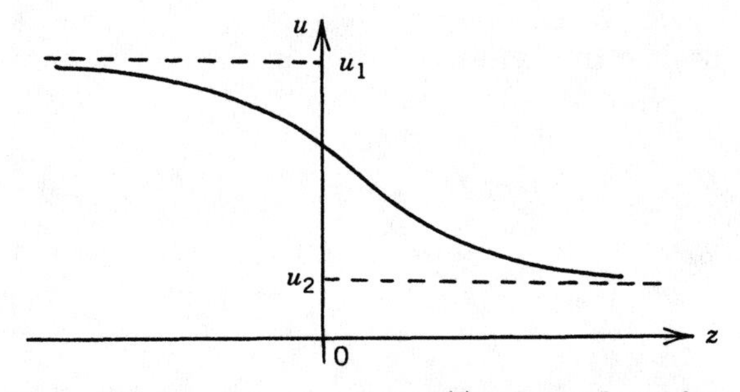

Figure 4.10. Traveling wave solution (7) to Burgers' equation.

actual, physical shock wave (not the mathematical shock wave defined by a discontinuity in Chapter 3), the tendency of a wave to break is balanced by viscous effects in the region where the steep gradients occur. We shall discuss the limiting behavior as $D \to 0$ in the sequel.

Initial Value Problem

Now we change gears and address the initial value problem for Burgers' equation:

$$u_t + uu_x - Du_{xx} = 0, \qquad x \in R, \quad t > 0 \tag{8}$$

$$u(x,0) = u_0(x), \qquad x \in R \tag{9}$$

Amazingly enough, this problem can be reduced to the initial value problem for the linear diffusion equation and thus can be solved analytically, in closed form. The reduction is accomplished by the Cole–Hopf transformation, a change of variables that was discovered independently by J. Cole and E. Hopf in the early 1950s. First, let us introduce the (potential) function w defined by the equation

$$u = w_x \tag{10}$$

Then (8) becomes

$$w_{xt} + w_x w_{xx} - Dw_{xxx} = 0$$

which can be integrated (with respect to x) immediately to obtain

$$w_t + \tfrac{1}{2}w_x^2 - Dw_{xx} = 0 \tag{11}$$

Now introduce the dependent function v defined by

$$w = -2D \ln v \tag{12}$$

It is easy to calculate the derivatives of w and substitute them into (11) to obtain

$$0 = w_t + \tfrac{1}{2}w_x^2 - Dw_{xx} = -2Dv^{-1}(v_t - Dv_{xx})$$

Therefore, the transformation

$$u = -\frac{2Dv_x}{v} \tag{13}$$

which is a combination of (10) and (12), reduces (8) to

$$v_t - Dv_{xx} = 0 \tag{14}$$

which is the diffusion equation. Equation (13) is the *Cole–Hopf transformation*. Also, via (13), the initial condition (9) on u transforms into an initial condition on v. We have

$$u(x,0) = u_0(x) = -\frac{2Dv_x(x,0)}{v(x,0)}$$

Integrating both sides of this equation yields

$$v(x,0) = v_0(x) = \exp\left(-\frac{1}{2D}\int_0^x u_0(y)\,dy\right) \tag{15}$$

Now, the solution to the initial value problem (14)–(15) for the diffusion equation is

$$v(x,t) = \left(\frac{1}{4\pi Dt}\right)^{1/2}\int_R v_0(\xi)\exp\left[-\frac{(x-\xi)^2}{4Dt}\right]d\xi$$

Therefore,

$$v_x(x,t) = -\left(\frac{1}{4\pi Dt}\right)^{1/2}\int_R v_0(\xi)\frac{x-\xi}{2Dt}\exp\left[-\frac{(x-\xi)^2}{4Dt}\right]d\xi$$

Consequently, from (13),

$$u = -\frac{2Dv_x}{v} = \frac{\int_R[(x-\xi)/t]v_0(\xi)\exp\left[-(x-\xi)^2/4Dt\right]dx}{\int_R v_0(\xi)\exp\left[-(x-\xi)^2/4Dt\right]d\xi} \tag{16}$$

where $v_0(\xi)$ is given by (15). It is now straightforward to see that the solution (16) of the initial value problem (8)–(9) can be written as

$$u(x,t) = \frac{\int_R[(x-\xi)/t]e^{-G(\xi,x,t)/2D}\,d\xi}{\int_R e^{-G(\xi,x,t)/2D}\,d\xi} \tag{17}$$

where

$$G(\xi,x,t) = \frac{(x-\xi)^2}{2t} + \int_0^\xi u_0(y)\,dy \tag{18}$$

In summary, we have found an analytic expression for the solution of the initial value problem associated with Burgers' equation.

EXERCISES

1. Show that the solution to the initial value problem

$$u_t + uu_x - Du_{xx} = 0, \qquad x \in R, \quad t > 0$$

$$u(x,0) = U \quad \text{if } x < 0; \qquad u(x,0) = 0 \quad \text{if } x > 0$$

can be written

$$Uu^{-1} = 1 + \frac{\exp[U(x - Ut/2)/2D]\operatorname{erfc}[-x/(2\sqrt{Dt})]}{\operatorname{erfc}[(x - Ut)/(2\sqrt{Dt})]}$$

where erfc = 1 − erf is the complementary error function. Discuss the behavior of the solution for small D. Sketch some time snapshots of the solution.

2. (a) Find an equilibrium solution $u = u^*(x)$ to Burgers' equation $u_t + uu_x = u_{xx}$ on the interval $0 < x < a$ subject to the boundary conditions $u = 0$ at $x = 0$ and $u = 1$ at $x = a$.

(b) Let $u(x, t) = u^*(x) + U(x, t)$, where U is a small perturbation of the equilibrium solution, and where $U = 0$ at $x = 0, a$. Show that the linearized perturbation equation for U is given by

$$U_t = U_{xx} - (u^*U)_x, \qquad 0 < x < a, \quad t > 0$$

(c) By assuming solutions of the form $U = \phi(x)e^{\lambda t}$, $\phi(0) = \phi(a) = 0$, show that λ must be negative and therefore u^* is linearly asymptotically stable to small perturbations. Hint: multiply the ϕ-equation by ϕ and integrate from $x = 0$ to $x = a$.

4.6 ASYMPTOTIC SOLUTIONS TO BURGERS' EQUATION

In the preceding section we obtained a formula [equations (17) and (18) in Section 4.5] for the solution to the pure initial value problem for Burgers' equation. Now we want to understand certain aspects of this solution. It is common practice in the subject of nonlinear partial differential equations to examine problems, or solutions to problems, in various limits; for example, if a parameter is present, one can inquire about the behavior in the limit of small or large values of the parameter. Or, one may ask about the long-time behavior of the solution. The first problem we consider is the behavior of the solution to the initial value problem for Burgers' equation in the limit of small D (i.e., for small values of the diffusion constant). Intuitively, we expect the solution to approach the solution to the inviscid Burgers' equation in the limit of small D; we shall show that our intuition proves to be correct. Second, we investigate how Burgers' equation propagates a delta function (i.e., a point source), both in the limit of small D and in the limit of large D.

To determine the solution in specific cases it is necessary to approximate the integrals in formula (17) by asymptotic formulas. Such formulas are presented and rigorously proved in books on asymptotic analysis (see, e.g., Murray [1984] or Bender and Orszag [1978]). We now state one of the fundamental results, Laplace's theorem, and give a heuristic proof.

Theorem 1 (*Laplace*). Consider the integral

$$f(s) = \int_{-\infty}^{\infty} g(\xi) e^{sh(\xi)}\, d\xi \tag{1}$$

where g is continuous and h is in class C^2. Let a be a single stationary point of h [i.e., $h'(a) = 0$] and assume that $h''(a) < 0$. Then

$$f(s) \sim g(a) e^{sh(a)} \left[-\frac{2\pi}{sh''(a)} \right]^{1/2} \qquad \text{as } s \to \infty \tag{2}$$

Formula (2) is an estimate giving the dominant behavior of the integral in (1) for large values of the parameter s. We can observe instantly that the integrals in equation (17) of Section 4.5 are of the form (1) if we make the identification $s = 1/D$, $h(\xi) = -G(\xi, x, t)/2$, and $g(\xi) = (x - \xi)/t$ in one case [the numerator of (17)] and $g(\xi) = 1$ in the other case [the denominator of (17)]. Then (2) will give the behavior as $D \to 0$.

To demonstrate the validity of (2) in a special case we expand g and h about $\xi = a$, using Taylor's formula to get

$$\begin{aligned} g(\xi) &= g(a) + g'(a)(\xi - a) + \cdots \\ h(\xi) &= h(a) + h'(a)(\xi - a) + \tfrac{1}{2}h''(a)(\xi - a)^2 + \cdots \\ &= h(a) + \tfrac{1}{2}h''(a)(\xi - a)^2 + \cdots \end{aligned}$$

(Thus we are assuming a higher degree of smoothness than stated in the theorem; as stated, our argument contains a degree of hand-waving.) Then, keeping only leading order terms, we have

$$f(s) \sim g(a) e^{sh(a)} \int_{-\infty}^{\infty} e^{sh''(a)(\xi - a)^2/2}\, d\xi$$

Making the change of variables

$$-z^2 = \frac{sh''(a)(\xi - a)^2}{2}, \qquad dz = \left(\frac{-sh''(a)}{2}\right)^{1/2} d\xi$$

gives

$$f(s) \sim g(a)e^{sh(a)}\left[-\frac{2}{sh''(a)}\right]^{1/2}\int_{-\infty}^{\infty}\exp(-z^2)\,dz$$

$$= g(a)e^{sh(a)}\left[-\frac{2\pi}{sh''(a)}\right]^{1/2}$$

which is the approximation (2).

Using the result of Laplace's theorem, we may now estimate the integrals in equation (17) of Section 4.5. Making the identification mentioned above, we have

$$\int_{-\infty}^{\infty} e^{-G(\xi,x,t)/2D}\,d\xi \sim e^{-G(a,x,t)/2D}\left(\frac{4D\pi}{|G''(a,x,t)|}\right)^{1/2} \qquad \text{as } D \to 0$$

where $\xi = a$ is the assumed, single, stationary point (a maximum) of G, regarded as a function of ξ, with x and t as parameters. Later we deduce the consequences of the assumption that G has a single stationary point. Now, from (18) in Section 4.5,

$$G'(\xi,x,t) = u_0(\xi) - \frac{(x-\xi)}{t}, \qquad G''(\xi,x,t) = u_0'(\xi) + \frac{1}{t}$$

Furthermore,

$$\int_{-\infty}^{\infty}\frac{x-\xi}{t}e^{-G(\xi,x,t)/2D}\,d\xi \sim \frac{x-a}{t}e^{-G(a,x,t)/2D}$$

$$\times\left[\frac{4\pi D}{|G''(a,x,t)|}\right]^{1/2} \qquad \text{as } D \to 0$$

Therefore, from (17) in Section 4.5, we conclude that

$$u(x,t) \sim \frac{x-a}{t} \qquad \text{as } D \to 0$$

where a is the single stationary point of G and is given by the equation

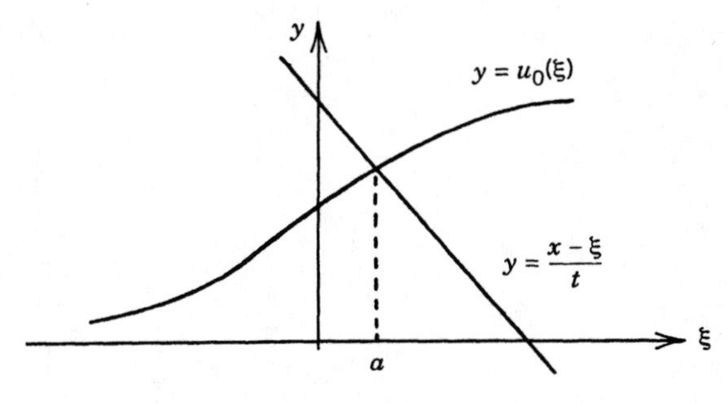

Figure 4.11

$u_0(a) = (x - a)/t$. Therefore, in this case we can write the solution to the initial value problem to Burgers' equation as

$$u(x,t) \sim u_0(a) \quad \text{as } D \to 0, \qquad \text{where } a \text{ is the single root of } x = a + tu_0(a)$$

We notice that this is the same solution that we obtained for the initial value problem for the inviscid Burgers' equation (with $D = 0$); therefore, we have recovered the inviscid solution in the limit as $D \to 0$.

We remark that if our assumption that $\xi = a$ is a single stationary point of G holds true, the graphs of $u_0(\xi)$ and $(x - \xi)/t$ must have a single intersection point. Figure 4.11 shows how the graph of the initial data u_0 may appear in this case, namely as an increasing function of ξ. We recall that shocks will not develop for such initial data when propagated by the inviscid Burgers' equation.

Evolution of a Point Source

For the linear diffusion equation we have already noted that a point source (a delta function) at $x = 0$, $t = 0$ evolves according to the fundamental solution of the diffusion equation. Further, in Section 4.3 we calculated the evolution of such a source for a nonlinear diffusion equation. It is interesting to ask how such a point source is propagated by Burgers' equation, which is a nonlinear convection–diffusion equation. Consequently, we consider the initial value problem

$$u_t + uu_x = Du_{xx}, \qquad x \in R, \quad t > 0 \tag{3}$$

$$u(x,0) = \delta(x), \qquad x \in R \tag{4}$$

We assume that the source is located at $x = 0-$, that is, just to the left of the origin. We apply the solution (17)–(18) of Section 4.5. The quantity G in

(18) is, in the present case,

$$G(\xi, x, t) = \frac{(x-\xi)^2}{2t} + \int_0^{\xi} \delta(\hat{\xi})\, d\hat{\xi}$$

Consequently,

$$G(\xi, x, t) = \frac{(x-\xi)^2}{2t} \quad \text{if } \xi > 0; \qquad G(\xi, x, t) = \frac{(x-\xi)^2}{2t} - 1 \quad \text{if } \xi < 0$$

Letting N and Δ denote the numerator and denominator in (17), we have

$$\Delta = e^{1/2D} \int_{-\infty}^{0} \exp\left[-\frac{(x-\xi)^2}{4tD}\right] d\xi + \int_0^{\infty} \exp\left[-\frac{(x-\xi)^2}{4Dt}\right] d\xi$$

Now make the substitution $z = (x - \xi)/\sqrt{4Dt}$ to obtain, after some calculus and algebra,

$$\Delta = (4Dt)^{1/2}\left[\pi + (e^{1/2D} - 1)\int_{x/\sqrt{4Dt}}^{\infty} e^{-z^2}\, dz\right] \tag{5}$$

The numerator can be calculated in nearly the same manner to obtain

$$N = 2De^{-x^2/4Dt}(e^{1/2D} - 1) \tag{6}$$

Therefore, the solution to the initial value problem (3)–(4) can be written

$$u(x,t) = \frac{N}{\Delta} = \left(\frac{D}{t}\right)^{1/2}(e^r - 1)\frac{e^{-x^2/4Dt}}{\left[\sqrt{\pi} + (e^r - 1)\int_{x/\sqrt{4Dt}}^{\infty} e^{-z^2}\, dz\right]},$$

$$r = \frac{1}{2D} \tag{7}$$

Remark. We observe that the solution in (7) has the form of a similarity solution $u = (D/t)^{1/2} f(r, z)$, where $z = x/\sqrt{Dt}$ (see Sections 4.2 and 4.3).

There are two different limits in which (7) could be examined. In the limit of small r (or large D) we expect diffusion to dominate, and in the limit of large r (or small D) we expect convection to dominate.

1. *Limit of Large D.* In this case we consider the ratio $u(x,t)/K(x,t)$, where u is given by (7) and K is the fundamental solution to the linear

diffusion equation. We shall show that this ratio tends to unity as D gets large, thus proving the expected result that u behaves asymptotically like the solution to the linear diffusion equation. To this end, it is straightforward to see that we may write u/K in terms of r as

$$\frac{u}{K} = \frac{r^{-1}(e^r - 1)}{1 + \pi^{-1/2}(e^r - 1)\int_{x\sqrt{r/2t}}^{\infty} e^{-z^2}\,dz} \tag{8}$$

The limit of the right side of (8), as $r \to 0$, is easily calculated to be 1 (see Exercise 3), thereby showing that

$$u(x,t) \sim K(x,t) \qquad \text{as } D \to \infty \tag{9}$$

2. *Limit of Small D.* This case is a little more difficult, and also more interesting. Let us first write the solution (7) of the initial value problem (3)–(4) as

$$u(x,t) = \left(\frac{2}{t}\right)^{1/2} F(z,r) \tag{10}$$

where $z = x/(2t)^{1/2}$ and

$$F(z,r) = \frac{1}{2\sqrt{r}}\,\frac{(e^r - 1)e^{-rz^2}}{\sqrt{\pi} + (e^r - 1)\int_{z\sqrt{r}}^{\infty} e^{-\zeta^2}\,d\zeta}$$

Clearly, we may replace $e^r - 1$ by e^r for large r and write

$$F(z,r) \sim \frac{1}{2\sqrt{r}}\,\frac{e^{r(1-z^2)}}{\sqrt{\pi} + e^r\int_{z\sqrt{r}}^{\infty} e^{-\zeta^2}\,d\zeta} \qquad \text{as } r \to \infty \tag{11}$$

We now determine the behavior of $F(z,r)$ for large r over different ranges of z.

If $z < 0$, the integral in the denominator of (11) approaches $\sqrt{\pi}$, and it is routine to show that

$$F(z,r) \sim \frac{e^{-rz^2}}{2(\pi r)^{1/2}} \qquad \text{as } r \to \infty$$

Therefore,

$$F(z,r) \sim 0 \qquad \text{as } r \to \infty \quad \text{for } z < 0 \tag{12}$$

If $z > 0$ we require the asymptotic approximation $\int_\eta^\infty e^{-\zeta}\,d\zeta \sim e^{-\eta^2}/2\eta$ as $\eta \to \infty$, which is the leading order approximation of the complementary error function erfc for large η (see, e.g., Abramowitz and Stegun [1964] or other handbooks of mathematical functions). Then (11) becomes

$$F(z,r) \sim \frac{z}{1 + 2\sqrt{\pi r}\, z e^{r(z^2-1)}} \qquad \text{as } r \to \infty \quad \text{for } z > 0 \tag{13}$$

Now, if $z > 1$, then $F(z,r) \sim 0$ as $r \to \infty$, and if $0 < z < 1$, then $F(z,r) \sim z$ as $r \to \infty$.

In summary, we have shown that $F(z,r) \sim 0$ as $r \to \infty$ for $z < 0$ and $z > 1$. In the interval $0 < z < 1$ we have $F(z,r) \sim z$. Translating this information back to the solution u given by (10), we have, as $D \to 0$,

$$u(x,t) \sim \frac{x}{t} \quad \text{for } 0 < x < \sqrt{2t}\,; \qquad u(x,t) \sim 0 \quad \text{otherwise} \tag{14}$$

Figure 4.12 shows the limiting solution (14) for a small diffusion constant D; as expected, we obtain a shocklike structure where diffusion plays a small role compared to nonlinear convection. Near $z = 1$, that is, near the wave front $x = \sqrt{2t}$, it can be shown that there is a steep transition region of order $O(D)$ (see Whitham [1974], p. 105).

The preceding analysis for Burgers' equation can be carried out because a closed-form solution to the initial value problem is available. For other nonlinear equations, like Fisher's equation, the analysis is often more difficult because such analytic solutions are not known.

In the next chapter (Section 5.4) we discuss the origins of Burgers' equation in the realm of gas dynamics.

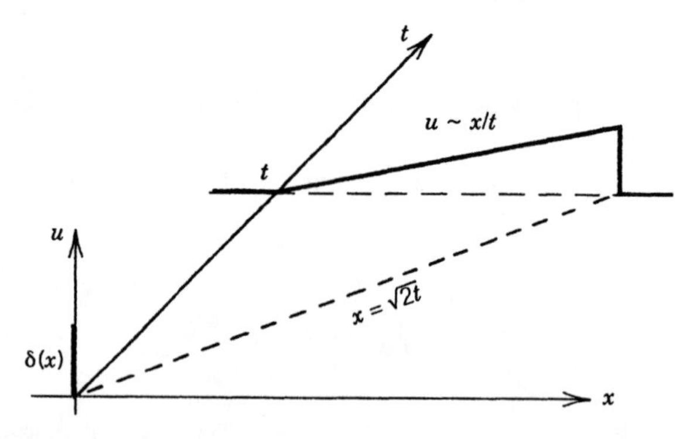

Figure 4.12. Evolution of a delta function initial condition, propagated by Burgers' equation, in the case of a small diffusion constant D.

EXERCISES

1. Verify (5) and (6).
2. In Laplace's theorem let $h(\xi) = 1 - \xi^2$ and sketch the graph of $e^{s(1-\xi^2)}$ for various values of s. Note that for large s the main contribution to the integral $f(s)$ comes from the neighborhood of $\xi = 0$, where h has a local maximum. Estimate the integral

$$f(s) = \int_{-\infty}^{\infty} \cos^2 \xi e^{s(1-\xi^2)}\, d\xi$$

for large s.
3. Show that the limit as $r \to 0$ of the right side of (8) is unity. *Suggestion:* Expand e^r in a Taylor series and also note that the integral $\int_{x\sqrt{r/2t}}^{\infty} e^{-z^2}\, dz$ tends to a constant.
4. Verify (12) and (13).
5. Show that near the wavefront ($z = 1$), formula (10) for $F(z, r)$ becomes

$$F(z,r) \sim \frac{1}{1 + 2\sqrt{\pi r}\, e^{2r(z-1)}} \qquad \text{as } r \to \infty$$

Hint: Note that $z^2 - 1 \sim 2(z - 1)$.
6. Derive the solution (7) using similarity methods.

APPENDIX: PHASE PLANE

In this appendix we give a brief presentation of the rudiments of phase-plane analysis. This treatment is meant to be a tool that will permit quick reference when reading certain parts of the book. So our interest is in reviewing terminology and basic results rather than developing phase-plane phenomena from scratch. For a detailed discussion the reader may consult one of the many excellent books on the subject; for example, see the recent text by Hale and Kocak [1991].

Our discussion will center on the nonlinear autonomous system

$$\frac{dx}{dt} = p(x, y), \qquad \frac{dy}{dt} = q(x, y) \tag{1}$$

where p and q are given functions that are assumed to have continuous derivatives of all order. By a *solution* of (1) we mean a pair of smooth functions $x = x(t)$, $y = y(t)$ that satisfy the differential equations (1) for all t in some interval I. The interval I is usually the whole real line. Graphically, we represent the solution parametrically as a curve in the xy-plane, which is called the *phase plane*. Such a solution curve is called an *orbit*, *path*, or *trajectory* of (1). The independent variable t is regarded as a parameter along the curve and is interpreted as time. The orbits have a natural positive direction to them, namely the direction that they are traced out as the time parameter t increases; to indicate this direction an arrow is always placed on a given orbit. Because the system is autonomous (t does not appear on the right sides), the solution is invariant under a time translation; therefore, the time t may be shifted along any orbit. A constant solution $x(t) = x_0$, $y(t) = y_0$ to (1) is called an *equilibrium solution* and its orbit is represented as a single point (x_0, y_0) in the phase plane. Clearly, such points must satisfy the algebraic relations

$$p(x, y) = 0 \qquad q(x, y) = 0 \tag{2}$$

The points that satisfy (2) are called *critical points* (also, *rest points* and *equilibrium points*), and each such point represents an equilibrium solution. It is evident that no solution curve or orbit can pass through a critical point at finite time t; otherwise, uniqueness would be violated. The totality of all the orbits of (1) and critical points, graphed in the phase plane, is called the *phase portrait* of (1). The qualitative behavior of the phase portrait is determined to a large extent by the location of the critical points and the local behavior of the orbits near those points. The *Poincaré–Bendixson theorem* in two dimensions characterizes the behavior of the possible orbits of (1):

(a) An orbit cannot approach a critical point in finite time; that is, if an orbit approaches a critical point, then, necessarily, $t \to \pm\infty$.

(b) As $t \to \pm\infty$, an orbit either approaches a critical point, moves on a closed path, approaches a closed path, or leaves every bounded set. A closed orbit is a *periodic solution*.

Because of our smoothness assumptions on p and q it follows that the initial value problem, consisting of (1) and initial conditions $x = x_i$, $y = y_i$ at $t = 0$, has a unique solution. It follows that at most one orbit passes through each point of the plane and all of the orbits of (1) cover the entire phase plane without intersecting. In principle, the orbits can be found by integrating the differential relationship

$$\frac{dy}{dx} = \frac{q(x, y)}{p(x, y)} \tag{3}$$

which comes from dividing the two equations in (1). When this is done, information about how a given orbit depends on the time parameter t is lost; however, this is often not crucial.

The right sides of (1) define a vector field $\langle p, q \rangle$ in the phase plane; the orbits are the curves that have this vector field as their tangents. Thus the orbits are the integral curves of (3). The loci $p(x, y) = 0$ and $q(x, y) = 0$, where the vector field is vertical and horizontal, respectively, are called the *nullclines*. The basic problem of determining the phase portrait of (1) is generally facilitated by graphing the nullclines and determining the direction of the vector field $\langle p, q \rangle$ in the various regions separated by the nullclines in the xy-plane.

The following two results of Bendixson and Poincaré, respectively, are also helpful.

(a) If $p_x + q_y$ is of one sign in a region of the phase plane, the system (1) cannot have a closed orbit in that region.
(b) A closed orbit of (1) must surround at least one critical point.

Critical points are also classified as to their stability. A critical point is *stable* if every orbit sufficiently near the point at some time t_0 remains in a prescribed circle about the point for all $t > t_0$. A critical point is *unstable* if it is not stable. A critical point is *asymptotically stable* if it is stable and every orbit sufficiently near the point at some time t_0 approaches the point as $t \to \infty$. An asymptotically stable critical point is called an *attractor*.

Linear Systems

Linear systems are of course the easiest to study. A linear system has the form

$$x' = ax + by \tag{4}$$

$$y' = cx + dy \tag{5}$$

Hereafter, we use a prime to denote d/dt. Let us assume up front that $ad - bc$ is nonzero; then the only critical point of (4)–(5) is the origin $x = 0$, $y = 0$. We may write (4)–(5) in matrix form as

$$\mathbf{u}' = A\mathbf{u} \tag{6}$$

where $\mathbf{u} = (x, y)^T$ is the vector of unknowns and A is the coefficient matrix given by

$$A = \begin{pmatrix} a & b \\ c & d \end{pmatrix}$$

By assumption, $\det A \neq 0$. Solutions of (6) are obtained by assuming that

$$\mathbf{u}(t) = \mathbf{v}e^{\lambda t} \tag{7}$$

where $\mathbf{v}$ is a constant vector and λ is a constant, both to be determined. Substituting this form into (6) yields the algebraic eigenvalue problem

$$A\mathbf{v} = \lambda\mathbf{v} \tag{8}$$

Any eigenpair $(\lambda, \mathbf{v})$ of (8) gives a solution of (6) of the form (7). Therefore, if $(\lambda_1, \mathbf{v}_1)$ and $(\lambda_2, \mathbf{v}_2)$ are two eigenpairs with λ_1 and λ_2 distinct, all solutions of (6) are given by the linear combination

$$\mathbf{u}(t) = c_1\mathbf{v}_1 \exp(\lambda_1 t) + c_2\mathbf{v}_2 \exp(\lambda_2 t) \tag{9}$$

where c_1 and c_2 are arbitrary constants. This includes, of course, the case when λ_1 and λ_2 are complex conjugates; then real solutions may be found by taking the real and imaginary parts of (9). If $\lambda_1 = \lambda_2$, there may not be two linear independent eigenvectors $\mathbf{v}_1$ and $\mathbf{v}_2$; if two independent eigenvectors exist, (9) remains valid; if not, the general solution to (6) is

$$\mathbf{u}(t) = c_1\mathbf{v}_1 \exp(\lambda_1 t) + c_2(\mathbf{w} + \mathbf{v}_1 t)\exp(\lambda_1 t) \tag{10}$$

for some constant vector $\mathbf{w}$ that must be determined.

Therefore, we may catalog the different types of solutions of the linear system (5) or (6), depending on the eigenvalues and eigenvectors of the coefficient matrix A. The results in the following summary come directly from the forms of the general solution (9) or (10).

CASE 1. If the eigenvalues are real and have opposite signs, the critical point $(0, 0)$ is classified as a *saddle point*, and a generic phase portrait is shown in Figure 4A.1. The two orbits entering the origin are the two *stable manifolds*, and the two orbits leaving the origin are the two *unstable manifolds*; the directions of these manifolds are determined by the two eigenvectors. The

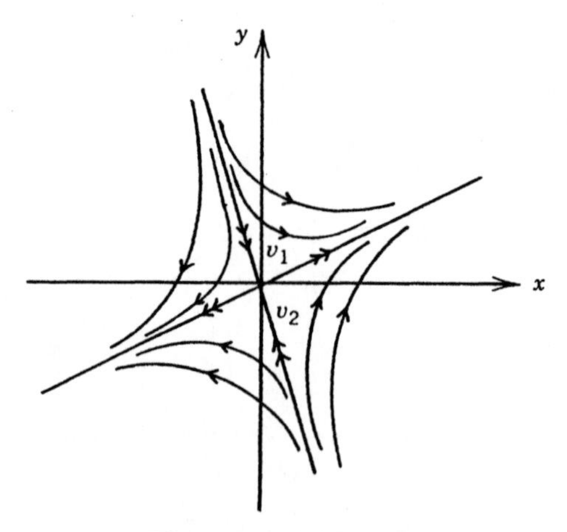

Figure 4A.1. Saddle point.

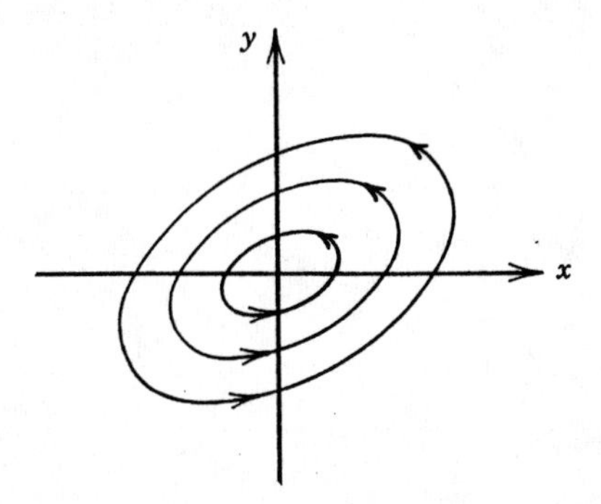

Figure 4A.2. Center.

stable manifold corresponds to the negative eigenvalue and the unstable manifold corresponds to the positive eigenvalue. These special manifolds are called *separatrices*. A saddle point is unstable.

CASE 2. If the eigenvalues are purely imaginary, the orbits form closed curves (ellipses) representing periodic solutions, and the origin is classified as a *center* (see Figure 4A.2). A center is stable.

CASE 3. If the eigenvalues are complex (conjugates) and not purely imaginary, the orbits spiral into, or out of, the origin, depending on whether the real part of the eigenvalues is negative or positive, respectively. In this case the origin is called a *spiral point* or a focus (see Figure 4A.3). If the real part of the eigenvalues is negative, the spiral point is an attractor; if the real part is positive, the point is unstable.

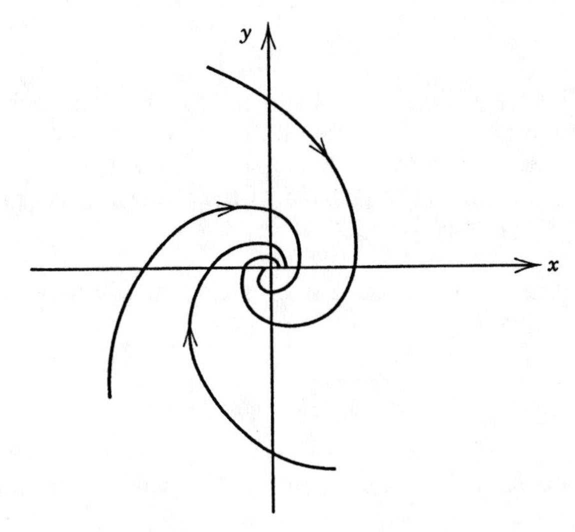

Figure 4A.3. Asymptotically stable spiral.

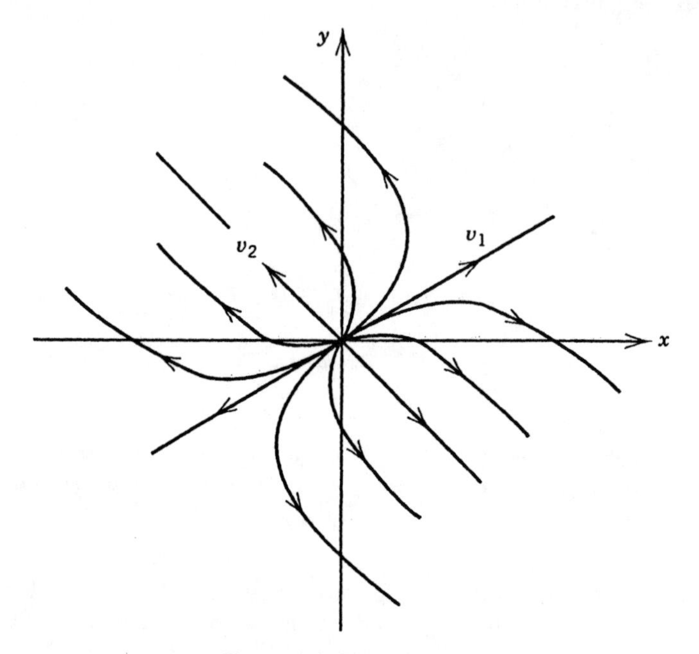

Figure 4A.4. Unstable node.

CASE 4. If the eigenvalues are real, distinct, and have the same sign, the origin is classified as a *node*. In this case all the orbits enter the origin if the eigenvalues are negative (giving an attractor), and all the orbits exit the origin if the eigenvalues are positive (giving an unstable node). The detailed orbital structure near the origin is determined by the eigenvectors. One of them defines the direction in which orbits enter (or exit) the origin, and the other defines the direction approached by the orbits as t approaches minus, or plus, infinity (see Figure 4A.4).

CASE 5. If the eigenvalues are real and equal, the critical point at the origin is still classified as a node, but it is of different type than in case iv since there may be only one eigenvector. In this case the node is called *degenerate* (see Figure 4A.5). The node is an attractor if the eigenvalue is negative, and it is unstable if the eigenvalue is positive.

We recall that the eigenvalues of A are easily calculated as the roots of the quadratic equation

$$\lambda^2 - \operatorname{tr} A\lambda + \det A = 0$$

where $\operatorname{tr} A$ denotes the trace of A and $\det A$ is the determinant of A. The

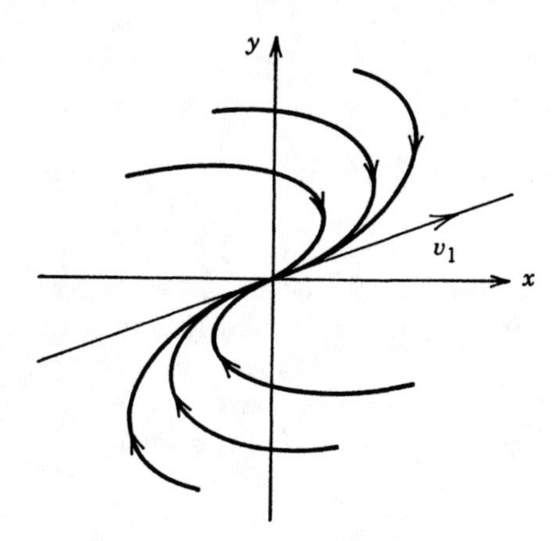

Figure 4A.5. Degenerate stable node.

corresponding eigenvectors are the solution to the homogeneous equation

$$(A - \lambda I)\mathbf{v} = \mathbf{0}$$

Nonlinear Systems. Now we return to the nonlinear system (1). The principal idea here is that (1) can be studied by examining a linear system. Let us first assume that $x = 0$, $y = 0$ is an isolated critical point of (1) [i.e., there is a neighborhood of $(0, 0)$ that contains no other critical points]. However, (1) may have other critical points spread throughout the plane. The local structure of the orbits near the origin for the nonlinear system (1) can be determined, under fairly broad conditions, by examining the linearized system obtained by neglecting the higher order terms (quadratic, cubic, etc.) in the Taylor series expansion of p and q about $(0, 0)$. More precisely, (1) can be written

$$x' = p(0,0) + p_x(0,0)x + p_y(0,0)y + \text{higher order terms}$$

$$y' = q(0,0) + q_x(0,0)x + q_y(0,0)y + \text{higher order terms}$$

But $p(0,0) = q(0,0) = 0$, so to leading order we have the linear system

$$\begin{pmatrix} x' \\ y' \end{pmatrix} = \begin{pmatrix} p_x(0,0) & p_y(0,0) \\ q_x(0,0) & q_y(0,0) \end{pmatrix} \begin{pmatrix} x \\ y \end{pmatrix} \tag{11}$$

Because x and y are small, this linearized system is a good approximation of the nonlinear system near $(0, 0)$, so the orbits of the nonlinear system (1) near the critical point $(0, 0)$ might be expected to behave like the orbits of the linear system (11). The latter, of course, are determined by the eigenvalues and eigenvectors of the coefficient matrix

$$J(0,0) = \begin{pmatrix} p_x(0,0) & p_y(0,0) \\ q_x(0,0) & q_y(0,0) \end{pmatrix}$$

which is called the *Jacobi matrix* of (1) at the critical point $(0,0)$. In the case that (x_0, y_0) is a nonzero isolated critical point of (1), the change of variables $\bar{x} = x - x_0$, $\bar{y} = y - y_0$, translates the critical point to the origin in the $\overline{xy}$-coordinate system. Therefore, without loss of generality we can analyze the critical point $(0,0)$. When a nonzero critical point is translated to the origin, it is easy to see that the Jacobi matrix (i.e., the matrix of the linearized system) is

$$J(x_0, y_0) = \begin{pmatrix} p_x(x_0, y_0) & p_y(x_0, y_0) \\ q_x(x_0, y_0) & q_y(x_0, y_0) \end{pmatrix}$$

We now can state the main result: *Let* $(0,0)$ *be an isolated critical point of* (1). *If the real parts of the eigenvalues of the Jacobi matrix* $J(0,0)$ *are not zero, then* (1) *and* (11) *have the same qualitative structure in a neighborhood of* $(0,0)$. By qualitative structure, we mean the same stability characteristics and the same nature of critical point (saddle, node, or focus). Notice that the case that does not extend to the nonlinear system is that of a center for the linear system. In this case one must examine the higher order terms to determine the nature of the critical point of the nonlinear system. If the Jacobi matrix has a zero eigenvalue, $(0,0)$ is a higher order critical point; in this case the higher order terms play a crucial role and the nature of the critical point may differ from a node, saddle, focus, or center. For example, it may have nodal structure on one side and a saddle structure on the other. A specific analysis is required in the case of higher order critical points. Similar remarks apply to any nonzero critical point of (1).

In summary, if the Jacobi matrix at an isolated critical point has nonzero eigenvalues, the nonlinear system (1) may be analyzed by examining the associated linearized system (11) in a neighborhood of the critical point. The phase portrait of (1) is determined by finding all the critical points, analyzing their nature and stability, and then examining the global behavior and structure of the orbits. For example, it is an interesting problem to determine if a separatrix connects two critical points in the system; such connections are called *heteroclinic orbits*. An orbit connecting a critical point to itself is called a *homoclinic orbit*.

Example 1. Consider the nonlinear system

$$\begin{aligned} x' &= y \\ y' &= x - y + x^2 - 2xy \end{aligned} \tag{12}$$

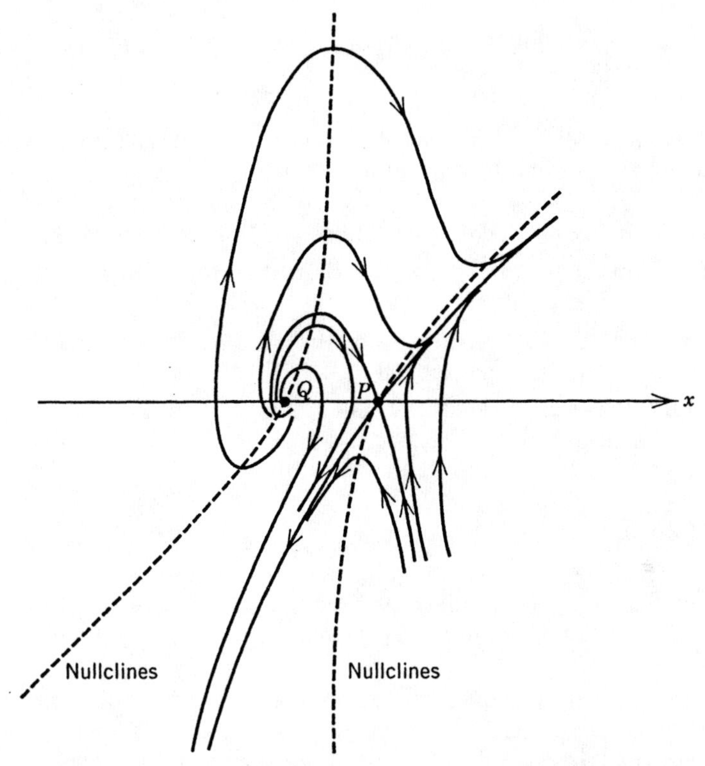

Figure 4A.6. Phase portrait of the nonlinear system (12). $P:(0,0)$ and $Q:(-1,0)$ are the two critical points, a saddle and an unstable spiral, respectively. The nullclines $y' = 0$ are dashed.

There are two critical points $P:(0,0)$ and $Q:(-1,0)$. The Jacobi matrix $J(0,0)$ at P is

$$\begin{pmatrix} 0 & 1 \\ 1 & -1 \end{pmatrix}$$

which has eigenvalues $(-1 \pm \sqrt{5})/2$, which are real and of opposite sign. Thus P is a saddle point. The Jacobi matrix $J(-1,0)$ at Q is

$$\begin{pmatrix} 0 & 1 \\ 1 - 2x - 2y & -1 - 2x \end{pmatrix}_{(0,0)} = \begin{pmatrix} 0 & 1 \\ -1 & 0 \end{pmatrix}$$

The eigenvalues are $(1 \pm i\sqrt{3})/2$, which are complex with positive real part. Thus Q is an unstable spiral point. The nullclines are $y = 0$ (where $x' = 0$ and the vector field is vertical) and $y = x(1 + x)/(1 + 2x)$ (where $y' = 0$ and the vector field is horizontal). A phase diagram is shown in Figure 4A.6. The nullclines $y' = 0$ are shown dotted. We observe that there is a heteroclinic orbit exiting the spiral point Q and entering the saddle point P along its separatrix.

There are several software packages that can easily graph phase portraits in two dimensions. Four such programs are:

1. PHASER (by H. Kocak, 1989), Springer-Verlag, New York.
2. PHASEPLANE version 3.0 (by B. Ermentrout, 1990), Brooks/Cole, Pacific Grove, Calif.
3. MATHEMATICA version 2.0 (by S. Wolfram, 1991), Wolfram Research, Inc., Champaign, Ill.
4. PLOD/PC, version 6.0 (by D. Kahaner and D. Barnett), NIST, Gaithersburg, Md.

REFERENCES

A good introduction to the diffusion equation and random walks is given in the first chapter of Zauderer [1989]. See also Murray [1989] and Okubo [1980].

An elementary discussion of the general similarity method is presented in Logan [1987]. A standard reference on this topic is Bluman and Kumei [1989]. See also Dresner [1983] and Rogers and Ames [1989]; the latter contains an extensive bibliography as well as numerous recent nontrivial applications. The similarity method as it applies to problems in the calculus of variations is discussed in Logan [1977, 1987].

For a discussion of Fisher's equation, see Murray [1977, 1989]. A good source for Burgers' equation is Whitham [1974]; Kriess and Lorenz [1989] discuss questions of uniqueness and existence of solutions to boundary value problems associated with Burgers' equation.

Abramowitz, M., and I. A. Stegun, 1964. *Handbook of Mathematical Functions*, National Bureau of Standards, Washington, D.C.

Bender, C. M., and S. A. Orszag, 1978. *Advanced Mathematical Methods for Scientists and Engineers*, McGraw-Hill, New York.

Bluman, G. W., and S. Kumei, 1989. *Symmetries and Differential Equations*, Springer-Verlag, New York.

Dresner, L., 1983, *Similarity Solutions of Nonlinear Partial Differential Equations*, Pitman Books, London.

Hale, J., and H. Kocak, 1991. *Dynamics and Bifurcations*, Springer-Verlag, New York.

Hartman, P., 1964. *Ordinary Differential Equations*, Wiley, New York.

Hochstadt, H., 1975. *Differential Equations: A Modern Approach*, Dover, New York.

Kreiss, H. O., and J. Lorenz, 1989. *Initial-Boundary Value Problems and the Navier–Stokes Equations*, Academic Press, New York.

Lin, C. C., and L. A. Segel, 1974. *Mathematics Applied to Deterministic Problems in the Natural Sciences*, Macmillan, New York. Reprinted by the Society for Industrial and Applied Mathematics, Philadelphia.

Logan, J. D., 1977, *Invariant Variational Principles*, Academic Press, New York.

Logan, J. D., 1987. *Applied Mathematics: A Contemporary Approach*, Wiley-Interscience, New York.

Murray, J. D., 1977. *Lectures on Nonlinear-Differential-Equation Models in Biology*, Oxford University Press, Oxford.

Murray, J. D., 1984. *Asymptotic Analysis*, Springer-Verlag, New York.

Murray, J. D., 1989. *Mathematical Biology*, Springer-Verlag, New York.

Okubo, A., 1980. *Diffusion and Ecological Problems: Mathematical Models*, Springer-Verlag, New York.

Rogers, C., and W. F. Ames, 1989. *Nonlinear Boundary Value Problems in Science and Engineering*, Academic Press, New York.

Whitham, G. B., 1974. *Linear and Nonlinear Waves*, Wiley-Interscience, New York.

Zauderer, E., 1989. *Partial Differential Equations of Applied Mathematics*, 2nd ed., Wiley-Interscience, New York.

HYPERBOLIC SYSTEMS

Most physical systems involve several unknown functions and therefore require several (usually, the same number as unknowns) PDEs for the governing equations. For example, the complete description of a fluid mechanical system might require knowledge of the density, pressure, temperature, and the particle velocity, so we would expect to formulate a system of equations to describe the flow. The central idea for hyperbolic systems, as was the case for a single equation, is that of characteristics, and it is this thread that weaves through the entire chapter.

In Section 5.1 we develop, from first principles, two physical models that will serve to illustrate the concepts developed subsequently. First, we derive the PDEs that govern waves in shallow water; this derivation is simple and it serves as a precursor to the formulation of the conservation laws of gas dynamics, the latter being the paradigm of applied nonlinear hyperbolic PDEs. In Section 5.2 we extend the notion of characteristics to systems of equations, and out of this development we are able to give a formal classification of hyperbolicity. Section 5.3 contains several examples of Riemann's method, or the method of characteristics, applied to the shallow water equations with various boundary conditions. In Sections 5.4 and 5.5 we take up the theory of weakly nonlinear waves and determine conditions along a wavefront propagating into a region of spacetime. Because Burgers' equation is so fundamental in PDEs, we give its derivation from the equations of gas dynamics in a weakly nonlinear limit.

5.1 WAVES IN SHALLOW WATER; GAS DYNAMICS

Shallow Water Waves

The partial differential equations governing waves in shallow water can be obtained from a limiting form of the general equations of fluid mechanics when the ratio of the water depth to the wavelength of a typical surface wave is small. In the present treatment, however, we obtain these equations a

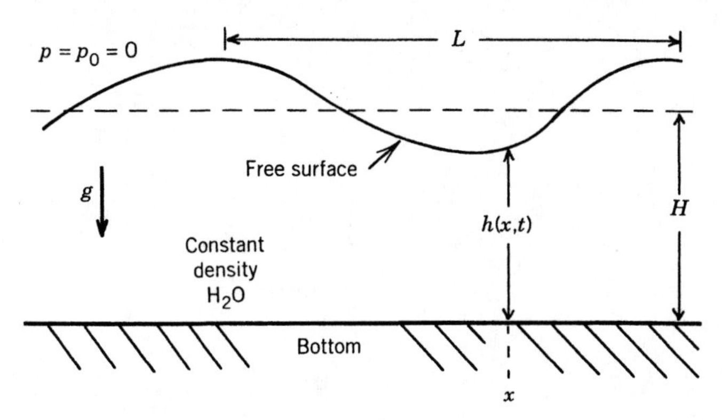

Figure 5.1

priori without the benefit of knowing the conservation laws of general fluid flow. The shallow water equations, which represent mass and momentum balance in the fluid, are highly typical of other hyperbolic systems (e.g., the equations of gas dynamics), and thus they provide a solid example of the methods of nonlinear analysis in a simple, understandable context.

Figure 5.1 shows the geometry. We assume that the water, which has constant density ρ, lies above a flat bottom. We measure the distance along the bottom by a coordinate x, and the coordinate y measures the height above the bottom. There is assumed to be no variation in the z direction, which is into the paper. Let H be the height of the undisturbed surface, and assume that the height of the free surface is given at any time t and at any location x by the function $y = h(x, t)$. At the free surface we take the pressure to be p_0, the ambient pressure of air. With no loss of generality, we may assume that $p_0 = 0$. If a typical wave on the surface has wavelength L, the shallow water assumption is that H is small compared to L (i.e., $H \ll L$). In this case, we may ignore vertical motions and assume that there is a flow velocity $u(x, t)$ in the x direction which is an average velocity over the depth.

Since vertical motions are ignored, each volume $\Delta x \, \Delta y \, \Delta z$ of fluid has a hydrostatic balance in the vertical direction. That is, the upward pressure force on a fluid volume must equal the downward pressure force plus the force caused by gravity. Figure 5.2 shows a typical fluid element. If the pressure is denoted by $P(x, y, t)$, we may write the hydrostatic balance equation as

$$P(x, y, t) \, \Delta x \, \Delta z = P(x, y + \Delta y, t) \, \Delta x \, \Delta z + \rho g \, \Delta x \, \Delta y \, \Delta z$$

Dividing by $\Delta x \, \Delta y \, \Delta z$ and then taking the limit as $\Delta y \to 0$ yields

$$P_y(x, y, t) = -\rho g$$

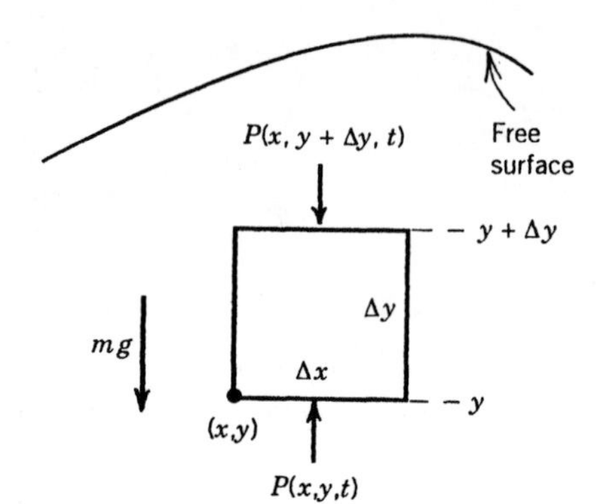

Figure 5.2

Now we can integrate this equation from a depth y to the free surface h to obtain the following equation for the pressure P:

$$P(x, y, t) = p_0 + \rho g(h(x, t) - y) \qquad (p_0 = 0) \tag{1}$$

Now we derive an equation for *mass balance*. Like all of the conservation laws obtained in earlier chapters, mass balance requires that the time rate of change of the total mass in a region between $x = a$ and $x = b$ must equal the mass flux into the region at $x = a$ minus the mass flux out of the region at $x = b$. In the present case, the mass flux at x is given by the density times the velocity u at x times the cross-sectional area at x, the latter being given by $h\,\Delta z$, where the width in the z direction is taken to be Δz. Thus *mass flux* $= \rho uh\,\Delta z$, in units of mass per unit time. Therefore, the mass balance law is, in integral form,

$$\frac{d}{dt}\int_a^b h(x, t)\,dx = u(a, t)h(a, t) - u(b, t)h(b, t) \tag{2}$$

where the constant quantity $\rho\,\Delta z$ was canceled from each term. As usual, if the functions u and h are smooth, then (as in Section 1.2) equation (2) may be cast into the PDE

$$h_t + (uh)_x = 0 \tag{3}$$

Equations (2) and (3) represent mass conservation in integral and differential form, respectively. We recall that the former must be used when the functions u and h have discontinuities.

The next equation is *momentum balance in the x direction* [recall that we assumed hydrostatic balance in the y direction, giving (1)]. Momentum balance requires that the time rate of change of the total momentum in the region between $x = a$ and $x = b$ must equal the net momentum flux (the rate that momentum flows into the region at $x = a$ minus the rate that momentum flows out of the region at $x = b$), plus the forces (pressure) acting on the region (at $x = a$ and $x = b$). Momentum is mass times velocity, so the total momentum in $a \le x \le b$ is given by

$$\text{total momentum} = \int_a^b \rho u(x,t) h(x,t)\, \Delta z\, dx$$

Momentum flux through a section at x is the momentum multiplied by the velocity at x, or $(\rho u h\, \Delta z)u$. Therefore, the net momentum flux is given by

$$\text{net momentum flux} = \rho h(a,t) u(a,t)^2\, \Delta z - \rho h(b,t) u(b,t)^2\, \Delta z$$

The force acting on the face of area $h(a,t)\,\Delta z$ is the pressure $P(a,y,t)$. But the pressure varies with depth according to (1), and therefore we must integrate over the depth to obtain the total force acting at $x = a$, which is

$$\begin{aligned}(\text{pressure force at } x = a) &= \int_0^{h(a,t)} P(a,y,t)\, \Delta z\, dy \\ &= \int_0^{h(a,t)} \rho g(h(a,t) - y)\, \Delta z\, dy \\ &= \frac{\rho g}{2} h(a,t)^2\, \Delta z\end{aligned}$$

Similarly,

$$(\text{pressure force at } x = b) = -\int_0^{h(b,t)} P(b,y,t)\, \Delta z\, dy = -\frac{\rho g}{2} h(b,t)^2\, \Delta z$$

Consequently, the integral form of the momentum balance equation is

$$\begin{aligned}&\frac{d}{dt}\int_a^b h(x,t) u(x,t)\, dx \\ &\quad = h(a,t)u(a,t)^2 - h(b,t)u(b,t)^2 + \frac{g}{2}\left[h(a,t)^2 - h(b,t)^2\right]\end{aligned} \tag{4}$$

where $\rho\,\Delta z$ has been canceled from each term. A differential form of this conservation law may be obtained in the usual manner if u and h have the requisite degree of smoothness. Then

$$(hu)_t + \left(hu^2 + \frac{g}{2}h^2\right)_x = 0 \tag{5}$$

Equations (3) and (5) are two coupled nonlinear PDEs for the velocity u and the height h of the free surface, and these equations are the governing evolution equations for shallow water waves. Clearly, equation (3) can be expanded to give

$$h_t + uh_x + hu_x = 0 \tag{6}$$

We leave it as an exercise to show that the momentum equation (5) can be written, with the aid of (6), simply as

$$u_t + uu_x + gh_x = 0 \tag{7}$$

Equations (6) and (7) are the model equations for shallow water theory. We note that they are quasilinear and strongly resemble the nonlinear model equations studied in Chapters 2 and 3.

Small-Amplitude Approximation

If we restrict the analysis to small-amplitude waves on the surface that do not deviate much from the undisturbed depth H, and the velocity is small, (6) and (7) may be simplified considerably; in fact, the equations can be linearized. To this end, let us assume that

$$h(x,t) = H + \eta(x,t), \qquad u(x,t) = 0 + v(x,t) \tag{8}$$

where h and v are small (η is assumed to be small compared to H), and the derivatives of η and v are assumed to be small as well. The small quantities η and v are called perturbations, and they represent quantities that deviate slightly from the equilibrium, constant state $h = H$, $u = 0$. Substituting (8) into (6) and (7) give nonlinear equations for these small deviations:

$$\eta_t + v\eta_x + (H + \eta)v_x = 0, \qquad v_t + vv_x + g\eta_x = 0 \tag{9}$$

These equations are called the *nonlinear perturbation equations*, and they govern the evolution of the small deviations from equilibrium, assuming that some initial and boundary conditions were specified. Equations (9) are no simpler than the original shallow water equations; however, if we retain only the linear terms in (9), we have

$$\eta_t + Hv_x = 0, \qquad v_t + g\eta_x = 0 \tag{10}$$

The reasoning here is that quadratic terms such as $v\eta_x$ are small compared to linear terms. Equations (10) are called the *linearized perturbation equations*. Now, it is easy to eliminate one of the unknowns in (10) and obtain a single, linear equation in one unknown. Differentiating the first equation in (10) with respect to t and the second with respect to x, and then using the equality of v_{xt} and v_{tx}, we obtain

$$\eta_{tt} - gH\eta_{xx} = 0 \tag{11}$$

We recognize this equation as the classical linear wave equation for η, which has general solution consisting of right and left traveling waves of the form

$$\eta(x, t) = F_1(x - \sqrt{gH}\,t) + F_2(x + \sqrt{gH}\,t)$$

where F_1 and F_2 are arbitrary functions. Thus small-amplitude surface waves are governed by the wave equation and they travel at speed $(gH)^{1/2}$. It is straightforward to observe that the velocity perturbation v also satisfies the wave equation

$$v_{tt} - gHv_{xx} = 0 \tag{12}$$

We emphasize that (11) and (12) are small-amplitude equations and hold only in the limit as the perturbations go to zero.

Gas Dynamics

The subject of gas dynamics is at the core of nonlinear PDEs, and it would be difficult to overstate its role in the history and development of the subject. Many of the analytical and numerical methods in nonlinear analysis were spawned by investigations in fluid mechanics, and in particular, in aerodynamics and the flow of gases. Any serious student of partial differential equations must be well grounded in the basic concepts of fluid flow.

The basic equations of gas dynamics consist of conservation laws, or balance laws, and constitutive relations that define the properties of gas. The conservation laws are nonlinear PDEs that express conservation of mass, momentum, and energy. For constitutive equations we consider the simplest case and assume an ideal gas law.

To fix the notion, we consider a tube of constant cross-sectional area A through which a gas is flowing in such a way that the physical variables are constant over any cross section. This assumption gives the motion its one-dimensional character. Further, we assume that the gas is a continuum; that is, the physical parameters may be regarded as point functions of time t and a fixed, laboratory, spatial coordinate x in the longitudinal direction. (see Figure 5.3). We let $u = u(x, t)$ denote the velocity of a cross section of gas, $\rho = \rho(x, t)$ the density of the gas, and $p = p(x, t)$ the pressure. By convention, the pressure $p(x, t)A$ is the force on the cross section at x caused by the gas to the left of x. We shall introduce additional parameters as needed. We emphasize that all physical quantities are measured in the fixed laboratory frame of reference; thus the velocity $u(x, t)$ is the velocity measured by an observer located at laboratory position x. This velocity will be the velocity of

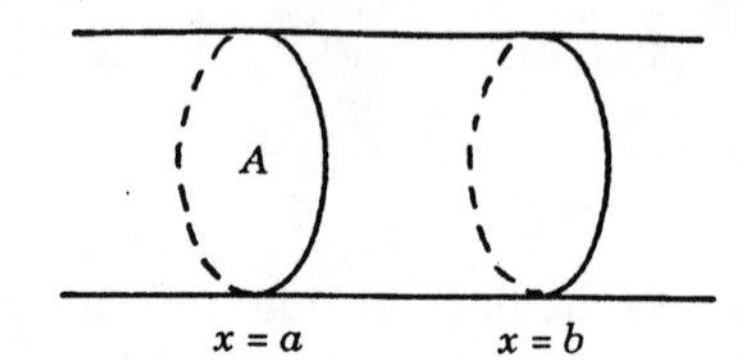

Figure 5.3. Cylindrical tube of cross-sectional area A.

different material as the gas passes by. Laboratory measurements of this type are called Eulerian measurements, and the laboratory coordinate x is called an Eulerian coordinate. There is an alternative description, the Lagrangian description, where all the physical quantities are measured with respect to an observer who is moving with the gas.

Now we consider an arbitrary section of the tube between $x = a$ and $x = b$. *Mass conservation* requires that the time rate of change of the total mass inside $[a, b]$ must equal the rate that mass flows in at $x = a$ minus the rate that mass flows out at $x = b$. In symbols,

$$\frac{d}{dt}\int_a^b \rho(x,t)A\,dx = u(a,t)\rho(a,t)A - u(b,t)\rho(b,t)A \tag{13}$$

which is the integral form of the mass conservation law. If u and ρ are smooth functions, we may deduce in the usual way the differential form of the conservation law,

$$\rho_t + (\rho u)_x = 0 \tag{14}$$

The partial differential equation (14) representing mass conservation is often called the *continuity equation*.

The *momentum balance* law states that the rate of change of momentum of the gas inside the region $[a, b]$ must equal the rate that momentum flows into the region at $x = a$, minus the rate that momentum flows out of the region at $x = b$, plus the net force on the region (the pressure at $x = a$ and $x = b$). We may express this balance law mathematically as

$$\frac{d}{dt}\int_a^b \rho(x,t)u(x,t)A\,dx = \rho(a,t)u(a,t)Au(a,t) - \rho(b,t)u(b,t)Au(b,t)$$
$$+\, p(a,t)A - p(b,t)A \tag{15}$$

This integral form may be expressed in differential form, again assuming smoothness, as

$$(\rho u)_t + (\rho u^2 + p)_x = 0 \tag{16}$$

We leave it as an exercise to show that with the aid of (14), equation (16) may be expressed as

$$\rho(u_t + uu_x) + p_x = 0 \tag{17}$$

Equations (14) and (17), expressing mass and momentum balance, are two quasilinear PDEs that govern gas flow in the tube. However, there are three unknowns, u, ρ, and p. Intuition dictates that an additional equation is required to have a well-determined system. We can turn, at this point, to a constitutive relation, or equation of state, which defines the material properties of the gas. The simplest assumption is to take the *barotropic equation of state*

$$p = f(\rho), \qquad f', f'' > 0 \tag{18}$$

where f is a given function. For example, if there are no temperature changes, some gases can be modeled by the equation $p = k\rho^\gamma$, where k is a positive constant and $\gamma > 1$ (for air, $\gamma = 1.4$). Equations (14), (17), and (18) then form a well-determined system and govern what is termed *barotropic flow*.

In general, however, there may be temperature changes in a system and the pressure p can depend on both the density ρ and the temperature $T = T(x, t)$. For example, an ideal gas satisfies the well-known equation

$$p = R\rho T \tag{19}$$

where R is the universal gas constant. Such an equation of state introduces another unknown, the temperature T, and therefore an additional equation is needed. This additional equation is the conservation of energy law, which must be considered when temperature changes occur.

We may write down an *energy balance* law in the same manner as we did the mass and momentum balance laws. There are two kinds of energy in a system, the kinetic energy of motion and the internal energy, the latter due to molecular movement, and so on. We denote the specific internal energy function by $e = e(x, t)$, given in energy units per unit mass. Thus the total energy in the region $[a, b]$ is given by

$$\begin{aligned}\text{total energy in } [a,b] &= \text{kinetic energy of } [a,b] + \text{internal energy inside } [a,b] \\ &= \int_a^b \tfrac{1}{2}\rho(x,t)u(x,t)^2 A\,dx + \int_a^b \rho(x,t)e(x,t)A\,dx\end{aligned}$$

How can the total energy be changed? Energy (both kinetic and internal) can flow in and out of the region, and energy can also change by doing work. It is the forces caused by the pressure that do the work, and the rate that a force does work is the force times the velocity. Therefore, in words, we may express the balance of energy as: The rate of change of total energy in the region

$[a, b]$ must equal the rate at which total energy flows into the region at $x = a$, minus the rate at which energy flows out of the region at $x = b$, plus the rate at which the pressure force does work at the face $x = a$, minus the rate at which pressure does work at the face $x = b$. Expressed symbolically, we have

$$\frac{d}{dt}\int_a^b \rho\left(\frac{u^2}{2} + e\right)A\,dx = \left[\frac{1}{2}\rho(a,t)u(a,t)^2 A + \rho(a,t)e(a,t)A\right]u(a,t)$$
$$-\left[\frac{1}{2}\rho(b,t)u(b,t)^2 A + \rho(b,t)e(b,t)A\right]u(b,t)$$
$$+ p(a,t)u(a,t)A - p(b,t)u(b,t)A \tag{20}$$

which is the integral form of the conservation or balance of energy law. Assuming smoothness, we then conclude that

$$\left[\rho\left(\frac{u^2}{2} + e\right)\right]_t + \left[\rho u\left(\frac{u^2}{2} + e\right) + pu\right]_x = 0 \tag{21}$$

which is the *balance of energy equation*.

Equations (14), (17), and (21) are the general equations of gas dynamics, expressing conservation of mass, and balance of momentum and energy. There are four unknowns: the density ρ, the velocity u, the pressure p, and the specific internal energy e. We point out that there are some physical effects that have been neglected in these equations. In the momentum equation we have assumed that there are no external forces present (e.g., gravity or an electromagnetic field), and there are no viscous forces present; both of these types of forces can change the momentum and could be included. Flows without viscous effects are called *inviscid*. Second, in the energy balance equation we neglected diffusion effects (heat transport) resulting from temperature gradients; if viscous forces were present, they would also do work and thus would have to be included as well.

Generally, equations (14), (17), and (21) are supplemented by two equations of state of the form

$$p = p(\rho, T) \tag{22}$$

and

$$e = e(\rho, T) \tag{23}$$

called a *thermal* equation of state and a *caloric* equation of state, respectively. Therefore, under these constitutive assumptions, there are five equations in five unknowns: ρ, u, p, T, and e. A special case is the *ideal gas law*,

$$p = R\rho T, \qquad e = c_v T \tag{24}$$

where c_v is the constant specific heat at constant volume. Under the assumption (24) of an ideal gas, the energy equation (21) can be written in several ways, depending on the choice of variables. For example, we leave it to the reader to show that the energy equation can be written

$$\rho c_v (T_t + uT_x) + pu_x = 0 \tag{25}$$

In the same way that linearized equations for small disturbances were obtained from the fully nonlinear shallow water equations, we may obtain linearized equations governing small-amplitude waves in gas dynamics. In this case the linearized theory is called *acoustics*, and the resulting equations govern ordinary sound waves in a gas. As in the case of shallow water waves, the linearized equations of acoustics are linear wave equations, and we leave their derivation to the exercises.

EXERCISES

1. Show that (7) follows from (6) and (5).
2. Derive (12).
3. Derive (25) under the ideal gas assumption (24).
4. In the case of an ideal gas [equations (24)], show that the energy balance equation can be written

$$p_t + up_x - \frac{p}{\rho}(\rho_t + u\rho_x) + (\gamma - 1)pu_x = 0$$

where $\gamma = 1 + R/c_v$.

5. *Acoustics* Consider the barotropic flow of a gas, which is governed by equations (14), (17), and (18). Let $c^2 = f'(\rho)$. Assume that the gas is in a constant-equilibrium state $u = 0$, $\rho = \rho_0$, $p = p_0$, and let $c_0^2 = f'(\rho_0)$. Let

$$u = 0 + \tilde{u}(x,t), \qquad \rho = \rho_0 + \tilde{\rho}(x,t)$$

where $\tilde{u}$ is a small velocity perturbation and $\tilde{\rho}$ is a small (compared to ρ_0) density perturbation, and derivatives of $\tilde{u}$ and $\tilde{\rho}$ are small as well. Show that under the assumption that quadratic terms in the perturbations may be ignored compared to linear terms, deviations $\tilde{u}$ and $\tilde{\rho}$ from the equilibrium state each satisfy the wave equation

$$\psi_{tt} - c_0^2 \psi_{xx} = 0$$

and therefore acoustic signals travel at speed c_0.

5.2 HYPERBOLIC SYSTEMS AND CHARACTERISTICS

We stated in Chapters 2 and 3 that a single first order PDE was essentially a wavelike equation (a fact we verify below), and we were able to identify a family of spacetime curves along which the PDE reduced to an ordinary differential equation. These curves, called the characteristics, were also the spacetime loci along which signals were transmitted in the problem. Now we ask if such curves, or special directions, exist for a *system* of first order PDEs. The answer to this question will lead to a general classification of systems of PDEs and give a precise definition of hyperbolicity, a concept that, until now, has only been intuitive.

To fix the idea and motivate the general definition and procedure, we analyze the shallow water equations

$$h_t + uh_x + hu_x = 0 \tag{1}$$

$$u_t + uu_x + gh_x = 0 \tag{2}$$

The technique that we apply in the subsequent calculation is applicable to other systems of PDEs. We are searching for a direction in spacetime in which both equations, simultaneously, will contain directional derivatives of h and u in that direction. To begin, we take a linear combination of (1) and (2) to get

$$\gamma_1(h_t + uh_x + hu_x) + \gamma_2(u_t + uu_x + gh_x) = 0$$

or, upon rearranging terms to get the derivatives of h and of u together,

$$\gamma_1\left[h_t + \left(u + \frac{g\gamma_2}{\gamma_1}\right)h_x\right] + \gamma_2\left[u_t + \left(u + \frac{\gamma_1}{\gamma_2}hu_x\right)\right] = 0 \tag{3}$$

where γ_1 and γ_2 (not necessarily constant) are to be determined. Now we observe from (3) that the total derivative of h and the total derivative of u will be in the same direction if we force

$$\frac{g\gamma_2}{\gamma_1} = \frac{h\gamma_1}{\gamma_2}$$

or

$$\frac{\gamma_2}{\gamma_1} = \pm\sqrt{\frac{h}{g}} \tag{4}$$

Therefore, let us take $\gamma_1 = 1$ and $\gamma_2 = \sqrt{h/g}$. Then (3) becomes

$$h_t + \left(u + \sqrt{gh}\,\right)h_x + \sqrt{\frac{h}{g}}\left[u_t + \left(u + \sqrt{gh}\,\right)u_x\right] = 0 \tag{5}$$

If $\gamma_1 = 1$ and $\gamma_2 = -\sqrt{h/g}$, then (3) becomes

$$h_t + \left(u - \sqrt{gh}\,\right)h_x - \sqrt{\frac{h}{g}}\left[u_t + \left(u - \sqrt{gh}\,\right)u_x\right] = 0 \tag{6}$$

Consequently, we have replaced the two original shallow water equations by two alternative equations, obtained by taking linear combinations, in which the derivatives of h and u occur in the same direction, namely $u + \sqrt{gh}$ in (5) and $u - \sqrt{gh}$ in (6). That is, along the curves defined by

$$\frac{dx}{dt} = u + \sqrt{gh} \tag{7}$$

the PDE (5) reduces to

$$\frac{dh}{dt} + \sqrt{\frac{h}{g}}\,\frac{du}{dt} = 0 \tag{8}$$

which is an ordinary differential equation. And along the curves defined by

$$\frac{dx}{dt} = u - \sqrt{gh} \tag{9}$$

the PDE (6) reduces to

$$\frac{dh}{dt} - \sqrt{\frac{h}{g}}\,\frac{du}{dt} = 0 \tag{10}$$

We can summarize (7)–(10) by writing

$$\frac{dh}{dt} \pm \sqrt{\frac{h}{g}}\,\frac{du}{dt} = 0 \qquad \text{along} \qquad \frac{dx}{dt} = u \pm \sqrt{gh} \tag{11}$$

Equations (11) are called the *characteristic equations* corresponding to (1) and (2), or the *characteristic form* of (1) and (2). Therefore, for the system of two PDEs (1) and (2) there are *two* families of characteristic curves and they are defined by equations (7) and (9). Along these curves the PDEs can be reduced to ordinary differential equations.

Classification

The classification of a system of n first order PDEs is based on whether or not there exist n directions along which the PDEs reduce to n ordinary differential equations. To be more precise, assume that we are given a system of n equations in n unknowns $u_1, \ldots, u_n$, which we write in matrix form as

$$\mathbf{u}_t + A(x, t, \mathbf{u})\mathbf{u}_x = \mathbf{b}(x, t, \mathbf{u}) \tag{12}$$

where $\mathbf{u} = (u_1, \ldots, u_n)^t$, $\mathbf{b} = (b_1, \ldots, b_n)^t$, and $A = (a_{ij}(x, t, \mathbf{u}))$ is an n by n matrix. The superscript t denotes the transpose operation; thus boldface letters denote column vectors. (In the subsequent analysis we prefer matrix notation rather than the more cumbersome index notation.) For example, the shallow water equations are, in matrix form,

$$\begin{pmatrix} h \\ u \end{pmatrix}_t + \begin{pmatrix} u & h \\ g & u \end{pmatrix}\begin{pmatrix} h \\ u \end{pmatrix}_x = \begin{pmatrix} 0 \\ 0 \end{pmatrix}$$

In this case $\mathbf{u} = (h, u)^t$, $\mathbf{b} = (0, 0)^t$, and the matrix A is given by

$$A = \begin{pmatrix} u & h \\ g & u \end{pmatrix} \tag{13}$$

Now we ask if there is a family of curves along which the PDEs (12) reduce to a system of ordinary differential equations, that is, in which the directional derivative of each u_i occurs in the same direction. We proceed as in the example of the shallow water equations and take a linear combination of the n equations in (12). This amounts to multiplying (12) on the left by a row vector $\boldsymbol{\gamma}^t = (\gamma_1, \ldots, \gamma_n)$ which is to be determined. Thus

$$\boldsymbol{\gamma}^t\mathbf{u}_t + \boldsymbol{\gamma}^t A\mathbf{u}_x = \boldsymbol{\gamma}^t\mathbf{b} \tag{14}$$

We want (14) to have the form of a linear combination of total derivatives of the u_i in the same direction λ; that is, we want (14) to have the form

$$\mathbf{m}^t(\mathbf{u}_t + \lambda\mathbf{u}_x) = \boldsymbol{\gamma}^t\mathbf{b} \tag{15}$$

for some $\mathbf{m}$. Consequently, we require that

$$\mathbf{m} = \boldsymbol{\gamma}, \qquad \mathbf{m}^t\lambda = \boldsymbol{\gamma}^t A$$

or

$$\boldsymbol{\gamma}^t A = \lambda \boldsymbol{\gamma}^t \tag{16}$$

Therefore, λ must be an eigenvalue of A and $\boldsymbol{\gamma}^t$ must be a corresponding left eigenvector. Note here that λ as well as $\boldsymbol{\gamma}$ can depend on x, t, and $\mathbf{u}$. So if $(\lambda, \boldsymbol{\gamma}^t)$ is an eigenpair, then

$$\boldsymbol{\gamma}^t \frac{d\mathbf{u}}{dt} = \boldsymbol{\gamma}^t \mathbf{b} \quad \text{along} \quad \frac{dx}{dt} = \lambda(x, t, \mathbf{u}) \tag{17}$$

and the system of PDEs (12) has been reduced to a single ordinary differential equation along the family of curves, called *characteristics*, defined by $dx/dt = \lambda$. The eigenvalue λ is called the *characteristic direction*. Clearly, since there are n unknowns, it would appear that n ordinary differential equations are required; but if A has n distinct real eigenvalues, we will have n ordinary differential equations, each holding along a characteristic direction defined by an eigenvalue. In this case we say that the system is hyperbolic. More precisely, we make the following definition.

Definition 1. Consider the quasilinear system (12). We say that (12) is *hyperbolic* if A has n real eigenvalues and n linearly independent left eigenvectors. If (12) is hyperbolic and the eigenvalues are distinct, (12) is said to be *strictly hyperbolic*. We say that (12) is *elliptic* if A has no real eigenvalues, and we say that (12) is *parabolic* if A has n real eigenvalues but fewer than n independent left eigenvectors. In the hyperbolic case, the system of n equations (17), obtained by selecting each of the n distinct eigenpairs, is called the characteristic form of (12).

No exhaustive classification is made in the case that A has both real and complex eigenvalues. We remind the reader that if a matrix has n distinct, real eigenvalues, it has n independent left eigenvectors, since distinct eigenvalues must have independent eigenvectors. It is, of course, possible for a matrix to have fewer than n distinct eigenvalues (multiplicities can occur), yet have a full complement of n independent eigenvectors. Finally, we recall that the eigenvalues can be calculated directly from the equation

$$\det(A - \lambda I) = 0$$

a condition that follows immediately from the stipulation that the homogeneous linear system of equations (16) must have nontrivial solutions.

We remark that more general systems of the form

$$B(x,t,\mathbf{u})\mathbf{u}_t + A(x,t,\mathbf{u})\mathbf{u}_x = \mathbf{b}(x,t,\mathbf{u}) \tag{18}$$

can be considered as well. If B is nonsingular, (18) can be clearly transformed into a system of the form (12) by multiplying through by B-inverse. However, let us follow through with the analysis of (18) as it stands. We ask if there is a vector $\boldsymbol{\gamma}$ such that the equation

$$\boldsymbol{\gamma}^t(B\mathbf{u}_t + A\mathbf{u}_x) = \boldsymbol{\gamma}^t\mathbf{b} \tag{19}$$

can take the form

$$\mathbf{m}^t(\beta\mathbf{u}_t + \alpha\mathbf{u}_x) = \boldsymbol{\gamma}^t\mathbf{b} \tag{20}$$

where α and β are scalar functions. If so, there is a single direction (α, β) where the directional derivative of each u_i in the equation is in that same direction. To elaborate, let $x = x(s)$, $y = y(s)$ be a curve Γ such that $dx/ds = \alpha$ and $dx/ds = \beta$. Then along this curve $d\mathbf{u}/ds = \alpha\mathbf{u}_x + \beta\mathbf{u}_t$, so (20) may be written

$$\mathbf{m}^t\frac{d\mathbf{u}}{ds} = \boldsymbol{\gamma}^t\mathbf{b} \tag{21}$$

which is a differential equation along Γ. Now, the conditions that (19) and (20) are equivalent are

$$\boldsymbol{\gamma}^t B = \mathbf{m}^t\beta, \qquad \boldsymbol{\gamma}^t A = \mathbf{m}^t\alpha \tag{22}$$

or, upon eliminating $\mathbf{m}$,

$$\boldsymbol{\gamma}^t(B\alpha - A\beta) = \mathbf{0} \tag{23}$$

A necessary and sufficient condition that the homogeneous equation (23) has a nontrivial solution $\boldsymbol{\gamma}^t$ is

$$\det(B\alpha - A\beta) = 0 \tag{24}$$

Equation (24) is a condition on the direction (α, β) of the curve Γ. Such a curve is said to be *characteristic*, and (21) is said to be in *characteristic form*. As before, we say that a system (18) satisfying (24) is *hyperbolic* if the linear algebraic system (23) has n linearly independent solutions $\boldsymbol{\gamma}^t$, where the directions (α, β) are real and not both α and β are zero. The directions, of course, need not be distinct. It may be the case that either B or A is singular, but we assume that A and B are not both singular so that the system becomes degenerate. If $\det B = 0$, then $\beta = 0$ is a solution of (24) for any α, and therefore the x direction is characteristic (i.e., the horizontal lines $t =$ constant are characteristics); similarly, if $\det A = 0$, the vertical coordinate lines $x =$ constant are characteristics. If no real directions (α, β) exist, the system (18) is called *elliptic*; the system (18) is called *parabolic* if there are n real directions defined by (24) but fewer than n linearly independent solutions of (23).

Example 1. For the shallow water equations (1) and (2) have the coefficient matrix A given by (13). The eigenvalues of (13) are found from the algebraic eigenvalue equation $\det(A - \lambda I) = 0$, or in this case,

$$\det\begin{pmatrix} u-\lambda & h \\ g & u-\lambda \end{pmatrix} = (u-\lambda)^2 - gh = 0$$

Therefore, the eigenvalues are $\lambda = u \pm \sqrt{gh}$, which are real and distinct. Thus the shallow water equations are hyperbolic, and in fact, strictly hyperbolic. Notice that the eigenvalues define the characteristic directions [compare (11)]. The eigenvectors are easily seen to be $(1, \sqrt{h/g})$ and $(1, -\sqrt{h/g})$, which define the appropriate linear combination of the PDEs.

Remark. Now we can see why a single first order quasilinear PDE is hyperbolic. The coefficient matrix for the equation

$$u_t + c(x,t,u)u_x = b(x,t,u)$$

is just the real scalar function $c(x,t,u)$, which has the single eigenvalue $c(x,t,u)$. In this direction [i.e., if $dx/dt = c(x,t,u)$], the PDE reduces to the ordinary differential equation $du/dt = b(x,t,u)$.

Example 2. Consider the system

$$u_t - v_x = 0$$

$$v_t - cu_x = 0 \tag{25}$$

where c is a constant. The equivalent matrix form is

$$\begin{pmatrix} u \\ v \end{pmatrix}_t + \begin{pmatrix} 0 & -1 \\ -c & 0 \end{pmatrix}\begin{pmatrix} u \\ v \end{pmatrix}_x = \begin{pmatrix} 0 \\ 0 \end{pmatrix}$$

The coefficient matrix is

$$\begin{pmatrix} 0 & -1 \\ -c & 0 \end{pmatrix}$$

which is easily seen to have eigenvalues $\lambda = \pm\sqrt{c}$. Therefore, if $c > 0$, the system is hyperbolic, and if $c < 0$, the system is elliptic. It is straightforward to eliminate the unknown v from the two equations (25) and obtain the single second order PDE

$$u_{tt} - cu_{xx} = 0 \tag{26}$$

for u. If $c > 0$, we recognize (26) as the wave equation, which is hyperbolic. If $c < 0$, then (26) is elliptic [if $c = -1$, then (26) is Laplace's equation]. Therefore, our classification scheme for systems is consistent with our classification in Chapter 1 of second order linear equations. For the system (25) in the hyperbolic case ($c > 0$), the characteristic directions are $\pm\sqrt{c}$, and the characteristic curves are given by

$$\frac{dx}{dt} = \pm\sqrt{c} \quad \text{or} \quad x = \pm\sqrt{c}\,t + \text{constant}$$

These two families of characteristics are straight lines of speed $\sqrt{c}$ and $-\sqrt{c}$. To find the characteristic form of the PDEs (25), we proceed as in the example of the shallow water equations and take a linear combination of the two equations in (25), to obtain

$$\gamma_1(u_t - v_x) + \gamma_2(v_t - cu_x) = 0$$

Rearranging to put the derivatives of u and the derivatives of v in different terms, we have

$$\gamma_1\left(u_t - \frac{c\gamma_2}{\gamma_1}u_x\right) + \gamma_2\left(v_t - \frac{\gamma_1}{\gamma_2}v_x\right) = 0$$

Thus the derivatives of u and v will be in the same direction if $c\gamma_2/\gamma_1 = \gamma_1/\gamma_2$, or

$$\frac{\gamma_1}{\gamma_2} = \pm\sqrt{c}$$

Therefore, we take $\gamma_1 = \pm\sqrt{c}$ and $\gamma_2 = 1$. Consequently, the characteristic form of the PDEs (25) is

$$\pm\sqrt{c}\left(u_t \mp \sqrt{c}\,u_x\right) + \left(v_t \mp \sqrt{c}\,v_x\right) = 0$$

Hence we may write

$$\pm\sqrt{c}\,\frac{du}{dt} + \frac{dv}{dt} = 0 \qquad \text{along} \quad \frac{dx}{dt} = \mp\sqrt{c}$$

or

$$\pm\sqrt{c}\,u + v = \text{constant} \qquad \text{along} \quad x = \mp\sqrt{c}\,t + \text{constant}$$

The expressions $\pm\sqrt{c}\,u + v$ are the *Riemann invariants* for this problem; they are expressions that are constant along the characteristic curves. It follows immediately that

$$\sqrt{c}\,u + v = f(x + \sqrt{c}\,t), \qquad -\sqrt{c}\,u + v = g(x - \sqrt{c}\,t)$$

where f and g are arbitrary functions. We can easily obtain formulas for u and v by solving these two equations simultaneously. Thus we have obtained the general solution of (25) in terms of arbitrary functions; the latter may be determined from initial or boundary data.

There are basically two ways to find the characteristic form of a system of hyperbolic equations. We may calculate the eigenvalues and eigenvectors directly and then use (17); or we may proceed directly and take linear combinations of the equations as we did for the shallow water equations at the beginning of this section and for the wave equation in Example 2. For a small number of equations ($n \le 3$) the latter method may be preferable. We leave further examples to the exercises.

In summary, a system of hyperbolic equations can be changed into a system of equations where each equation in the new system involves directional derivatives in only one direction, that is, in the direction of the characteristics. This transformed set of equations is called the set of characteristic equations, and they instantly reduce to ordinary differential equations along the characteristic curves. Often, these differential equations can be solved and one can deduce the solution to the original system of hyperbolic equations. This method is called the *method of characteristics*, and it is fundamental in the treatment of hyperbolic problems.

Example 3. The characteristic form of the shallow water equations (1)–(2) is given by (11). We may write (11) as

$$\sqrt{\frac{g}{h}}\,\frac{dh}{dt} \pm \frac{du}{dt} = 0 \qquad \text{on} \qquad \frac{dx}{dt} = u \pm \sqrt{gh}$$

The differential equations on the characteristic curves may be integrated directly to obtain

$$R_{\pm} = 2\sqrt{gh} \pm u = \text{constant} \qquad \text{on} \qquad \frac{dx}{dt} = u \pm \sqrt{gh}$$

The two quantities $R_{\pm}$ are called *Riemann invariants*, and they are constant along characteristics; R_{+} is constant along a characteristic of the family $dx/dt = u + \sqrt{gh}$, and R_{-} is constant along a characteristic of the family

$dx/dt = u - \sqrt{gh}$. The constant may vary from one characteristic to another. In the shallow water equations, the characteristics with direction $u + \sqrt{gh}$ are called the *positive characteristics* (or C^+ characteristics), and the characteristics with direction $u - \sqrt{gh}$ are called the *negative characteristics* or (C^- characteristics). Consequently, as was the case for a single PDE, characteristics carry signals or information; the C^+ characteristics carry the constancy of R_+, and the C^- characteristics carry the constancy of R_-. Armed with this fact, we shall be able to solve the shallow water equations for some types of initial and boundary data; we carry out these calculations in Section 5.3, where we discuss the Riemann method.

Example 4. It is clear that the diffusion equation $u_t - u_{xx} = 0$ may be written as the first order system

$$u_t = v_x, \qquad u_x = v \tag{27}$$

which in matrix form is

$$\begin{pmatrix} 1 & 0 \\ 0 & 0 \end{pmatrix}\begin{pmatrix} u \\ v \end{pmatrix}_t + \begin{pmatrix} 0 & -1 \\ 1 & 0 \end{pmatrix}\begin{pmatrix} u \\ v \end{pmatrix}_x = \begin{pmatrix} 0 \\ v \end{pmatrix}$$

This is in the form (18) with det $B = 0$. Here $\det(B\alpha - A\beta) = \beta^2 = 0$ forces $\beta = 0$, so the characteristic direction $(\alpha, 0)$ is in the direction of the x-axis, confirming our previous remarks. The eigenvalue equation $\boldsymbol{\gamma}^t(B\alpha - A\beta) = \mathbf{0}$ becomes

$$(\gamma_1 \quad \gamma_2)\begin{pmatrix} \alpha & 0 \\ 0 & 0 \end{pmatrix} = (0 \quad 0)$$

which has only one solution, namely $\boldsymbol{\gamma}^t = (0, c)$, where c is a constant. Therefore, the system (27) is parabolic, consistent with the classification of the second order diffusion equation. If we interpret the characteristics, here horizontal straight lines with speed infinity, as curves in spacetime that carry signals, then in this case of the diffusion equation, signals travel at infinite speed. Recall that the diffusion equation with an initial point source (zero everywhere but at a single point) has a solution that is nonzero for all real x for $t > 0$. Thus we can imagine that signals travel instantly when governed by the linear diffusion equation.

EXERCISES

1. The current $i = i(x,t)$ and voltage $v = v(x,t)$ at a position x at time t along a transmission line satisfy the first order system

$$Cv_t + i_x = -Gv, \qquad Li_t + v_x = -Ri$$

where C, G, L, and R are positive constants denoting the capacitance, leakage, inductance, and resistance, respectively, all per unit length in the line. Show that this system is hyperbolic and write it in characteristic form. Sketch the two families of characteristics on an xt-diagram.

2. (*Gas Dynamics*). From Section 5.1, the conservation equations governing barotropic flow are given by

$$\rho_t + (\rho u)_x = 0, \qquad (\rho u)_t + (\rho u^2 + p)_x = 0, \qquad p = f(\rho)$$

where $f', f'' > 0$.

(a) Show that these equations may be written

$$p_t + up_x + c^2 \rho u_x = 0, \qquad \rho u_t + \rho u u_x + p_x = 0$$

where $c^2 = f'(\rho)$.

(b) Derive the characteristic form of the equations in part (a). You should get

$$\frac{dp}{dt} \pm \rho c \frac{du}{dt} = 0 \qquad \text{on} \quad C^{\pm}: \frac{dx}{dt} = u \pm c$$

(c) Prove that

$$R_{\pm} = \int \frac{c(\rho)}{\rho} \, d\rho \pm u = \text{constant} \quad \text{on} \quad C^{\pm}$$

(d) Determine the Riemann invariants $R_{\pm}$ in part (c) if the equation of state is $f(\rho) = k\rho^{\gamma}$, where $k > 0$ and $\gamma > 1$ are constants. *Solution:*

$$R_{\pm} = \frac{2c}{\gamma - 1} \pm u$$

3. Consider the linear nonhomogeneous system

$$\begin{pmatrix} -1 & a^2 \\ 1 & -1 \end{pmatrix} \begin{pmatrix} u \\ v \end{pmatrix}_t + \begin{pmatrix} u \\ v \end{pmatrix}_x = \begin{pmatrix} f(t) \\ g(t) \end{pmatrix}$$

on $x > 0$, $t > 0$, where $0 < a < 1$, and where f and g are given continuous functions.

(a) Show that the system is hyperbolic and find the characteristic form of the equations.

(b) Sketch the characteristics on an xt-diagram.

(c) Determine expressions that are constants on the characteristics.

(d) To the system of PDEs, append the initial and boundary conditions

$$u(0,t) = 0, \quad t > 0;$$
$$u(x,0) = \alpha v(x,0), \quad x > 0, \qquad \text{where } \alpha > 0$$

For any $t > 0$, prove that the boundary condition on v is given by

$$v(0,t) = \delta v(0,rt) + \frac{\delta[F(rt) + aG(rt)]}{a(a-1)} + \frac{F(t) - G(t)}{a(1-a)}$$

where F and G are the antiderivatives of f and g, and

$$r = (1-a)/(1+a), \delta = (a - \alpha)/(a + \alpha)$$

[See J. D. Logan, *Appl. Anal.* **33**, 255–266 (1989).]

4. Consider the linear, strictly hyperbolic system

$$\mathbf{u}_t + A(x,t)\mathbf{u}_x = \mathbf{0}$$

Show that the transformation $\mathbf{u} = P\mathbf{w}$, where $P^{-1}AP = D$, and D is a diagonal matrix having the eigenvalues of A on its diagonal, transforms the system into a linear system

$$\mathbf{w}_t + D(x,t)\mathbf{w}_x = C(x,t)\mathbf{w}$$

with the derivatives of the components of $\mathbf{w}$ decoupled. What are the characteristic directions? Why does such a matrix P exist? Write the system in characteristic form.

5. Solve the linear initial value problem

$$u_t + v_t + 3u_x + 2v_x = 0$$
$$-u_t + v_t + 5u_x + 2v_x = 0, \qquad x \in R, \quad t > 0$$
$$u(x,0) = \sin x, \qquad v(x,0) = e^x, \quad x \in R$$

Sketch the characteristics in the xt-plane. *Solution:* $u = \sin(x+t)$, $v = [8\sin(x-2t) + 6\exp(x-2t) - 8\sin(x+t)]/6$.

6. Consider the system

$$u_t + 4u_x - 6v_x = 0$$
$$v_t + u_x - 3v_x = 0$$

(a) Show that the system is strictly hyperbolic.
(b) Transform the system to characteristic form by finding the eigenvalues and left eigenvectors.
(c) Find the general solution in terms of arbitrary functions.
(d) What are the characteristics? What are the Riemann invariants?

7. For a simple analog of detonation, W. Fickett [*Am. J. Phys.* **47**, 1050–1059 (1979)] posed the equations

$$u_t + f(u, z)_x = 0, \qquad z_t = -r(u, z)$$

where $u = u(x, t)$ is a temperature-like quantity and $z = z(x, t)$ is the mass fraction of the reactant A in an exothermic, irreversible, chemical reaction A → B (i.e., z is the mass of A divided by the sum of the masses of A and B); $f(u, z)$ defines an equation of state and $r(u, z)$ defines the chemical reaction rate. It is assumed that $f_u, f_{uu} > 0$, and $f_z < 0$. The equations above represent an idealized for model chemical–fluid mechanic interactions in a chemically reactive medium. Show that this system is hyperbolic, find the characteristic directions, and transform the system to characteristic form.

5.3 RIEMANN'S METHOD

In this section, using the shallow water equations as the model, we illustrate how to solve a variety of initial boundary value problems by the method of characteristics, or *Riemann's method*. Riemann, along with Stokes, was one of the early investigators of wave phenomena and he pioneered some of the techniques and methods for nonlinear equations and shock waves, particularly in the area of fluid dynamics. The Riemann method uses to advantage the characteristic form of a system of hyperbolic equations. As we observed in Section 5.2, under some circumstances it is possible to determine expressions, called Riemann invariants, that are constant along the characteristic curves. Knowing these invariants and relating them back to the initial and boundary conditions permits the determination of their constant values on the characteristics, and using this information then points the way toward the solution of the problem.

Jump Conditions for Systems

Before proceeding with some representative problems, we pause to consider the shock relations for a system of hyperbolic equations in conservation form. The situation is much the same as in the single-equation case discussed in Section 3.2. We assume that the equations governing some physical system are integral conservation laws of the form

$$\frac{d}{dt}\int_a^b \Psi_k \, dx = -\Phi_k\Big|_{x=a}^{x=b} + \int_a^b f_k \, dx \qquad (k = 1, \ldots, n) \tag{1}$$

where the Ψ_k, Φ_k, and f_k are physical quantities of interest, depending on x, t, and n unknown functions, or densities, $u_1(x, t), \ldots, u_n(x, t)$. We interpret the Φ_k as flux terms and the f_k as local source terms. Under the hypothesis that the u_k are continuously differentiable, we could proceed in the usual way to obtain a system of PDEs of the form

$$(\Psi_k(x,t,\mathbf{u}))_t + (\Phi_k(x,t,\mathbf{u}))_x = f_k(x,t,\mathbf{u}) \qquad (k=1,\dots,n) \quad (2)$$

where $\mathbf{u} = (u_1,\dots,u_n)$. We say that the system (2) is in conservation form. If the functions u_k have simple discontinuities along a smooth curve $x = s(t)$ in spacetime, the same argument as in Section 3.2 can be applied to obtain the *n jump conditions*

$$-s'[\Psi_k] + [\Phi_k] = 0 \qquad (k = 1,\dots,n) \quad (3)$$

where $s' = ds/dt$, and where $[Q] = Q^- - Q^+$ denotes the jump in the quantity Q across $x = s(t)$. The curve $x = s(t)$ is the *shock path*, s' is the *shock velocity*, and the waveforms u_k themselves, as their discontinuities propagate along $x = s(t)$, are collectively called a *shock wave*. The same proviso holds as for the case of the single scalar conservation law: namely, the correct jump conditions can be obtained from integral forms of the conservation laws since there is no unique way to obtain PDEs in conservation form from a system of arbitrary PDEs.

Example 1. The shallow water equations (see Section 5.1) were obtained from integral conservation laws, which led, in turn, to a system of PDEs in conservation form

$$h_t + (hu)_x = 0 \quad (4)$$

$$(hu)_t + \left(hu^2 + \tfrac{1}{2}h^2\right)_x = 0 \quad (5)$$

where we have set $g = 1$. (The fact that the shallow water equations can be rescaled to set the acceleration due to gravity equal to unity is explored in Exercise 1.) Thus, according to (3), the jump conditions are given by

$$s' = \frac{[hu]}{[h]} \quad \text{and} \quad s' = \frac{\left[hu^2 + h^2/2\right]}{[hu]} \quad (6)$$

We can think of (6) as two equations relating the five quantities consisting of the values of h and of u ahead and behind the shock, and the shock velocity s'. In the context of shallow water theory, a propagating shock is often called a *bore*.

Dam Breaking Problem

The first problem that we consider is what is commonly called the dam breaking problem. We imagine water at height 1 held motionless in $x < 0$ by a dam placed at $x = 0$. Ahead of the dam ($x > 0$) there is no water. At time $t = 0$ the dam is suddenly removed, and the problem is to determine the height and velocity of the water for all $t > 0$ and $x \in R$, according to the shallow water theory. Consequently, the governing equations are (4) and (5) subject to the initial conditions

$$u(x,0) = 0 \quad \text{for } x \in R; \qquad h(x,0) = 1 \quad \text{for } x < 0,$$
$$\text{and} \qquad h(x,0) = 0 \quad \text{for } x > 0 \tag{7}$$

The solution of this problem, using Riemann's method, is a blend of physical intuition and analytic calculations, the former guiding the latter. We prefer this approach, which interrelates the physics and the mathematics, rather than purely analytic reasoning that ignores the origins of the problem. From our calculation in Section 5.2 we know that the shallow water equations (4) and (5) may be put in the characteristic form

$$\frac{dh}{dt} \pm \sqrt{h}\,\frac{du}{dt} = 0 \qquad \text{on} \quad C^{\pm}: \frac{dx}{dt} = u \pm \sqrt{h} \tag{8}$$

Direct integration then gives

$$R^{\pm} = 2\sqrt{h} \pm u = \text{constant} \qquad \text{on} \quad C^{\pm}: \frac{dx}{dt} = u \pm \sqrt{h} \tag{9}$$

where R^+ and R^- are the two Riemann invariants, which are expressions that are constant on the C^+ and C^- characteristics, respectively. Let us first examine the C^- curves emanating from the negative x-axis, where the water is located. The speed of the negative characteristics is $u - \sqrt{h}$, and therefore the C^- curves leave the negative x-axis (where $u = 0$ and $h = 1$) with speed -1. Along a C^- characteristic we have

$$R^- = 2\sqrt{h} - u = 2\sqrt{h(x,0)} - u(x,0) = 2 \tag{10}$$

Similarly, a C^+ characteristic emanating from the negative x-axis has initial speed $+1$, and on such a characteristic we have

$$R^+ = 2\sqrt{h} + u = 2 \tag{11}$$

Therefore, at a point P in the region $x < -t$, both (10) and (11) hold, thereby giving $u = 0$ and $h = 1$ in the domain $x < -t$. Since h and u are constant in this entire region, the speeds of the $C+$ and $C-$ characteristics are constants ($+1$ and -1, respectively), so these characteristics are straight lines in the region $x < -t$. This portion of the characteristic diagram is shown in Figure 5.4. The region $x < -t$ is the region of spacetime that cannot be affected by the removal of the dam—so in that region the initial

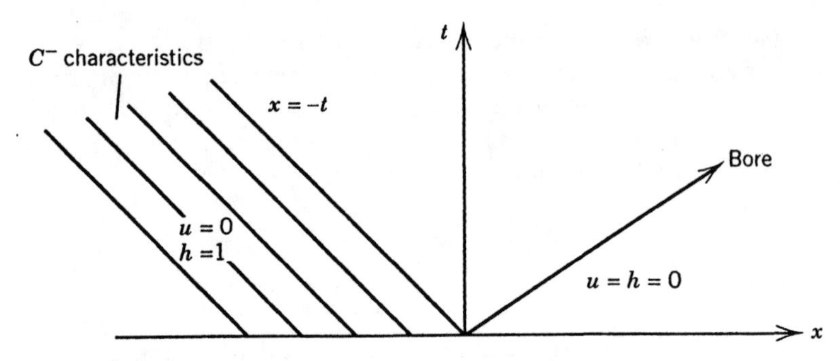

Figure 5.4. Negative (C^-) characteristics for the dam breaking problem.

conditions $u = 0$ and $h = 1$ remain valid. The first signal back into the water is the leading edge of rarefaction wave that releases the lower water height ahead of the wave. This signal travels along $x = -t$ at speed -1.

Now, we also expect the water to move into $x > 0$ as a bore, subject to the jump conditions (6). Ahead of the bore (see Figure 5.4) we have $u = h = 0$, and therefore the jump conditions (6) become

$$s' = u_-, \qquad s' = \frac{u_-^2 + h_-/2}{u_-} \tag{12}$$

where h_- and u_- are the values of h and u just behind the bore. It follows immediately from (12) that $h_- = 0$; that is, there is no jump in height across the bore. To determine u_-, and hence the bore velocity s', we need to know the solution $u(x, t)$ in the triangular region behind the bore and in front of the leading edge of the rarefaction $x = -t$.

One way to proceed is to note that the C^+ characteristics carry information forward in time; they originate on the negative x axis and end on the bore. Because equation (11) holds on *every* C^+ characteristic, it must in fact hold *everywhere* behind the bore (this is because the C^+ come from a constant state, $u = 0$ and $h = 1$, in this problem; if the initial state were not constant, this conclusion could not be made). Hence (11) must hold just behind the bore, or

$$2\sqrt{h_-} + u_- = 2$$

Consequently, since $h_- = 0$, we must have $u_- = 2$ and $s' = 2$. We have concluded that the bore is a straight line $x = 2t$; the jump in the height h across the bore is zero, and the jump in the velocity u is 2.

To find the solution $u(x, t)$ and $h(x, t)$ in the region behind the bore we expect, from our earlier experience with rarefactions occurring in traffic flow and in other problems, to fit in a fan of C^- characteristics connecting the solution along $x = -t$ to the solution along $x = 2t$. Therefore, we insert straight-line C^- characteristics into this region, passing through the origin,

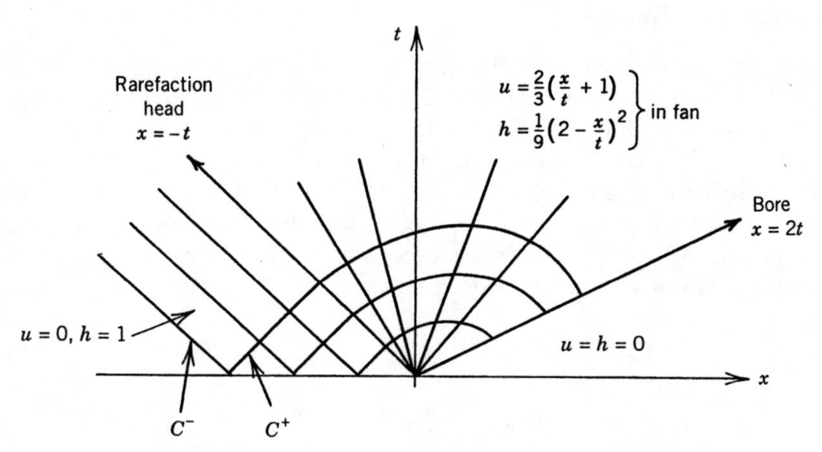

Figure 5.5. Characteristic diagram for the dam breaking problem.

with equations

$$x = (u - \sqrt{h})t \tag{13}$$

where $u - \sqrt{h}$ is the speed of a C^- characteristic. Then, in this triangular region,

$$u = \frac{x}{t} + \sqrt{h} \tag{14}$$

Using (14) along with (11), the latter holding at all points of the region, gives

$$u(x,t) = \frac{2(x/t + 1)}{3}, \qquad h(x,t) = \frac{(2 - x/t)^2}{9} \qquad (-t < x < 2t) \tag{15}$$

We already know that $u = h = 0$ for $x > 2t$, and $u = 0$, $h = 1$ for $x < -t$. Thus we have a complete solution to the problem. Figure 5.5 shows a complete characteristic diagram of the solution, and Figure 5.6 shows typical time snapshots of the velocity and height.

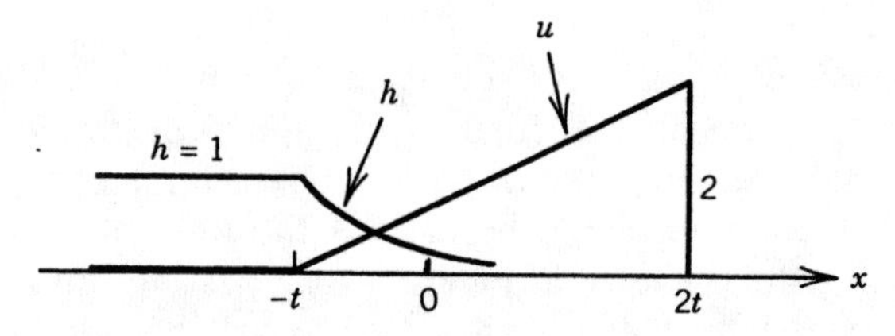

Figure 5.6. Velocity and water height profiles for the dam breaking problem.

Retracting Wall Problem

We now imagine water in $x > 0$ being held motionless at height $h = 1$ by a wall at $x = 0$. At time $t = 0$ the wall is pulled back along a given spacetime path $x = X(t)$, $t > 0$, where X satisfies the conditions $X'(0) = 0$, $X'(t) < 0$, and $X''(t) < 0$. Thus the wall has negative velocity and it continues to accelerate backward as time increases. The problem is to determine the velocity and height of the water for $X(t) < x$ and $t > 0$. The shallow water equations are assumed to govern the evolution of the system. The initial and boundary conditions are expressed analytically by

$$u(x,0) = 0, \qquad h(x,0) = 1 \quad \text{for } x > 0 \tag{16}$$

and

$$u(X(t),t) = X'(t), \qquad t > 0 \tag{17}$$

Equation (17) expresses the fact that the velocity of the water adjacent to the wall is the same as the velocity of the wall. As it turns out, we cannot independently impose a condition on height of the water at the wall; intuitively, we should be able to determine h at the wall as a part of the solution to the problem.

Again we resort to the characteristic form of the shallow water equations,

$$R^+ = 2\sqrt{h} + u = \text{constant} \quad \text{on} \quad C^+ : \frac{dx}{dt} = u + \sqrt{h} \tag{18}$$

$$R^- = 2\sqrt{h} - u = \text{constant} \quad \text{on} \quad C^- : \frac{dx}{dt} = u - \sqrt{h} \tag{19}$$

in terms of the Riemann invariants. There is enough information in this characteristic form of the equations to solve the problem. The basic idea is to let the C^+ and C^- characteristics carry signals (the constancy of R^+ and R^-) from the boundaries into the region of interest. First we consider the C^- characteristics emanating from the positive x-axis. Since $u = 0$ and $h = 1$ along the x-axis, the C^- characteristics leave the x-axis with speed -1. Along each of these characteristics R^- has the same constant value, namely

$$2\sqrt{h} - u = 2 \tag{20}$$

But since (20) holds along every C^- characteristic, *it must hold everywhere* in $x > X(t)$. That is, the negative Riemann invariant is constant throughout the region of interest. Since the speed of the C^- characteristics is more negative than the speed of the wall along the path of the wall, the C^- curves must intersect the wall path as shown in Figure 5.7. In passing, we notice that (20) permits us to calculate the water height h at the wall, because u at the wall is

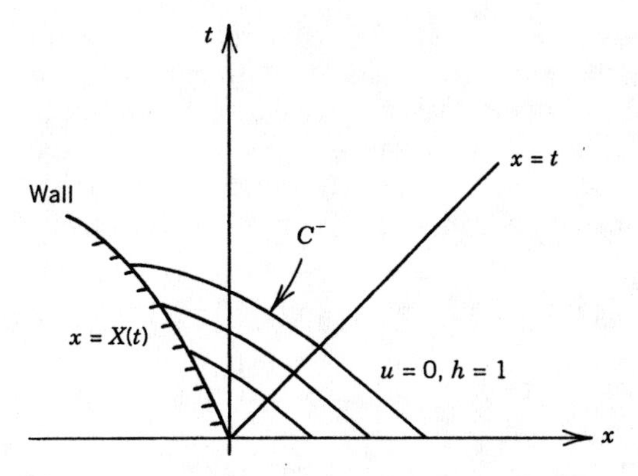

Figure 5.7. Negative characteristics for the retracting wall problem.

known. Thus the height of the water at the wall is given by

$$h(X(t), t) = \left(\frac{2 + X'(t)}{2} \right)^2 \tag{21}$$

The C^+ characteristics emanating from the x-axis have initial speed 1, and it is easy to see that in the region $x > t$ we have the constant state $u = 0$ and $h = 1$; this is the region of spacetime that is not influenced by the motion of the wall. The leading signal into the water ahead travels at speed 1 along the limiting C^+ characteristic $x = t$. Thus, since both u and h are constant ahead of this initial signal, both sets of characteristics are straight lines in this region. Now we consider a C^+ characteristic emanating from the wall, as shown in Figure 5.8. It is straightforward to see that such a C^+ must be a straight line; along a C^+ we have

$$2\sqrt{h} + u = \text{constant} \tag{22}$$

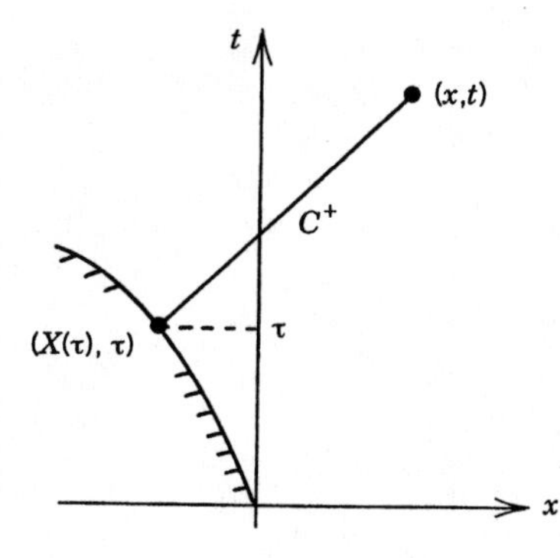

Figure 5.8. C^+ characteristic emanating from the retracting wall.

(this constant, of course, will vary from one C^+ to another since u is changing along the wall). Adding and subtracting (22) and (20), the latter holding everywhere, shows that on a specific C^+ characteristic we must have both u and h constant. Hence the speed $u + \sqrt{h}$ of a specific C^+ must be constant, and the characteristic is therefore a straight line. The C^+ characteristic emanating from $(X(\tau), \tau)$ has equation given by

$$
\begin{aligned}
x - X(\tau) &= (u + \sqrt{h})(t - \tau) \\
&= \left[1 + \frac{3X'(\tau)}{2}\right](t - \tau)
\end{aligned} \tag{23}
$$

where we have used (21) and (17), and the fact that u and h are constant on the characteristic.

Therefore, we may determine the solution h and u at an arbitrary point (x, t) in the region $X(t) < x < t$. Again using the constancy of h and u on a C^+, we have

$$u(x, t) = u(X(\tau), \tau) = X'(\tau) \tag{24}$$

and

$$h(x, t) = h(X(\tau), \tau) = \left[\frac{2 + X'(\tau)}{2}\right]^2 \tag{25}$$

where $\tau = \tau(x, t)$ is given implicitly by (23). The characteristic diagram is shown in Figure 5.9.

We have noted in the two preceding problems that one of the Riemann invariants is constant throughout the region of interest. Whenever this occurs, we say that the solution in the nonuniform region is a *simple wave*. One can show that a simple wave solution always exists adjacent to a uniform

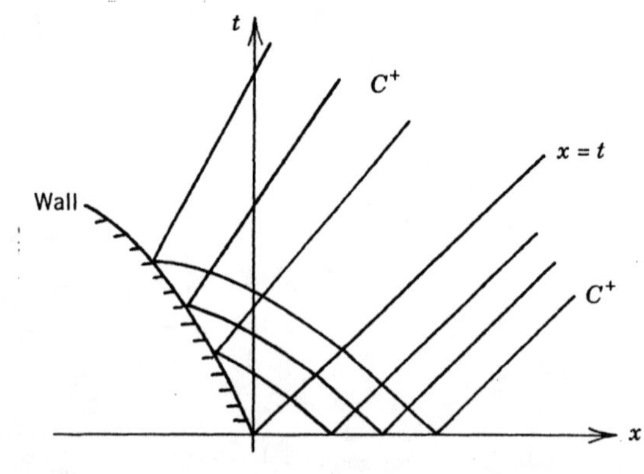

Figure 5.9. Characteristic diagram for the retracting wall problem.

state provided that the solution remains smooth. In the two examples we considered, the constancy of one of the Riemann invariants arose because all of the characteristics originated from a constant state; if the initial conditions were not constant, the simple wave argument could not be made.

Formation of a Bore

Now we examine the preceding problem in the case that the wall is pushed forward instead of being retracted. If we draw a typical wall path in this case, it is easy to see that two C^+ characteristics will collide, indicating the formation of a shock wave or bore (see Figure 5.10). Assume that the wall path is $x = X(t)$, where $X'(t) > 0$ and $X''(t) > 0$, and let τ_1 and τ_2 be two values of t with $\tau_1 < \tau_2$. The C^+ characteristics emanating from $(X(\tau_1), \tau_1)$ and $(X(\tau_2), \tau_2)$ have speeds [see (23)] given by

$$1 + \frac{3X(\tau_1)}{2} \quad \text{and} \quad 1 + \frac{3X(\tau_2)}{2}$$

respectively. Because X' is increasing by assumption, $X'(\tau_2) > X'(\tau_1)$ and it follows that the characteristic at τ_2 is faster than the characteristic τ_1. Thus the two characteristics must intersect.

Let us now consider a special case and take the wall path to be

$$x = X(t) = at^2, \qquad t > 0$$

where a is a positive constant. Then, from (23), the C^+ characteristics coming from the wall have the equation

$$x - a\tau^2 = (1 + 3a\tau)(t - \tau) \tag{26}$$

Equation (26) defines τ implicitly as a function of x and t. Therefore, given x and t, we may ask when it is possible to solve (26) uniquely for τ

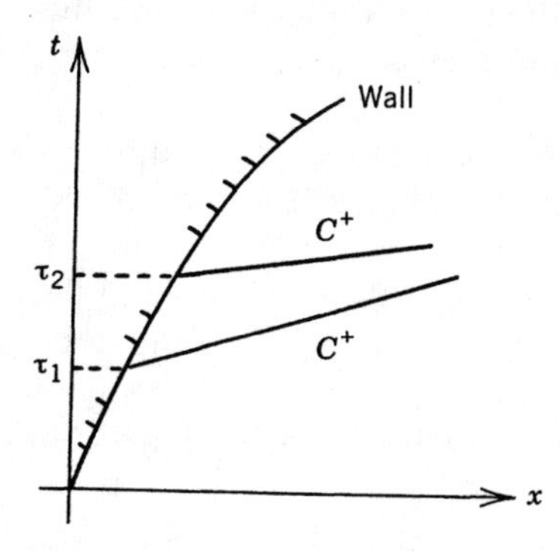

Figure 5.10. Two colliding C^+ characteristics emanating from an accelerating wall.

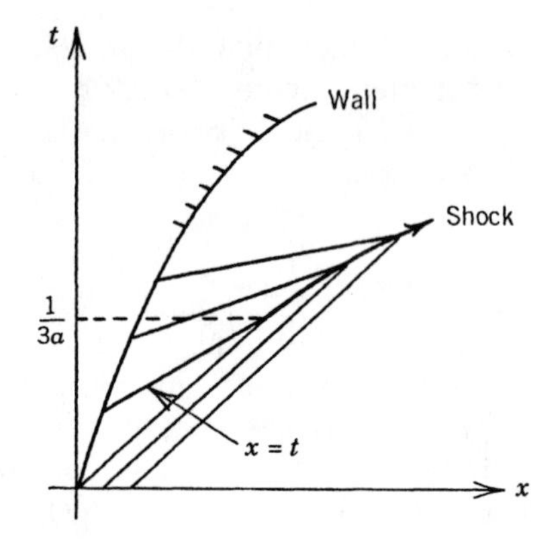

Figure 5.11. Characteristics intersecting to form a shock wave at time $t = 1/3a$.

[i.e., determine a unique C^+ characteristic passing through (x, t)]. Writing (26) as

$$F(x,t,\tau) = x - a\tau^2 + (1 + 3a\tau)(\tau - t) = 0 \tag{27}$$

we know from analysis that it is possible to solve $F(x, t, \tau) = 0$ for τ (at least locally) if $F_\tau(x, t, \tau) \neq 0$. In the present case

$$F_\tau = 1 + 4a\tau - 3at$$

Therefore, the formation of the bore will occur at the first time t_b when $F_\tau = 0$, that is,

$$t_b = \min\left\{\frac{1 + 4a\tau}{3a} : \tau > 0\right\} = \frac{1}{3a}$$

Thus the breaking time is $1/(3a)$ and occurs along the $\tau = 0$ characteristic, or $x = t$. Figure 5.11 shows the characteristic diagram for this problem. The bore, after it forms, moves into a constant state $u = 0$, $h = 1$. The speed of the bore must be determined by the jump conditions (6) and the state u and h behind the bore. Like most nonlinear initial boundary value problems, this problem is computationally complex and cannot be resolved analytically, and one must resort to numerical methods.

We consider a simpler problem, where the wall moves into the motionless water with depth 1 at a *constant*, positive speed V; that is, the path of the wall is $x = X(t) = Vt$, $t > 0$. We expect a bore to form immediately and move with constant velocity into the uniform state $u = 0$, $h = 1$. Behind the bore we also expect that there is a uniform state $u = V$, with h still to be determined (note that the shallow water equations can have constant solutions). The speed of the bore and the depth of water h_- behind the bore

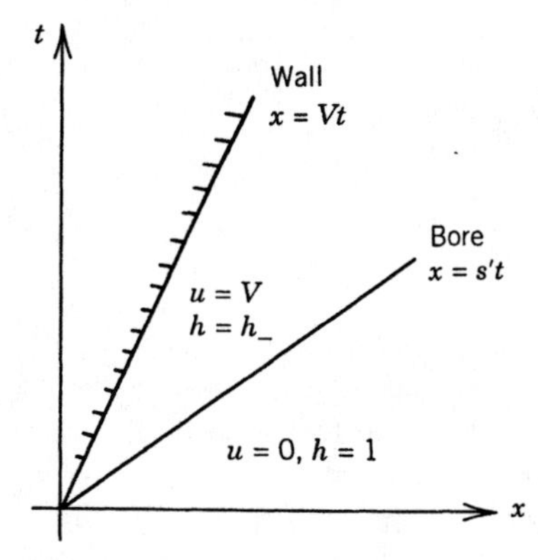

Figure 5.12. Bore with speed s' caused by a constant-velocity wall.

can be determined by the jump conditions (6). We have, from those conditions,

$$s' = \frac{h_- V}{h_- - 1}, \qquad s' = \frac{V^2 h_- + h_-^2/2 - 1/2}{V h_-} \tag{28}$$

The bore speed s' can be eliminated to obtain

$$h_-^3 - h_-^2 - (1 + 2V^2)h_- + 1 = 0 \tag{29}$$

which is a cubic equation for h_-. It is easy to see, using calculus, that there is a single negative root and two positive roots, one less that 1 and one greater than 1. We reject the negative root because the depth must be nonnegative, and we reject the smaller positive root because the bore speed [from (28)] would be negative. Therefore, h_- is the largest positive root of (29). The solution that we have derived is shown on the diagram in Figure 5.12.

Gas Dynamics

The equations governing one-dimensional gas dynamics, derived in Section 5.1, are given in conservation form by

$$\rho_t + (\rho u)_x = 0, \qquad (\rho u)_t + (\rho u^2 + p)_x = 0 \tag{30}$$

where ρ is the density, u the velocity, and p the pressure given by the equation of state $p = k\rho^\gamma$, where $k > 0$ and $\gamma > 1$. It was shown in Exercise 2 of Section 5.2 that the characteristic form of (30) is

$$\frac{2c}{\gamma - 1} \pm u = \text{constant} \qquad \text{on} \quad \frac{dx}{dt} = u \pm c \tag{31}$$

where $c^2 = p'(\rho)$. The form of the PDEs (30) and the Riemann invariants in

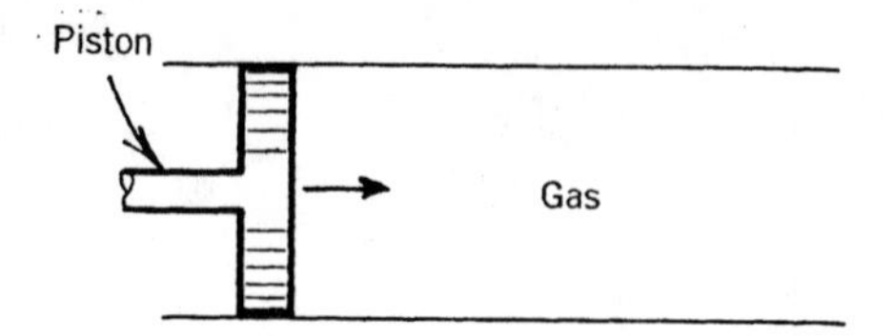

Figure 5.13. Piston moving into a gas.

(31) bear a strong resemblance to the equations from shallow water theory that were discussed above. We have chosen to analyze the shallow water equations, believing that the reader will have a better feeling for the underlying physical problem. We could just as well have studied the so-called *piston problem* in gas dynamics. In this problem we imagine that the gas in a tube (see Figure 5.13) is set into motion by a piston at one end (the piston plays the role of the wall or wavemaker in the shallow water equations). A device of a piston is not so unrealistic as it may first appear. For example, in aerodynamics the piston may model a blunt object moving into a gas; or, the piston may represent the fluid on one side of a valve after it is opened, or it may represent a detonator in an explosion process. The piston problem, that is, the problem of determining the motion of the gas for a given piston movement, is fundamental in nonlinear PDEs and gas dynamics. We leave some of these problems to the exercises.

EXERCISES

1. The shallow water equations (4) and (5) are conservation laws in dimensioned variables. Select new dimensionless independent and dependent variables according to

$$\bar{x} = \frac{x}{L}, \qquad \bar{t} = \frac{t}{T}, \qquad \bar{h} = \frac{h}{H}, \qquad \bar{u} = \frac{u}{\sqrt{gH}}$$

   where $T = L/\sqrt{gH}$, and where H is the undisturbed water height and L is a typical wavelength; rewrite the shallow water equations in terms of the dimensionless variables.

2. Solve the retracting wall problem in the case that the wall path is given by $x = -Vt$, where V is a positive constant.

3. Verify that the cubic equation (29) always has one negative root and two positive roots.

4. Consider a gas at rest ($u = 0$, $\rho = \rho_0$) in a cylindrical tube $x > 0$, and suppose that the governing equations are given by (30), or (31). At time $t = 0$ a piston located at $x = 0$ begins to move along the spacetime path

$x = X(t)$, $t > 0$. Discuss the resulting motion of the gas in the following cases:

(a) $X(t)$ satisfies the conditions $X(0) = 0$, $X'(t) < 0$, and $X''(t) < 0$.
(b) $X(t) = Vt$, where $V > 0$.
(c) If $X(t) = at^2$, where $a > 0$, at what time does a shock wave form?

5. A metallic rod of constant cross section initially at rest and occupying $x > 0$ undergoes longitudinal vibrations. The governing equations are

$$v_h - e_t = 0 \qquad v_t - \frac{2Y}{\rho_0} e e_h = 0$$

where t is time, h is a spatial coordinate attached to a fixed cross section (i.e., a material coordinate, with $h = x$ at $t = 0$), $v = v(h, t)$ is the velocity of the section h, $e = e(h, t)$ is the strain (the lowest order approximation of the distortion, or the relative change in length) at h, $Y > 0$ is the stiffness (Young's modulus), and ρ_0 is the initial constant density. Beginning at $t = 0$ the back boundary ($h = 0$) is moved with velocity $-t$ [i.e., $v(0, t) = -t$, $t > 0$]. Write this problem in characteristic form, identifying the Riemann invariants and characteristics. Can this initial boundary value problem be solved analytically by the Riemann method? Use the similarity method (the method of stretchings) introduced in Chapter 4 to determine the solution and the shock path.

5.4 HODOGRAPHS; WAVEFRONTS

In this section, using the shallow water equations as our model, we illustrate some basic procedures to analyze and understand nonlinear hyperbolic systems. These are standard techniques that are applicable to most problems.

Hodograph Transformation

The hodograph transformation is a nonlinear transformation that changes a system of quasilinear equations to a linear system. The basic idea, due originally to Riemann, is to interchange the dependent and independent variables. To illustrate the method, let us consider the shallow water equations, which we write in the form

$$h_t + uh_x + hu_x = 0 \tag{1}$$

$$u_t + uu_x + h_x = 0 \tag{2}$$

Here we are using the dimensionless form of the equations (see Exercise 1, Section 5.3). Both the depth h and the velocity u are functions of x and t, that is,

$$h = h(x,t), \qquad u = u(x,t) \tag{3}$$

Regarding (3) as a transformation from xt-space to hu-space, let us assume for the moment that the Jacobian of the transformation, namely

$$J = \begin{vmatrix} h_x & h_t \\ u_x & u_t \end{vmatrix} = h_x u_t - u_x h_t$$

is never zero. Then the transformation (3) is invertible, and we can solve (3) for x and t in terms of h and u to obtain $x = x(h,u)$, $t = t(h,u)$. The derivatives of the various quantities are related by the chain rule. We have

$$x_t = x_h h_t + x_u u_t = 0, \qquad x_x = x_h h_x + x_u u_x = 1$$

and similarly for t_t and t_x. From these equations it follows immediately that

$$x_u = -\frac{h_t}{J}, \qquad x_h = \frac{u_t}{J}, \qquad t_u = \frac{h_x}{J}, \qquad t_h = -\frac{u_x}{J}$$

The shallow water equations (1) and (2) may then be written

$$-x_u + u t_u - h t_h = 0 \tag{4}$$

$$x_h - u t_h + t_u = 0 \tag{5}$$

which is a linear system of equations for $x = x(h,u)$ and $t = t(h,u)$. We may eliminate x by cross differentiation to obtain a single second order equation for $t = t(h, \text{u})$, that is,

$$t_{uu} = h^{-1}(h^2 t_h)_h \tag{6}$$

This linear hyperbolic equation can be reduced to canonical form by the standard method of introducing characteristic coordinates ξ and η defined by

$$\xi = 2h^{1/2} - u, \qquad \eta = 2h^{1/2} + u$$

Then it easily follows that (6) becomes, in terms of canonical coordinates,

$$-4t_{\xi\eta} = 0$$

which has the general solution

$$t = f(\xi) + g(\eta)$$

where f and g are arbitrary functions. Therefore, equation (6) for t has general solution

$$t(h,u) = f(2h^{1/2} - u) + g(2h^{1/2} + u) \tag{7}$$

In principle, equation (4) can then be solved for x (by integrating with respect to h) to obtain $x = x(h,u)$.

Of course, to obtain the solution to an initial–boundary value problem one would have to determine the arbitrary functions from boundary or initial data and then invert the equations $t = t(h,u)$, $x = x(h,u)$ to obtain the desired h and u. The analytical procedure can be outlined as follows. For example, suppose that initial conditions of the form

$$u(x,0) = u_0(x), \qquad h(x,0) = h_0(x), \quad x \in R \tag{8}$$

are appended to the shallow water equations (1) and (2). Then, in principle, the variable x could be eliminated from (8) to obtain a locus $\mathscr{C}$ in the hu-plane, along which $t = 0$ and x is given. Thus we would have reduced the initial value problem (1), (2), and (8) to a Cauchy problem for (4) and (5). For all practical purposes this procedure is usually impossible to carry out, and it is better to perform numerical calculations on the original nonlinear problem; thus, as noted by Whitham [1974, p. 184], the hodograph analysis seems to be mainly of academic interest.

Wavefront Expansions

In Section 2.4 we introduced the idea of an expansion near a wavefront for a single first order PDE. Now we wish to show that the same procedure works for systems of PDEs as well. The basic problem is to determine how a continuous waveform with a derivative discontinuity propagates in time. In particular, along what path in spacetime do derivative discontinuities propagate, and how do the magnitudes of the jumps in the derivatives evolve? We expect, of course, that discontinuities in the derivatives will propagate along characteristics in the case of systems of PDEs. One can validate this expectation by an argument similar to that presented in Section 2.4. Our main interest in this subsection is to determine how the magnitude of the jumps propagate in the special case that a wavefront is moving into a constant state. In this case an expansion in a neighborhood of the wavefront proves to be a convenient procedure, and we illustrate it with a specific example.

At $t = 0$ we imagine a water wave, governed by the shallow water theory, whose depth profile $h(x,0)$ and accompanying velocity profile $u(x,0)$ are continuous functions with a simple jump discontinuities in their spatial derivatives at $x = 0$, and smooth otherwise. We assume that ahead of the wave ($x > 0$) there is a uniform state $h = h_0$ and $u = u_0$. Then, for $t > 0$, we

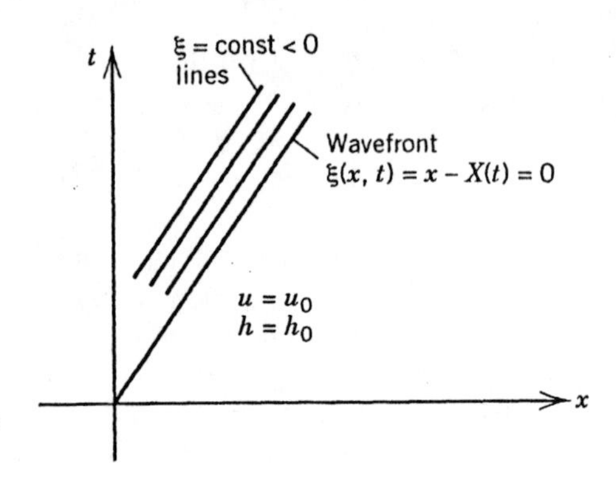

Figure 5.14. Wavefront moving into a constant medium.

assume that the discontinuity propagates along a curve $x = X(t)$ in spacetime, called a wavefront, with both h and u being smooth on each side of the curve. We introduce a coordinate ξ whose coordinate lines are parallel to the curve $x = X(t)$; in particular, we define

$$\xi = x - X(t)$$

Then the wavefront is given by $\xi = 0$; ahead of the wavefront we have $\xi > 0$, and $\xi < 0$ behind the wavefront (see Figure 5.14). We are interested in the behavior of the wave in a neighborhood of the wavefront, that is, for ξ small and negative. Ahead of the wave we assume a uniform state, and behind the wave we assume that h and u have expansions as power series in ξ. Hence we make the Ansatz

$$h = h_0, \qquad u = u_0 \quad \text{for } \xi > 0 \tag{9}$$

$$h = h_0 + h_1(t)\xi + \tfrac{1}{2}h_2(t)\xi^2 + \cdots \tag{10}$$

$$u = u_0 + u_1(t)\xi + \tfrac{1}{2}u_2(t)\xi^2 + \cdots \qquad \text{for } \xi < 0$$

Then for $\xi < 0$ we have

$$h_t = h_1'\xi - h_1X' - h_2X'\xi + O(\xi^2), \qquad h_x = h_1 + h_2x + O(\xi^2) \tag{11}$$

and similarly for u_t and u_x. Substituting all these expressions into the governing PDEs (1) and (2), and then setting the coefficients of the powers of ξ equal to zero, gives, to leading order,

$$(u_0 - X')h_1 + h_0u_1 = 0 \tag{12}$$

$$h_1 + (u_0 - X')u_1 = 0 \tag{13}$$

and at order $O(\xi)$

$$h_1' - X'h_2 + u_0 h_2 + 2u_1 h_1 + h_0 u_2 = 0 \tag{14}$$

$$u_1' - X'u_2 + u_1^2 + u_0 u_2 + h_2 = 0 \tag{15}$$

The leading order equations (12) and (13) are two equations for u_1 and h_1, which according to (10) define the corrections to u_0 and h_0 in the expansions for u and h. Notice, by (11), that h_1 gives the leading order approximation for h_x behind the wavefront, and since h is constant ahead of the front, h_1 therefore provides the leading order approximation for the jump in h_x across the front. A similar remark holds true for u_1. Now, the system (12) and (13) is homogeneous and will have a nontrivial solution u_1 and h_1 if, and only if, the determinant of the coefficient matrix vanishes, that is,

$$X' = u_0 \pm h_0^{1/2} \tag{16}$$

Equation (16) defines the wavefront speed, which is constant. Consequently, the wavefront, to leading order, is a straight line. There are two possibilities, depending on whether the plus or minus sign is chosen in (16). If the plus sign is selected, we refer to the wavefront as a *downstream* wave, and if the minus sign is chosen (giving a slower speed), we refer to the wavefront as an *upstream* wave. In either case, the solution to the homogeneous system (12)–(13) can be written

$$u_1 = \frac{(X' - u_0)h_1}{h_0} \tag{17}$$

Therefore, u_1 and h_1 are linearly related and hence we need only determine one of these quantities.

We first examine the case of downstream waves where $X' = u_0 + h_0^{1/2}$. Substituting into the $O(\xi)$ equations (14) and (15), we obtain

$$h_1' - h_0^{1/2} h_2 + 2u_1 h_1 + h_0 u_2 = 0 \tag{18}$$

$$u_1' - h_0^{1/2} u_2 + u_1^2 + h_2 = 0 \tag{19}$$

Our goal is to obtain an equation that will determine h_1 or u_1. At first observation we have a difficulty since (18) and (19) contain both h_2 and u_2, neither of which is known. However, further observation indicates that, remarkably, h_2 and u_2 may be eliminated from (18) and (19) by multiplying (19) by $h_0^{1/2}$ and then adding the result to (18). This yields

$$h_1' + 2u_1 h_1 + h_0^{1/2} u_1' + h_0^{1/2} u_1^2 = 0 \tag{20}$$

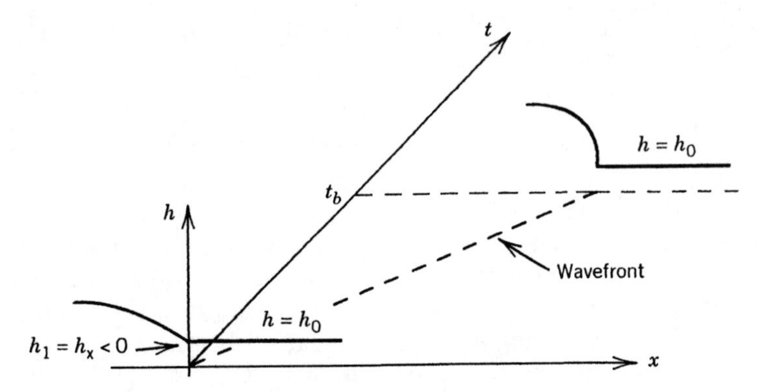

Figure 5.15. Wave with a negative gradient breaking downstream.

Now we may use (17) to eliminate u_1 from (20) and obtain

$$h_1' + \frac{3}{2h_0^{1/2}}h_1^2 = 0 \tag{21}$$

which is a single ordinary differential equation for h_1. The variables in (21) separate and it is easy to find the solution of (21) as

$$h_1 = \frac{2h_0^{1/2}}{3t + 2h_0^{1/2}/h_1(0)} \tag{22}$$

where $h_1(0)$ denotes the initial value of h_1 (i.e., the initial value of the jump in the derivative h_x at $x = t = 0$).

Now, using (22), we can make some interesting observations regarding the evolution of a jump discontinuity. If $h_1(0)$ is negative, as shown in Figure 5.15, the wave will eventually break downstream. This is because the denominator of (22) will vanish at finite time, giving an infinite gradient just behind the wavefront. Waves with positive gradients will not break downstream.

In the case of *upstream waves*, where the minus sign is chosen in (16), a calculation similar to the one presented above leads to the following equation for h_1:

$$h_1' - \frac{3}{2h_0^{1/2}}h_1^2 = 0$$

Easily, this equation has solution

$$h_1 = \frac{2h_0^{1/2}}{2h_0^{1/2}/h_1(0) - 3t}$$

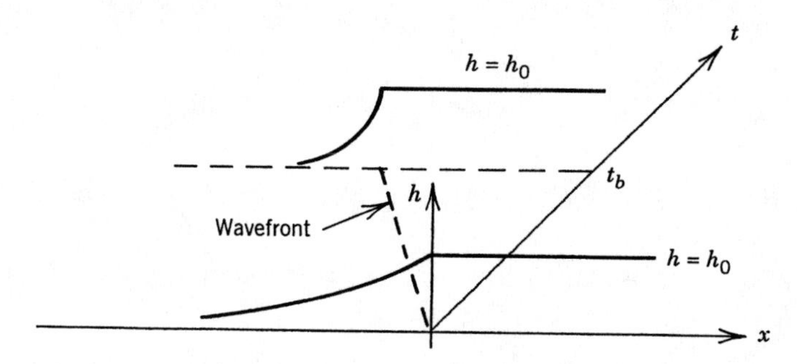

Figure 5.16. Tidal bore breaking upstream.

Consequently, if $h_1(0)$, the initial jump in h_x across the wavefront, is positive, the wave will break in finite time as shown in Figure 5.16. Such waves are called *tidal bores*.

We refer the reader to Whitham [1974] for a discussion of wavefront expansions for general hyperbolic systems and how those expansions relate to wavefronts in geometrical optics. In the next section we devise a procedure, called weakly nonlinear analysis, similar to the one discussed above to determine how special types of short-wavelength signals are propagated in a gas dynamic medium.

EXERCISES

1. Consider the gas dynamic equations

$$\rho_t + (\rho u)_x = 0, \qquad \rho(u_t + uu_x) + p_x = 0, \qquad p = k\rho^\gamma$$

Define c, the sound speed, by the equation $c^2 = p'(\rho)$.

(a) Show that the equations may be written in terms of c and the particle velocity u as

$$c_t + uc_x + \frac{\gamma - 1}{2} cu_x = 0$$

$$u_t + uu_x + \frac{2}{\gamma - 1} cc_x = 0$$

(b) Perform the hodograph transformation on the equations in part (a) to obtain the system

$$x_u - ut_u + \frac{\gamma - 1}{2} ct_c = 0, \qquad x_c - ut_c + \frac{2}{\gamma - 1} ct_u = 0$$

for $x = x(c, u)$ and $t = t(c, u)$.

(c) Show that t satisfies the second order wave equation

$$t_{bb} + \frac{2n}{b} t_b = t_{uu}, \qquad \text{where} \quad b = \frac{2c}{\gamma - 1}$$

and $n = (\gamma + 1)/[2(\gamma - 1)]$ and obtain the general solution in the case $n = 1$.

2. This exercise deals with the hodograph transformation and the *Born–Infeld equation*

$$(1 - \phi_t^2)\phi_{xx} + 2\phi_x\phi_t\phi_{xt} - (1 + \phi_x^2)\phi_{tt} = 0$$

(a) Introduce new independent and dependent variables via

$$\xi = x - t, \qquad \eta = x + t$$
$$u = \phi_\xi, \qquad v = \phi_\eta$$

and then apply the hodograph transformation to obtain the system

$$\xi_v - \eta_u = 0, \qquad v^2\eta_v + (1 + 2uv)\xi_v + u^2\xi_u = 0$$

(b) Reduce the system in part (a) to the single linear hyperbolic equation

$$u^2\xi_{uu} + (1 + 2uv)\xi_{uv} + v^2\xi_{vv} + 2u\xi_u + 2v\xi_v = 0$$

and then introduce characteristic coordinates

$$r = \frac{(1 + 4uv)^{1/2} - 1}{2v}, \qquad s = \frac{(1 + 4uv)^{1/2} - 1}{2u}$$

to obtain $\xi_{rs} = 0$.

(c) Prove that one may take

$$\xi = F(r) - \int s^2 G'(s)\, ds, \qquad \eta = G(s) - \int r^2 F'(r)\, dr$$

where F and G are arbitrary functions, and thus show that the solution ϕ is given by

$$\phi = \int rF'(r)\, dr + \int sG'(s)\, ds$$

(d) Finally, introduce $\rho = F(r)$ and $\sigma = G(s)$, with inverses $r = \Phi_1'(\rho)$ and $s = \Phi_2'(\sigma)$, and show that the solution may be written in the form

$$\phi = \Phi_1(\rho) + \Phi_2(\sigma)$$

where

$$\rho = x - t + \int_{-\infty}^{\sigma} \Phi_2'^2(\sigma)\, d\sigma, \qquad \sigma = x + t - \int_{\rho}^{\infty} \Phi_1'^2(\rho)\, d\rho$$

(See B. Barbishov and N. Chernikov, *Sov. Phys. JETP* **24**, 437–442 (1966); also, Whitham [1974, p. 617]).

3. The shallow water equations governing flood waves down a flat river of incline α are given by

$$h_t + uh_x + hu_x = 0, \qquad u_t + uu_x + g^* h_x = g^* S - \frac{Cu^2}{h}$$

where $S = \tan\alpha$, $g^* = g\cos\alpha$, and C is the coefficient of friction. Here h is the depth of the water measured perpendicular from the bottom and u is the velocity in the x direction along the bed (see Kevorkian [1990]). Consider a wavefront $x = X(t)$ propagating into a uniform state $u = u_0$, $h = h_0$, along which h and u are continuous but have discontinuous derivatives. Perform a wavefront expansion analysis and show that if

$$u_0 > 2(g^* h_0)^{1/2}$$

a wave will break regardless of the sign of $h_x(0^+, 0)$. If $u_0 < 2(g^* h_0)^{1/2}$, show that the jump in h_x will decay exponentially in the case $h_x(0+, 0) > 0$, but in the case

$$h_x(0^+, 0) < 2S(g^* h_0)^{1/2} \frac{1 - u_0/\left(2(g^* h_0)^{1/2}\right)}{3u_0}$$

the wave will break in finite time (i.e., a sufficiently strong downstream flood wave will be headed by a bore).

4. The small, longitudinal vibrations of a bar of density $\rho = \rho(x)$, stiffness $E = E(x)$, and cross-sectional area $\sigma = \sigma(x)$ are governed by the wave equation $\rho\sigma u_{tt} = (\sigma E u_x)_x$, where $u = u(x, t)$ is the displacement of a cross section (see Logan [1987] for a derivation). Observe that this equation can be transformed to a system of two first order equations by introducing the variables $v = u_t$ (the velocity) and $w = u_x$ (the stress). The purpose of this exercise is to determine how a wavefront (caused, say, by striking one end of the bar) propagates down the length of the bar.

(a) Assume a wavefront expansion

$$v = v_0 + v_1(x)\xi + \cdots, \qquad w = w_0 + w_1(x)\xi + \cdots \quad \text{for } \xi < 0$$

where $\xi = t - \psi(x) = 0$ is the wavefront and where $v = v_0 =$

constant, $w = w_0 = 0$ for $\xi > 0$. Show that to leading order, the wavefront is described by the differential equation

$$\psi'(x) = \pm\left[\frac{\rho(x)}{E(x)}\right]^{1/2}$$

(b) Show that

$$v_1'(x) = -\frac{(\sigma E\psi')'v_1(x)}{2\sigma E\psi'}$$

and therefore

$$v_1(x) = v_1(0)\left[\frac{\sigma(0)E(0)\psi'(0)}{\sigma(x)E(x)\psi'(x)}\right]^{1/2}$$

along the wavefront.

(c) Consider a uniform (ρ and E constant) bar in the shape of a wedge as shown in Figure 5.17. Show that jumps in v_x and w_x both increase as the area decreases. Interpret this result physically.

5. Consider the nonlinear system

$$u_t - v_x = 0, \qquad v_t - uu_x = 0$$

Assume that a wavefront is moving into a constant state. To leading order, determine the speed of the front and determine how a jump in u_x evolves along the front.

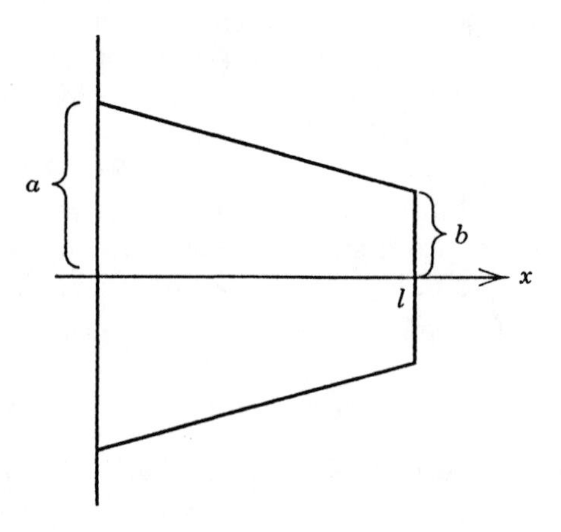

Figure 5.17. Wedge of unit thickness.

5.5 WEAKLY NONLINEAR APPROXIMATIONS

It should be evident at this point that general systems of nonlinear partial differential equations are complex and that no procedure, simple or otherwise, can be developed to solve most of these problems analytically. Therefore, much effort has gone into developing techniques to reduce the complexity of these systems. Often, this means examining a problem in some special limit whereby the system can be reduced to a simpler one, and progress can be made in elucidating some of the qualitative features of the original problem. One such method, which arose out of weakly nonlinear geometrical optics, is an asymptotic method that yields a problem in what is termed a *weakly nonlinear limit*. In geometrical optics one attempts to construct formal solutions that are high-frequency asymptotic solutions for bounded time intervals. The idea extends to other systems and can be explained simply as follows. Suppose that a medium exists in a rest state and that in some localized region of space a small-amplitude small-wavelength disturbance occurs. This event is then propagated to nearby regions by the equations of motion, which describe the dynamics of the problem. In this process we try to identify a small parameter and then we assume that the states can be represented as an expansion in that small parameter. The reader will recall that the acoustical approximation in gas dynamics was obtained in a similar manner, and the linear wave equation resulted in the special limit of waves of small amplitude. But now we wish to carry the analysis one step further and retain some of the nonlinear interactions that are present in the original system.

We illustrate the weakly nonlinear analysis by considering the gas dynamic equations. The central idea is to consider small-amplitude waves whose wavelength is small compared to the overall size of the spatial region where the problem is defined. We shall show that such disturbances are propagated by Burgers' equation, thereby giving a derivation of one of the important model equations of applied analysis.

Derivation of Burgers' Equation

We consider the one-dimensional flow of a barotropic fluid that is governed by the gas dynamic equations

$$\rho_t + u\rho_x + \rho u_x = 0 \tag{1}$$

$$\rho u_t + \rho u u_x + p_x = \tfrac{4}{3}\mu u_{xx} \tag{2}$$

$$p = k\rho^\gamma \tag{3}$$

where ρ is the density, u the particle velocity, p the pressure, μ the constant viscosity, and k and γ positive constants. Equation (1) is mass conservation, and equation (2) is momentum balance, with the only two forces acting being

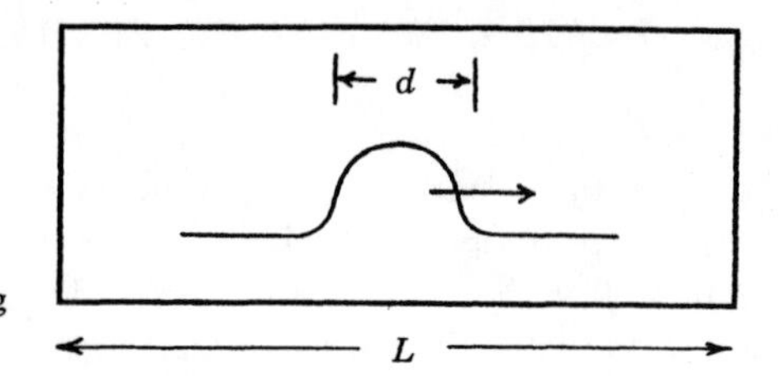

Figure 5.18. Wavelet of thickness d propagating in a container of characteristic length L.

the pressure and the viscous force. Equation (3) is the barotropic equation of state. In this model flow there is no heat diffusion and no heat generated or dissipated by mechanical or viscous forces. Our goal is to reduce the complexity of (1)–(3) by restricting the range of applicability, that is, by investigating a wave propagation problem in a special limit.

To this end, assume that a wave with representative thickness d is propagating to the right into a constant, undisturbed state $u = 0$, $p = p_0$, $\rho = \rho_0$, and let $x = X(t)$ represent the location of some reference point on the wave as it moves forward in time; moreover, let L be a length that is representative of the size of the region where the wave is propagating (e.g., the length of a tube or vessel) (see Figure 5.18). Our assumption will be that d is small compared to L, and therefore the parameter ε defined by

$$\varepsilon = \frac{d}{L}$$

is a small quantity that can serve as an expansion parameter in a perturbation series.

In this problem it is essential to introduce normalized variables and reformulate the equations in dimensionless terms; only then can we be certain of the order of magnitude of the various quantities in the equations. The sound speed in the region ahead of the wave is given by $c_0 = (\gamma p_0/\rho_0)^{1/2}$; therefore, the acoustic time t_a defined by $t_a = L/c_0$ is the time required for a signal traveling at speed c_0 to traverse a distance comparable to the magnitude of the length of the vessel. We take the acoustic time as a characteristic time for the problem and introduce the dimensionless time variable

$$\tau = \frac{t}{t_a}$$

For the characteristic length we select the representative wave width d and introduce the dimensionless spatial coordinate

$$\xi = \frac{x - X(t)}{d}$$

which measures the distance to the location of the wave on the scale of the wave. We choose ρ_0, p_0, and c_0 to be the density, pressure, and velocity scales, respectively, and we define the dimensionless dependent variables $\bar{\rho}$, $\bar{p}$, and $\bar{u}$ by

$$\bar{\rho} = \frac{\rho}{\rho_0}, \qquad \bar{p} = \frac{p}{p_0}, \qquad \bar{u} = \frac{u}{u_0}$$

Letting $D(t) = X'(t)$ denote the velocity of the reference point on the wave, we also introduce $D(t)$, defined by

$$\overline{D} = \frac{D}{c_0}$$

Then, in terms of the dimensionless variables, equations (1)–(3) become

$$\varepsilon\bar{\rho}_\tau - \overline{D}\bar{\rho}_\xi + \bar{u}\bar{\rho}_\xi + \bar{\rho}\bar{u}_\xi = 0 \tag{4}$$

$$\varepsilon\bar{\rho}\bar{u}_\tau - \bar{\rho}\overline{D}\bar{u}_\xi + \bar{\rho}\bar{u}\bar{u}_\xi + \gamma^{-1}\bar{p}_\xi = 2\varepsilon\nu\bar{u}_{\xi\xi} \tag{5}$$

$$\bar{p} = \bar{\rho}^\gamma \tag{6}$$

where

$$\nu = \frac{2\mu}{3\rho_0 c_0 \varepsilon d} \tag{7}$$

is a dimensionless constant and where $\varepsilon \ll 1$. The reader is asked to verify these equations in Exercise 1.

Now comes a crucial step, namely the assumption of the form of the expansion of the dependent variables in powers of the small parameter ε. We take

$$\bar{\rho}(\xi,\tau) = 1 + \varepsilon\rho_0(\xi,\tau) + \varepsilon^2\rho_1(\xi,\tau) + O(\varepsilon^3) \tag{8}$$

$$\bar{u}(\xi,\tau) = \varepsilon u_0(\xi,\tau) + \varepsilon^2 u_1(\xi,\tau) + O(\varepsilon^3) \tag{9}$$

$$\overline{D}(\tau) = D_0(\tau) + \varepsilon D_1(\tau) + O(\varepsilon^2) \tag{10}$$

Because of the equation of state (6) we automatically have, by substituting (8) into (6), the expansion

$$\bar{p}(\xi,\tau) = 1 + \varepsilon\gamma\rho_0(\xi,\tau) + \varepsilon^2\frac{\gamma(\gamma-1)}{2}\rho_0(\xi,\tau)^2 + O(\varepsilon^3) \tag{11}$$

Expression (11) is just the Taylor series expansion of $\bar{\rho}^\gamma$ about $\varepsilon = 0$. Thus our assumption is that the disturbance, whose density is $\bar{\rho}$, is of order 1 with small variations; the velocity D of the wave is, to leading order, on the same order as the speed of sound in the undisturbed medium ahead of the wave, and the particle velocity $\bar{u}$ within the wave is small in comparison. Such assumptions describe what is called a *weakly nonlinear model.*

The next step is to substitute the expansions (8)–(11) into the governing PDEs (4)–(5). It is straightforward to show that (4) becomes

$$\varepsilon^2\rho_{0\tau} - (D_0 + \varepsilon D_1)(\varepsilon\rho_{0\xi} + \varepsilon^2\rho_{1\xi}) + \varepsilon^2 u_0\rho_{0\xi} + (1 + \varepsilon\rho_0)(\varepsilon u_{0\xi} + \varepsilon^2 u_{1\xi}) = O(\varepsilon^3) \tag{12}$$

and (5) becomes

$$\varepsilon^2 u_{0\tau} - (1 + \varepsilon\rho_0)(D_0 + \varepsilon D_1)(\varepsilon u_{0\xi} + \varepsilon^2 u_{1\xi}) + \varepsilon^2 u_0 u_{0\xi} + \varepsilon\rho_{0\xi} + \varepsilon^2\rho_{1\xi} + \varepsilon^2(\gamma - 1)\rho_0\rho_{0\xi} = 2\varepsilon^2\nu u_{0\xi\xi} + O(\varepsilon^3) \tag{13}$$

Since these two equations should hold for any ε, we are free to set the coefficients of the various powers of ε equal to zero. The leading order contributions from equations (12) and (13) are

$$-D_0\rho_{0\xi} + u_{0\xi} = 0 \tag{14}$$

$$-D_0 u_{0\xi} + \rho_{0\xi} = 0 \tag{15}$$

and the next order contributions, which are the correction terms, are

$$-D_0\rho_{1\xi} + u_{1\xi} = -\rho_{0\tau} + D_1\rho_{0\xi} - u_0\rho_{0\xi} - \rho_0 u_{0\xi} \tag{16}$$

$$-D_0 u_{1\xi} + \rho_{1\xi} = -u_{0\tau} + D_1 u_{0\xi} + D_0\rho_0 u_{0\xi} - u_0 u_{0\xi} - (\gamma - 1)\rho_0\rho_{0\xi} + 2\nu u_{0\xi\xi} \tag{17}$$

Equations (14) and (15) are two homogeneous equations for u_0 and ρ_0; there will be nontrivial solution provided that the determinant of the coefficient matrix vanishes, that is,

$$D_0 = 1 \tag{18}$$

In this case both (14) and (15) become $\rho_{0\xi} = u_{0\xi}$. Since both u_0 and ρ_0 are zero ahead of the wave, we must have $u_0 = \rho_0$. We have therefore determined the leading order wave velocity D_0 in (10), and we have a relation between ρ_0 and u_0.

To find a single equation for ρ_0, we analyze the correction equations (16) and (17). Substituting (18) into (16) and (17) and using $u_0 = \rho_0$, we obtain

$$-\rho_{1\xi} + u_{1\xi} = -\rho_{0\tau} + D_1\rho_{0\xi} - 2\rho_0\rho_{0\xi} \tag{19}$$

$$\rho_{1\xi} - u_{1\xi} = -\rho_{0\tau} + D_1\rho_{0\xi} - (\gamma - 1)\rho_0\rho_{0\xi} + 2\nu\rho_{0\xi\xi} \tag{20}$$

Adding these two equations together gives a single equation for ρ_0. It is convenient to introduce the quantity $U(\xi, \tau)$ defined by

$$U = \frac{(\gamma + 1)\rho_0}{2} = \frac{(\gamma + 1)u_0}{2} \tag{21}$$

Then the sum of (19) and (20) gives

$$U_\tau - D_1 U_\xi + U U_\xi = \nu U_{\xi\xi} \tag{22}$$

which is similar to Burgers' equation. We may eliminate the term in (22) involving D_1 by defining a new spatial coordinate η by

$$\eta = \xi + \int_0^\tau D_1(y)\, dy \tag{23}$$

With this transformation, equation (22) becomes

$$u_\tau + u u_\eta = \nu u_{\eta\eta} \tag{24}$$

which is *Burgers' equation*. Here, we are using $u = u(\eta, \tau) = U(\xi(\eta, \tau), \tau)$; the lowercase letter u should not be confused with the velocity in the original equations (1) and (2).

Let us summarize. We have obtained Burgers' equation in a weakly nonlinear limit of the gas dynamic equations (1)–(3). The procedure was to look for equations that govern small deviations from a uniform state in acoustic time, on the scale of the width of the disturbance. The reader should contrast the situation in acoustics, where linear equations (in fact, the wave equation) govern the small perturbations; there, acoustic signals are small-amplitude disturbances on the time scale of acoustic time and on the length scale of the vessel. Burgers' equation, which followed from the weakly nonlinear limit, governs small-amplitude disturbances of short wavelength, compared to the size of the vessel, over the acoustical time scale.

It is common in applied mathematics to study complex problems on different time scales or length scales; generally, one obtains different governing equations in each limit, accompanied by an overall simplification. The solution of the problem in a specific limit often clarifies some aspects of complex physical systems, giving information about the behavior of the system under special conditions. A good example of the utility of scaling and examining a problem in special limits occurs in combustion theory, where there are several time scales: for example, the acoustic time scale, the chemical time scale, the time scale for diffusion, and the hydrodynamic time scale. By examining a problem on a rapid chemical time scale, for example, one may be able to neglect diffusion terms in the equations since diffusion may occur in some regimes on a much slower scale. In this manner one reduces the complexity of problems. Another illustration is water waves. Some of the important equations of applied mathematics, such as the Korteweg–deVries equation and the Boussinesc equations, result from studying the full nonlinear equations of hydrodynamics in a special limit (e.g., when the wavelength of a surface wave is long in comparison to the depth of the channel). In a later chapter the process of combustion is investigated in detail.

In Chapter 4 we showed the existence of traveling wave solutions to Burgers' equation. Now we relate that calculation to the discussion presented above. We recall that the determination of traveling waves for Burgers' equation involves a constant, unknown wave speed. This wave speed is related to the correction term D_1 in (10) and (23) in the following manner. The disturbance is centered along the coordinate $\xi = 0$ for all time. But in the $\eta\tau$-coordinate system of equation (24) the wave is on the path $\eta = \int_0^\tau D_1(y)\,dy$, and its speed is $D_1(\tau)$, which is unknown. So we are free to search for constant values D_1 for which Burgers' equation (24) admits traveling wave solutions of speed D_1. Generally, D_1 is not constant for wave profiles that change in time.

EXERCISES

1. Verify equations (4)–(6).
2. The purpose of this exercise is to perform a weakly nonlinear analysis on a simple nonlinear equation. Consider the scalar equation

$$u_t + \phi(u)_x = \mu u_{xx}$$

where t, x, and u are dimensionless variables of order 1, $\mu = \mu_1\varepsilon^2 + O(\varepsilon^3)$, $\varepsilon \ll 1$, and $\phi(u)$ can be expanded in a Taylor series about $u = u_0 =$ constant. Let $x = X(t)$ be a representative location on a wave that is propagating into the uniform state u_0. Introduce the variable $\xi = (x - X(t))/\varepsilon$ and assume that

$$u(\xi, t) = u_0 + \varepsilon u_1(\xi, t) + \varepsilon^2 u_2(\xi, t) + O(\varepsilon^3)$$

$$D(t) = D_0(t) + \varepsilon D_1(t) + O(\varepsilon^2)$$

where $D = X'$. To leading order, determine the speed of the wave, and show that u_1 satisfies the Burgers'-like equation

$$u_{1t} + \left(\frac{\alpha}{2}u_1^2\right)_\xi = \mu_1 u_{1\xi\xi}$$

where α is a constant.

3. Consider the PDE in Exercise 2 where $\mu = \mu(\varepsilon)$. In studying the long-time behavior of a system, one assumes the expansion

$$u(\eta, \tau) = u_0 + \varepsilon u_1(\eta, \tau) + \varepsilon^2 u_2(\eta, \tau) + O(\varepsilon^3)$$

where

$$\eta = x - X(t), \qquad \tau = \varepsilon t$$

where $x = X(t)$ is a reference location on a wave propagating into the constant state $u = u_0$. To leading order determine the speed of the wave and find an equation for the first correction $u_1(\eta, \tau)$. Discuss the two cases $\mu = O(\varepsilon)$ and $\mu = O(\varepsilon^2)$.

REFERENCES

Again, our policy is to give only a few basic references. Several of the references given below have extensive bibliographies that may be used as a guide to the literature. Parallel introductory treatments of gas dynamics and the shallow water equations can be found in Logan [1987] and Kevorkian [1990], respectively. Certainly, two of the outstanding works on gas dynamics are the classic treatises by Courant and Friedrichs [1948] and Whitham [1974]. Smoller [1983] addresses some of the mathematical questions in these areas. Another excellent complete reference is Rozdestvenskii and Janenko [1983]. An introduction to the theory of hyperbolic conservation laws is given in Majda [1986].

Courant, R., and K. O. Friedrichs, 1948. *Supersonic Flow and Shock Waves*, Wiley-Interscience, New York. Reprinted by Springer-Verlag, New York, 1976.

Kevorkian, J., 1990. *Partial Differential Equations*, Brooks/Cole, Pacific Grove, Calif.

Logan, J. D., 1987. *Applied Mathematics: A Contemporary Approach*, Wiley-Interscience, New York.

Majda, A. J., 1986. *Compressible Fluid Flow and Systems of Conservation Laws in Several Space Variables*, Springer-Verlag, New York.

Rozdestvenskii, B. L., and N. N. Janenko, 1983. *Systems of Quasilinear Equations and Their Applications to Gas Dynamics*, American Mathematical Society, Providence, R.I. Translated from the original 1968 Russian edition by J. R. Schulenberger.

Smoller, J., 1983. *Shock Waves and Reaction–Diffusion Equations*, Springer-Verlag, New York.

Whitham, G. B., 1974. *Linear and Nonlinear Waves*, Wiley-Interscience, New York.

6

REACTION–DIFFUSION EQUATIONS

During the last few decades there has been an intense research effort devoted to the study and application of reaction–diffusion equations. One of the motivating factors was a celebrated paper by A. Turing in 1952. He proposed that reaction–diffusion could serve as a model for the chemical basis of morphogenesis, or the development of form and pattern in the embryological stage of various organisms. His basic observation was that spatially inhomogeneous patterns can be generated by diffusion-driven instabilities. It is not surprising, therefore, that tremendous interest was spawned in the biological sciences in the general subject of reaction–diffusion equations. A parallel development has also taken place in combustion phenomena, where reaction–diffusion processes occur with equal importance. The interest in reaction–diffusion equations that these two areas, the biological sciences and combustion theory, generated in the mathematical community has been profound and a significant step has been taken in the understanding of nonlinear processes and nonlinear partial differential equations.

This chapter is an introduction to reaction–diffusion equations. In Chapter 4 we introduced the Fisher equation, which is a model for a diffusing population with a logistics growth law. Now we wish to study some simple examples where systems of such equations arise. In Section 6.1 we formulate models for competing populations, interacting chemical species, chemotaxis (the interaction of a diffusing population and a chemical attractant), and the deposition of a pollutant in groundwater flows. The latter process also includes nonlinear convection. In Section 6.2 we examine a basic problem for reaction–diffusion equations: the existence of traveling wave solutions. In Sections 6.3 and 6.4 we begin to investigate some of the theoretical questions associated with reaction–diffusion equations, and nonlinear parabolic equations in general. In particular, we ask about existence and uniqueness of solutions, and we formulate maximum principles and comparison theorems for such problems. In Section 6.5 we introduce the energy method, a technique for obtaining estimates and bounds on various quantities (e.g., the energy) associated with the solutions of reaction–diffusion or other evolution equations; we also study the blow-up problem and the method of invariant sets.

6.1 REACTION–DIFFUSION MODELS

In Chapter 4 we examined the Fisher equation

$$u_t - Du_{xx} = ru\left(1 - \frac{u}{K}\right)$$

which represents a prototype for a single reaction–diffusion equation. In this section we discuss the origin of *systems* of reaction–diffusion equations and we take up several mathematical models that illustrate some problems that are of interest. Let $u_i(x,t)$, $i = 1,\dots,n$, be smooth (C^2) functions representing n unknown concentrations or densities in a given one-dimensional system (say, in a given tube of unit cross-sectional area and either of infinite or finite length). These quantities may be interacting population densities, chemical concentrations, energy densities, or whatever, distributed over the length of the tube. Let $\phi_i(x,t)$, $i = 1,\dots,n$, denote the flux of the quantity u_i, with the usual convention that ϕ_i is positive if the net flow of u_i is to the right, and let f_i be the rate of production of u_i. As before, the dimensions of the quantities u_i, ϕ_i, and f_i are:

$$[u_i] = \frac{\text{amount}}{\text{volume}}, \qquad [\phi_i] = \frac{\text{amount}}{\text{time} \cdot \text{area}}, \qquad [f_i] = \frac{\text{amount}}{\text{time} \cdot \text{volume}}$$

Generally, the production rate f_i of u_i may depend on both space and time as well as all the densities, that is,

$$f_i = f_i(x,t,u_1,\dots,u_n)$$

The basic conservation law developed in Chapter 1 for a single density now applies to each density u_i. That is, if $I = [a,b]$ is an interval representing an arbitrary section of the tube, then

$$\frac{d}{dt}\int_a^b u_i\,dx = -\phi_i\Big|_{x=a}^{x=b} + \int_a^b f_i\,dx \qquad (i = 1,\dots,n) \tag{1}$$

These n equations express the fundamental principle that the rate of change of the total amount of the quantity u_i in I must be balanced by the net rate that u_i flows into I plus the rate that u_i is produced in I. In the usual way, assuming smoothness, the system of integral balance laws (1) can be reduced to a system of n partial differential equations

$$(u_i)_t + (\phi_i)_x = f_i \qquad (i = 1, \ldots, n) \tag{2}$$

In (2), the densities u_i and the fluxes ϕ_i are unknown, and the production rates (also called the source terms or the reaction terms) f_i are assumed to be given. It is clear that (2) may be written in vector form as

$$\mathbf{u}_t + \boldsymbol{\phi}_x = \mathbf{f}(x, t, \mathbf{u}) \tag{3}$$

where

$$\mathbf{u} = (u_1, \ldots, u_n)^T, \qquad \boldsymbol{\phi} = (\phi_1, \ldots, \phi_n)^T, \qquad \mathbf{f} = (f_1, \ldots, f_n)^T$$

We observe that (3) represents n equations with $2n$ unknowns ($\mathbf{u}$ and $\boldsymbol{\phi}$). Additional equations come from the assumption of constitutive relations that connect the flux to the density gradients. When appended to (3), such assumptions effectively eliminate the flux from the equations and we obtain a system of PDEs for the unknown densities alone.

The simplest constitutive assumption is to hypothesize that the flux ϕ_i depends only on the gradient of the ith density u_i, not upon the remaining densities. In symbols,

$$\phi_i = -D_i(u_i)_x \qquad (i = 1, \ldots, n) \tag{4}$$

where the D_i are diffusion constants. Here we do not assume that the diffusion constants are necessarily positive, which indicates that the flow is from regions of high density or concentrations to regions of low density or concentration; it may occur, for example, that organisms are attracted to their kind and movement is *up the concentration gradient*, that is, from regions of low concentrations to regions of high concentrations. In the latter case, the diffusion constant would be negative. In any case, substituting (4) into (2) gives a *system of reaction–diffusion equations*

$$(u_i)_t - D_i(u_i)_{xx} = f_i(x, t, u_1, \ldots, u_n) \qquad (i = 1, \ldots, n) \tag{5}$$

for the n unknown densities $u_1, \ldots, u_n$. In vector form, (5) can be written

$$\mathbf{u}_t - D\mathbf{u}_{xx} = \mathbf{f}(x, t, \mathbf{u}) \tag{6}$$

where $D = \text{diag}(D_1, \ldots, D_n)$ is an n by n diagonal matrix with the diffusion constants on the diagonal. In the system (6) the coupling is through the reaction terms f_i; the diffusion operators on the left side of the equations are decoupled because of the simple form of the constitutive relations (4).

A step up in difficulty is to assume that there is *cross diffusion* between the densities u_i. This means that the flux ϕ_i of the density u_i may depend on the gradients of all the densities, not just the gradient of u_i alone. Thus we assume that

$$\phi_i = -\sum D_{ij}(u_j)_x \qquad (i = 1, \ldots, n) \tag{7}$$

or in matrix form

$$\boldsymbol{\phi} = -D\mathbf{u}_x \tag{8}$$

where $D = (D_{ij})$ is an n by n matrix of diffusion coefficients. D_{ij} is a measure of how u_i diffuses into u_j. Therefore, substitution of (8) into (3) gives a system the same form as (6), but with D a nondiagonal matrix; thus, in the case of cross diffusion, the system (6) represents a system of reaction–diffusion equations that is coupled through both the reaction terms and the differential operators. It is, of course, possible that the diffusion coefficients depend on the densities u_i. Examples of this type of phenomenon were given in Chapter 4 in one-dimensional heat flow.

We now consider several situations where reaction–diffusion equations arise naturally. These examples come from a variety of physical and biological contexts, and they indicate the breadth of application of this important class of equations. In general, reaction–diffusion equations are of great interest in applied mathematics, and they have been the subject of much research during recent years.

Predator–Prey Model

We have already observed that the Fisher equation

$$u_t - Du_{xx} = ru\left(1 - \frac{u}{K}\right)$$

which is a prototype of a single reaction–diffusion equation, is a simple model of a population u that diffuses while it grows according to the logistics growth law. Now we wish to consider a one-dimensional model of two interacting populations; one is a predator, whose population density is denoted by v, and a prey, whose population density is denoted by u. The prey is assumed to grow at a rate proportional to its population (the Malthus model), and the predator, in the absence of prey, is assumed to die at a rate proportional to its population. The number of interactions between the predators and the prey is assumed to be proportional to the product uv; an interaction, of course, increases the number of predators while it reduces the number of prey. In summary, our assumptions

$$\text{Growth rate of the prey} = f_1(u, v) = au - buv$$

$$\text{Growth rate of the predator} = f_2(u, v) = -cv + buv$$

where a, b, and c are the positive proportionality constants. If ϕ_1 and ϕ_2 denote the fluxes of the prey and the predators, respectively, and we assume Fick's law for each, that is,

$$\phi_1 = -D_1 u_x, \qquad \phi_2 = -D_2 v_x \qquad (D_1 \text{ and } D_2 \text{ positive constants})$$

then we have the balance laws

$$u_t - D_1 u_{xx} = au - buv$$
$$v_t - D_2 v_{xx} = -cv + buv$$

which govern the two populations. The growth rates in these coupled reaction–diffusion equations come from laws originally proposed by V. Volterra in 1926 in an effort to model the dynamics of fish catches in the Adriatic Sea. Here we have added Fickian diffusion. We observe that the Volterra model provides for an interaction model where one growth rate is increased by the interaction, while the other is decreased. This type of model, in general, is called a *predator–prey* model. If both growth rates are decreased by an interaction, the model is called a *competition* model. If both are increased, the model is *symbionic*. In summary, population models in ecological problems is a rich source of reaction–diffusion equations. The consideration of different interaction terms, population-dependent diffusion coefficients, and even several species all lead to important models that have been studied extensively in population dynamics.

Combustion Phenomena

Another area where reaction–diffusion equations enjoy considerable application is in combustion theory. In Chapter 7 we take up the subject of combustion in a general setting; for the present we formulate a simple combustion model that illustrates the role of reaction–diffusion processes in combustion phenomena.

Quite generally, a combustion process involves mass, momentum, and energy transport in a chemically reacting fluid. Here we consider a simplified system where a substance in a tube undergoes an exothermic (heat releasing) chemical reaction $A \to B$, where A is the reactant and B is the product. As the reaction proceeds, temperature changes occur and both heat energy and matter diffuse through the medium. In this simple model, there is no transport. Now, let $a = a(x,t)$ be the concentration (in moles per volume) of the reactant A, and let $T = T(x,t)$ be the absolute temperature of the medium, in kelvin. We now write down the basic balance laws for mass and energy. Mass balance takes the form

$$a_t + \phi_x = f(a, T) \tag{9}$$

where ϕ is the flux of A and $f(a,T)$ is the production rate of A (here the rate that A is consumed by the reaction). We assume that the flux is given by Fick's law,

$$\phi = -Da_x \tag{10}$$

where D is the *mass diffusion coefficient*. Thus the flux is caused by concentration gradients and there is no transport. The production rate f is the reaction rate of the chemical reaction $A \to B$. From chemical kinetics theory it is known that the reaction rate has the form

$$f(a,T) = -sa^m \exp\left(-\frac{E}{RT}\right) \tag{11}$$

where $m > 0$ is the *order of the reaction*, E the *activation energy*, R the universal gas constant, and s the *preexponential factor*. Equation (11) is called the *Arrhenius equation*, after its originator. Finally, (9)–(11) give the *mass balance law*

$$a_t - Da_{xx} = -sa^m \exp\left(-\frac{E}{RT}\right) \tag{12}$$

Now energy balance. If c is the specific heat per unit volume (cal/deg · vol), cT represents the energy density in calories/volume. We are assuming that the entire chemical mixture has the same specific heat. The energy flux is assumed to be given by $-KT_x$ (Fourier's law), where K is the thermal conductivity and the energy (heat) production term is proportional to the reaction rate and is given by

$$Qsa^m \exp\left(-\frac{E}{RT}\right)$$

where $Q > 0$ is the heat of reaction. The energy balance equation is therefore

$$(cT)_t - (KT_x)_x = Qsa^m \exp\left(-\frac{E}{RT}\right)$$

or

$$T_t - kT_{xx} = c^{-1}Qsa^m \exp\left(-\frac{E}{RT}\right) \tag{13}$$

where $k = K/c$ is the diffusivity. Therefore, we have obtained a reactive–diffusive model (12)–(13) for the temperature and concentration of the reactant. The reader should observe that these equations conform with the general type of reaction–diffusion equations discussed at the beginning of this section.

For the model under discussion, equations (12) and (13) are usually supplemented by initial and boundary conditions. If we assume that the tube is of finite length ($0 \le x \le L$), at the ends of the tube we should expect that

$$a_x(0,t) = a_x(L,t) = 0, \qquad t > 0 \tag{14}$$

which means that the flux of A is zero at the ends, or that no chemical can escape the confines of the tube. If the ends of the tube are held at fixed temperatures, then

$$T(0,t) = T_1, \qquad T(L,t) = T_2, \quad t > 0 \tag{15}$$

The initial conditions are

$$T(x,0) = T_0(x), \qquad a(x,0) = a_0(x), \quad 0 < x < L \tag{16}$$

Conditions (14)–(16) give a well-posed reactive–diffusive system sometimes called the solid-fuel combustion problem.

Chemotaxis

Another reactive–diffusive process that has garnered attention in recent years is the motion of organisms under the influence of diffusion and chemotaxis, the latter being motion induced by variations in the concentration of chemicals produced by the organisms themselves. One such example can be described as follows. Slime mold amoebae feed on bacteria in the soil or in dung and are generally uniformly spatially distributed when the food supply is plentiful. However, as the food supply becomes depleted, the organisms begin to secrete a chemical (cyclic AMP) that acts as an attractant. It is observed that the amoebae move *up the concentration gradient* toward the high concentrations of the chemical and interesting wave patterns and aggregations form. These types of chemotactic motions occur in a variety of biological phenomena. Another example is the release of pheromones by some organisms; these chemicals act as sexual attractants or communication devices for predation.

Again we assume a one-dimensional model and let $u = u(x,t)$ denote the population density of an organism (say, amoebae) and $c = c(x,t)$ denote the concentration of a chemical attractant. We will assume that there is motion of the organism due to random movement (diffusion) and due to chemotaxis. The chemical is assumed to be produced by the organism and allowed to diffuse through the medium. The reaction–diffusion system is, by our previous account,

$$u_t + \phi_{1x} = 0 \tag{17}$$

$$c_t + \phi_{2x} = f(u, c) \tag{18}$$

where ϕ_1 and ϕ_2 are the fluxes of the organism and chemical, respectively, and f is the rate of production of the chemical. Now we make some specific constitutive assumptions. The chemical is assumed to move only by diffusion, and therefore we take

$$\phi_2 = -Dc_x \tag{19}$$

where D is its diffusion constant. The chemical is assumed to decay at a constant rate, and its production rate is assumed to be proportional to the population of the organism. That is,

$$f(u, c) = q_1 u - q_2 c \tag{20}$$

where q_1 is the rate of secretion and q_2 is the decay rate. Finally, the flux ϕ_1 of the organism consists of two parts: one contribution is from diffusion and the other is from chemotaxis. The hypothesis is

$$\phi_1 = \phi_{\text{diff}} + \phi_{\text{chem}} \tag{21}$$

We assume Fickian diffusion, or

$$\phi_{\text{diff}} = -ku_x$$

where $k > 0$ is the motility. For the chemotactic contribution we argue that ϕ_{chem} should be proportional to the gradient of the chemical concentration; as well, if the number of organisms is doubled for a given concentration gradient, the flux should be twice as great, and therefore ϕ_{chem} should also be proportional to u. Consequently, we assume that

$$\phi_{\text{chem}} = muc_x$$

where $m > 0$ is a proportionality constant that measures the strength of the chemotaxis. Note that there is no minus sign in the last constitutive equation because the organism is assumed to move toward the attractant, or up the gradient.

Combining the preceding equations [substituting (19)–(21) into (17) and (18)] gives a coupled system of reaction–diffusion equations

$$u_t - ku_{xx} + m(uc_x)_x = 0 \tag{22}$$

$$c_t - Dc_{xx} = q_1 u - q_2 c \tag{23}$$

which governs the concentration c of the chemical and population u of the organism. We observe that this system is strongly coupled in that the differential operator in (22) involves derivatives of both c and u. Further, the nonlinear chemotactic flux assumption leads to a more complicated nonlinear diffusion term in the governing equations than in the previous examples.

It is clear that different constitutive relations for f, ϕ_{chem}, and ϕ_{diff} could be considered, leading to still more complicated models. For example, D and k could depend on the concentrations themselves, or the source term f for the secreted chemical could have a nonlinear functional dependence on c and u. Some of these models have been examined in the literature. The reader can consult Murray [1989] for references and applications.

Traveling Reaction Fronts

Another source of reaction–diffusion systems is in the study of chemical kinetics and wavefronts that propagate through reacting flows. For example, let us consider a model sequence of reactions

$$A + Y \xrightarrow{k_1} Z + P, \qquad Z + Y \xrightarrow{k_2} 2P, \qquad A + Z \xrightarrow{k_3} 2Z, \qquad 2Z \xrightarrow{k_4} P + A \tag{24}$$

where A, Z, Y, and P are chemical species entering into the reactions. The single arrow indicates that the reaction can go only one way, and the constants $k_1, \ldots, k_4$ are constants indicating the rate of the reactions. In chemistry, the law of mass action states that the rate of a given reaction is proportional to the products of the concentrations of the reactants. Using lowercase letters to denote the concentrations of the various chemical species (e.g., a is the concentration of A, z is the concentration of Z, etc.), the law of mass action leads to a system of ordinary differential equations for the concentrations a, z, and so on, regarded as a function of time t. In the present case, let us assume that a and p are maintained at constant levels, and that only z and y vary. The chemical species Z enters into each of the four reactions in (24), and the law of mass action gives

$$\frac{dz}{dt} = \begin{cases} k_1 ay & \text{for the first reaction} \\ -k_2 zy & \text{for the second reaction} \\ k_3 az & \text{for the third reaction} \\ -k_4 z^2 & \text{for the fourth reaction} \end{cases}$$

The rate is positive when the net concentration of Z increases in a given reaction, and it is negative when the net concentration decreases.

Consequently, the total rate of production of Z is

$$\frac{dz}{dt} = k_1 ay - k_2 zy + k_3 az - k_4 z^2 \tag{25}$$

In the same manner we find that the rate of production of Y is

$$\frac{dy}{dt} = -k_1 ay - k_2 zy \tag{26}$$

Equations (25) and (26) form a system of two nonlinear ordinary differential equations for the concentrations z and y. If we now assume that the chemical species Z and Y can diffuse (say, with the same diffusion coefficients D), we can append diffusion terms to (25) and (26) and regard z and y as functions of both time t and a spatial coordinate x to obtain a system of reaction–diffusion equations

$$z_t - Dz_{xx} = k_1 ay - k_2 zy + k_3 az - k_4 z^2 \tag{27}$$

$$y_t - Dy_{xx} = -k_1 ay - k_2 zy \tag{28}$$

The reaction mechanism (24) is a special case of the celebrated *Belousov–Zhabotinskii reaction* (BZ reaction, for short), the classic example, and now the prototype, of an oscillating reaction. It is easy to see from this example that reaction–diffusion equations play an important role in the theory of chemical kinetics and waves in chemical and biochemical systems.

The four examples developed in the preceding discussion show the broad applicability of reaction–diffusion equations. We have done no analysis of these systems, but we can infer that there is a rich underlying theory that applies to reaction–diffusion equations in general and consequently, answers many questions that may arise in a variety of applications. For example, do solutions exist, and are they unique? What kinds of initial boundary value problems are well posed? Are there traveling wavefront solutions? What is the long-time behavior of solutions, and can spatial patterns form? Are wavefront solutions stable to small perturbations? In the sequel we address some fundamental notions regarding such equations.

EXERCISES

1. Consider the reactive–diffusive system

$$u_t - u_{xx} = u(1 - u - rv), \qquad v_t - v_{xx} = -buv \qquad (r, b > 0)$$

Show that in the special case $v = (1 - b)(1 - u)/r$ (b not equal to 1), the system reduces to Fisher's equation.

2. Let $u = u(x,t)$ denote the population of an organism and $n = n(x,t)$ denote the concentration of a nutrient to which the organisms are attracted. Discuss the origin and meaning of the model equations

$$u_t = \left(Du_x - \frac{au}{n}n_x\right)_x, \qquad n_t = -ku$$

where D, a, and k are positive constants. In the special case that $a = 2D$, find traveling wave solutions of the form $u = u(x - ct)$, $n = n(x - ct)$, where $u(\pm\infty) = 0$ and $n(-\infty) = 0$, $n(+\infty) = 1$. Sketch the solutions and interpret them biologically (see Murray [1989], p. 357).

3. Let $u = u_0$, $c = c_0$ be a uniform, constant solution to (22)–(23), and assume that $u = u_0 + U(x,t)$ and $c = c_0 + C(x,t)$, where U and C represent small perturbations from the uniform state.

(a) Show that the linearized perturbation equations for U and C are

$$U_t = kU_{xx} - mu_0C_{xx} \qquad C_t = DC_{xx} + q_1U - q_2C$$

(b) Assume a nontrivial solution of the form $(U, C) = (c_1, c_2)e^{\lambda t}e^{ibx}$ where c_1, c_2, b, and λ are constants. Show that a necessary and sufficient condition for linearized stability ($\lambda < 0$) of the uniform state is that $kDb^2 + q_2k > q_1mu_0$. Thus conclude that instabilities are more likely if the decay rate is small and the perturbation has a long wavelength. (This instability is a proposed mechanism for causing the aggregation of slime molds in the absence of a food supply.)

4. Consider the chemical reaction mechanism

$$A \to X, \qquad 2X + Y \to 3X, \qquad B + X \to Y + D, \qquad X \to E$$

where the concentrations a, b, d, and e of the chemical species A, B, D, and E, respectively, are held constant. Assume that the rate constants are unity and write down a system of two ordinary differential equations that govern the time evolution of the concentrations x and y of X and Y. If $a = 2$ and $b = 6$, describe how the concentrations x and y evolve. Suggestion: go to a phase plane.

6.2 TRAVELING WAVE SOLUTIONS

One of the fundamental questions regarding nonlinear partial differential equations is the existence of a special class of solutions called traveling waves. We have already introduced this concept in Chapters 1 and 4 for a single partial differential equation. In this section we study the existence of such solutions for systems of partial differential equations by considering two examples, the geographic spread of a disease, which is a basic reactive–diffusive system, and the flow and sedimentation of pollutants in groundwater, which is a reactive–convective–diffusive system.

It is difficult to formulate encompassing principles that apply to general classes of equations, and therefore the usual strategy is to examine each system separately. The procedure, however, is easy to follow: Assume a solution of the PDEs of the form of a traveling wave, $\mathbf{u} = \mathbf{u}(x - ct)$, where c is the wave speed, and thereby reduce the problem to a system of ordinary differential equations for the waveforms $\mathbf{u}$. The latter dynamical system is a problem in a two-, or more, dimensional phase space, and the question of existence of TWS is often reduced to the existence of an orbit of the dynamical system connecting two critical points (a heteroclinic orbit) or an orbit connecting a single critical point (a homoclinic orbit). The critical points usually represent the boundary conditions at infinity. Interpreted in a different way, the wave speed c acts as an unknown eigenvalue (i.e., a number for which there is a solution to a given boundary value problem). To show rigorously that such orbits exist in phase space is often a difficult problem and involves determining the local structure of the vector field near the critical points, as well as the global structure. Two examples will now illustrate this procedure in some simple cases.

Model for the Spread of a Disease

We now present an example discussed in several other sources: namely, the geographic spread (in one dimension) of a rabies epidemic in a population of foxes. Some of the same assumptions will apply to the spread of other epidemics among other organisms. We assume that the population of foxes is divided into two classes, the number of susceptibles, denoted by $S = S(x, t)$, and the number of infectives (which includes those in the incubation stage), denoted by $I = I(x, t)$. Our next assumption is that the infected class increases at a rate rSI, where $r > 0$ is the infection rate, and therefore the susceptible population decreases at the same rate. Further, the death rate of infectives is proportional to the number of infectives, or is equal to $-aI$, where $a > 0$ is the death rate. We assume that susceptibles become infectives with a very short incubation period. Finally, we postulate that infectives diffuse, possibly because of disorientation, but the susceptibles do not diffuse. The model equations are therefore

$$S_t = -rSI \tag{1}$$

$$I_t - DI_{xx} = rSI - aI \tag{2}$$

where D is the diffusion constant of the infectives. It is straightforward to normalize the reaction–diffusion system (1) and (2) to obtain dimensionless equations

$$S_t = -SI \tag{3}$$

$$I_t - I_{xx} = SI - bI \tag{4}$$

where b is a dimensionless parameter given by $b = a/rS_0$, representing the ratio of the death rate to the rate of infection. In the sequel we assume that $b < 1$. The quantity S_0 is the initial number of susceptibles [to nondimensionalize, we scaled both S and I by S_0, time by $(rS_0)^{-1}$, and x by $(D/rS_0)^{1/2}$].

We now seek traveling wave solutions (TWS) to the reactive–diffusive system (3)–(4). These are solutions of the form

$$S = S(z), \qquad I = I(z), \qquad z = x - ct \tag{5}$$

We anticipate that ahead of the wave we should have the boundary conditions

$$S(+\infty) = 1, \qquad I(+\infty) = 0 \tag{6}$$

which implies that the wavefront is moving into a state where there are no infectives. After the wave passes, we expect that again there are no infectives (rabies is almost always fatal). At the present time it is not possible to know the number of susceptibles, if any, after passage of the epidemic wave. Therefore, at $z = -\infty$ we assume the boundary conditions

$$S'(-\infty) = 0, \qquad I(-\infty) = 0 \tag{7}$$

where it is required that the derivative of S vanish. Substituting (5) into (3) and (4) gives the nonlinear system

$$cS' = SI, \qquad -cI' - I'' = SI - bI \tag{8}$$

The second equation in (8) becomes, upon using the first,

$$-cI' - I'' = cS' - \frac{cbS'}{S}$$

which integrates to

$$-cI + I' = cS - cb \ln S + A \tag{9}$$

where A is a constant of integration. Evaluating (9) at $z = +\infty$ gives $A = -c$, and therefore we have [from (8) and (9)] the first order dynamic

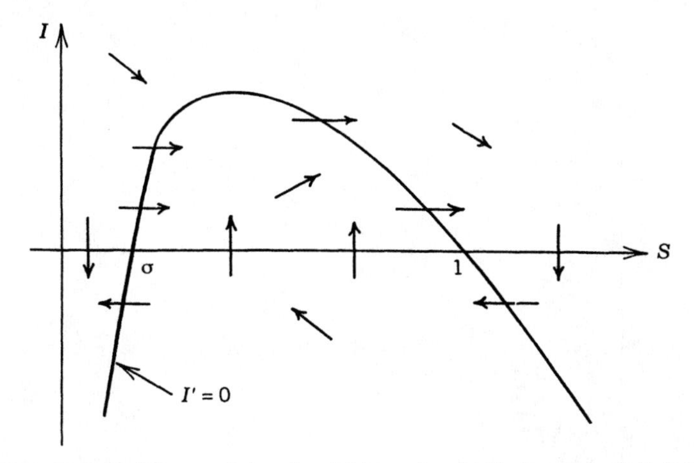

Figure 6.1. Nullclines and direction field associated with the system (10)–(11).

system

$$S' = c^{-1}SI \tag{10}$$

$$I' = -c(S + I) + bc \ln S + c \tag{11}$$

This system can be analyzed in the usual manner by phase-plane techniques (see the appendix to Chapter 4). The nullclines (where $S' = 0$ and where $I' = 0$) are easily determined. $S' = 0$ along $S = 0$ and $I = 0$, and $I' = 0$ along the locus

$$I = b \ln S - S + 1 \tag{12}$$

It is easily checked that this locus, when sketched in the SI-plane, has a positive maximum at $S = b$ and crosses the S-axis at $S = 1$ and $S = \sigma$, where σ satisfies the algebraic equation $b \ln \sigma - \sigma + 1 = 0$. Note that $\sigma < b < 1$. In the first quadrant we have $S' > 0$, while in the third quadrant $S' < 0$. Above the locus defined by (12) we have $I' < 0$, and below the locus $I' > 0$. The critical points are $(1, 0)$ and $(\sigma, 0)$, which are the intersections of the loci $S' = 0$ and $I' = 0$. A graph of the direction field is shown in Figure 6.1. The type and stability of the two critical points can be determined by linearization. The Jacobi matrix is

$$J(S, I) = \begin{pmatrix} \dfrac{I}{c} & \dfrac{S}{c} \\ -c + \dfrac{bc}{S} & -c \end{pmatrix}$$

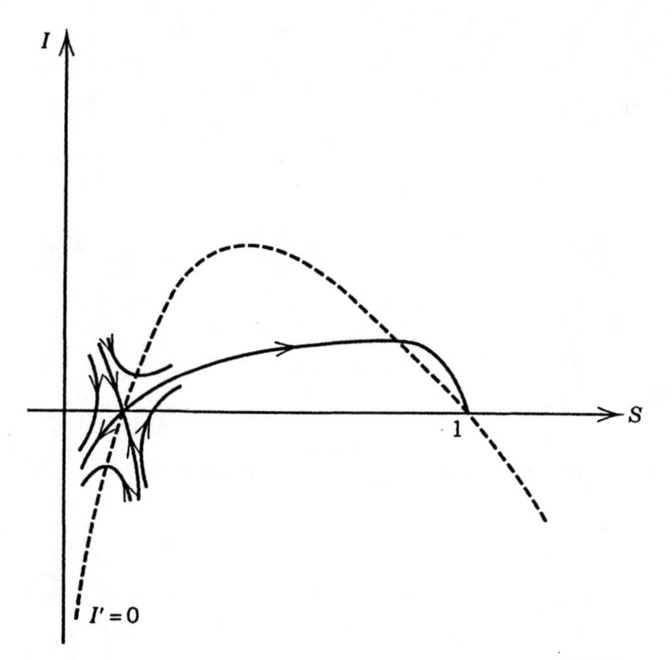

Figure 6.2. Phase portrait of (10)–(11) in the case $c > 2\sqrt{1-b}$.

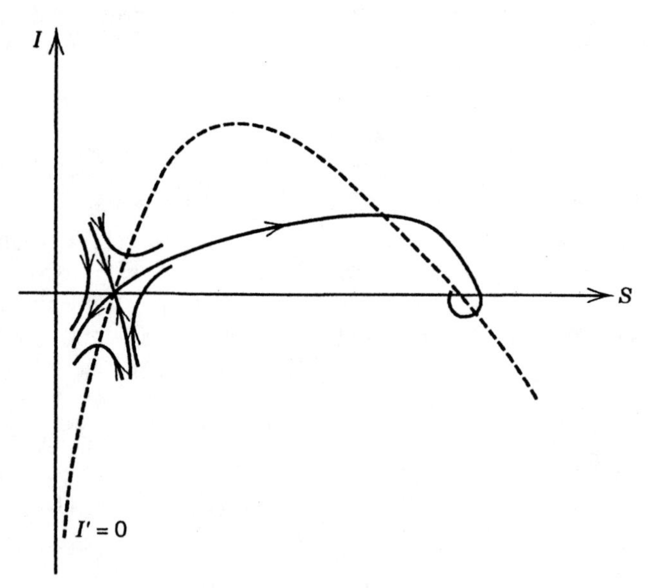

Figure 6.3. Phase portrait of (10)–(11) in the case $c < 2\sqrt{1-b}$.

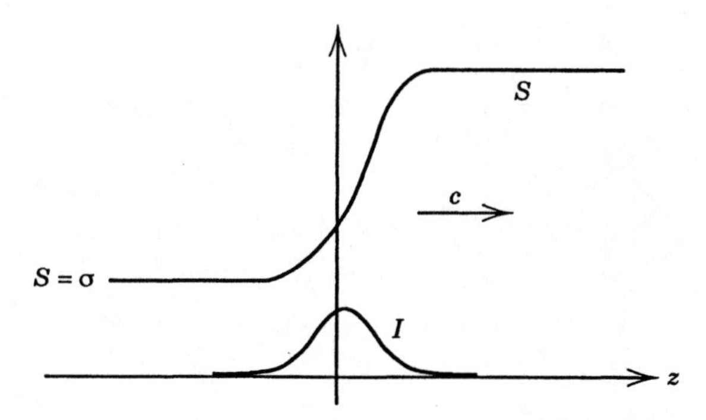

Figure 6.4. Traveling wave solution to the reactive–diffusive system (3)–(4).

At $S = \sigma$ and $I = 0$ the matrix $J(\sigma, 0)$ has real eigenvalues of opposite sign, so $(\sigma, 0)$ is a saddle point. At $S = 1$ and $I = 0$ the matrix $J(1, 0)$ has eigenvalues λ that satisfy the characteristic equation $\lambda^2 + c\lambda + (1 - b) = 0$. Clearly, therefore,

$$\lambda = \frac{1}{2}\left\{-c \pm \left[c^2 - 4(1 - b)\right]^{1/2}\right\}$$

If $c > 2(1 - b)^{1/2}$, the eigenvalues are both real and negative, and thus $(1, 0)$ is a stable node. If $c < 2(1 - b)^{1/2}$, the eigenvalues are complex with negative real parts, and thus $(1, 0)$ is a stable spiral. The phase diagram is shown in Figures 6.2 and 6.3 for both of these cases. In both cases there is a unique separatrix connecting the critical point $(\sigma, 0)$ at $z = -\infty$ to $(1, 0)$ at $z = +\infty$, and each corresponds to a traveling wave. On physical grounds we reject the case when $(1, 0)$ is a spiral because in that case the number of infectives oscillates around $I = 0$ for large z, giving negative values for the number of infectives. Consequently, if $c > 2(1 - b)^{1/2}$, there is traveling wave solution of the form shown in Figure 6.4 which corresponds to the heteroclinic orbit (the separatrix) connecting the saddle point $(\sigma, 0)$ to the stable node $(1, 0)$. This wave travels at speed c.

We remark that σ is the number of susceptibles that survive the epidemic, and σ is an increasing function of the parameter b; thus the smaller b, the fewer foxes survive the rabies epidemic. If $b = 0.5$, for example, σ is about 0.2 and so 20 percent of the foxes survive; the dimensionless wave speed is $c = 1.414$. The reader is referred to Murray [1989] for references to actual studies that compare the theoretical results to experimental observations.

Groundwater Model

In this example we formulate a nonlinear convection–reaction–diffusion model for the transport of solutes (e.g., pollutants) through a one-dimensional porous medium (e.g., soil). Such a model will be based on mass balance and

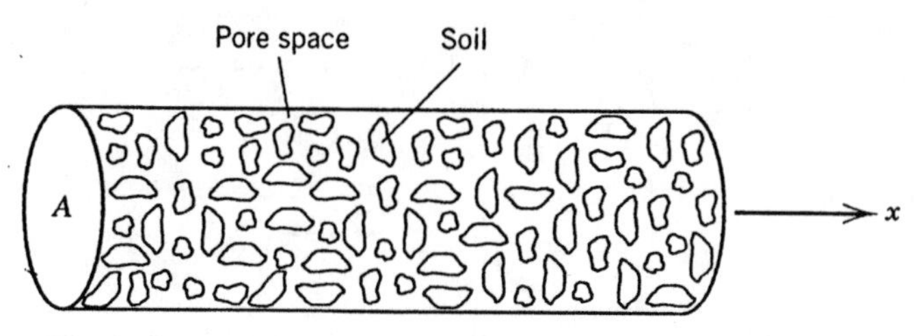

Figure 6.5. Porous medium of cross-sectional area A carrying a chemical dissolved in a liquid.

constitutive assumptions for the flux of the solute and the rate of sorption of the solute in the medium. The assumptions will lead to a coupled system of reaction–diffusion equations with a nonlinear convection term, and the question of existence of TWS will then be addressed.

We consider a one-dimensional underground flow in the horizontal direction (denoted by x) occurring in a tube of cross-sectional area A (see Figure 6.5). The tube is assumed to be a porous medium where the fluid (water) can occupy only a fraction ϕ of the total volume. The fraction ϕ is called the *porosity* of the medium, and we assume that it is constant. We let $C = C(x,t)$ denote the mass concentration of a chemical (the solute) dissolved in the water, and $N = N(x,t)$ denote the mass concentration of the chemical that is sorbed in the soil. By $q = q(x,t)$ we denote the flux of the solute. In the usual manner we can express mass balance in integral form as

$$\frac{d}{dt}\int_a^b \phi AC(x,t)\,dx = Aq(a,t) - Aq(b,t) - \int_a^b AN_t(x,t)\,dx \qquad (13)$$

where $[a,b]$ is an arbitrary interval representing a section of the tube. Equation (13) states that the time rate of change of the total amount of the solute in the section $[a,b]$ must be balanced by the net rate that the solute flows into the section minus the rate that is sorbed by the soil in that section. Assuming smoothness of the functions C, N, and q, we may proceed in the usual way and write (13) as a partial differential equation

$$\phi C_t + q_x + N_t = 0 \qquad (14)$$

To complete the formulation we append constitutive relations to (14). We assume that the flux q has two contributions, the first being due to Fickian diffusion, and the second a transport term caused by the bulk movement of the fluid through the medium. Thus we take

$$q = -DC_x + \frac{\phi BC^2}{2} \qquad (15)$$

where D is the diffusion constant and B is a bulk movement constant. An additional constitutive relation is needed to relate C and N. Such relations are usually determined empirically, the simplest being a linear relationship $N = \alpha C$. A better assumption, however, is to take $N_t = a(N_0 - N)$, where a is a positive constant and N_0 is the maximum concentration where the soil becomes saturated. This equation determines N as a function of time and it can be substituted, along with (15), into (14) to obtain a single equation for C. The growth rate a could also depend in some manner on the solute concentration C. Some of these assumptions are explored in the exercises, but for the present discussion we assume a constitutive equation for the rate of sorption of the form

$$N_t = KR(N, C) \tag{16}$$

where K is a rate constant and where R is a rate function having the form

$$R(N, C) = \frac{N_0 C^2}{C_0^2 + C^2} - N \tag{17}$$

where N_0 is a maximum, constant concentration where the soil is saturated with the chemical, and where C_0 is a threshold value of the solute concentration where the rate function switches on. Equations (16) and (17) imply that the rate of sorption increases with the concentration C up to some maximum rate, while it decreases with the concentration N; in other words, sorption decreases as the soil becomes more sorbed with the chemical pollutant. A graph of the equilibrium locus $R(N, C) = 0$ is shown in Figure 6.6. Above the equilibrium curve the sorption rate is negative and the sorbed chemical is returned to the water; below the equilibrium curve the sorption rate is positive and the solute is deposited in the soil. Consequently, in this model there is a tendency toward an equilibrium state. We remark that this

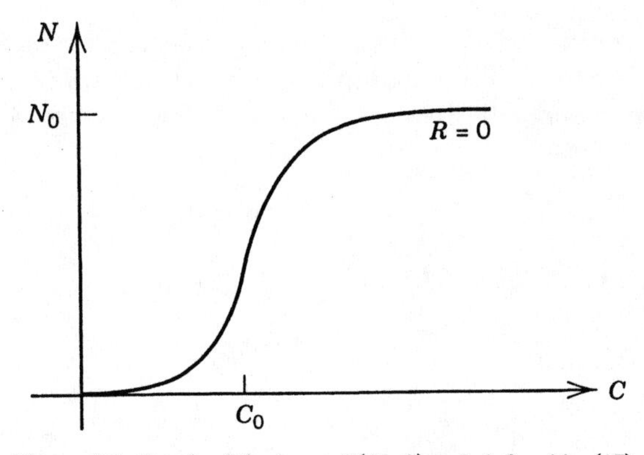

Figure 6.6. Graph of the locus $R(N, C) = 0$ defined by (17).

model is a generalization of some of the other models found in the literature; the reader can consult, for example, Guenther and Lee [1988] for an elementary discussion of some of the linear models.

With the assumptions (15) and (17), equations (14) and (16) become

$$\phi C_t + \phi B C C_x + N_t = D C_{xx} \tag{18}$$

$$N_t + K\left(N - \frac{N_0 C^2}{C_0^2 + C^2}\right) = 0 \tag{19}$$

These equations can be nondimensionalized by introducing the dimensionless parameters

$$c = \frac{C}{C_0}, \qquad n = \frac{N}{N_0}, \qquad \tau = \frac{C_0^2 B^2 \phi}{D} t, \qquad \xi = \frac{C_0 B \phi}{D} x$$

Then we have

$$c_\tau + c c_\xi + \beta n_\tau = c_{\xi\xi} \tag{20}$$

$$n_\tau + k\left(n - \frac{c^2}{1 + c^2}\right) = 0 \tag{21}$$

where β and k are dimensionless constants defined by

$$\beta = \frac{N_0}{\phi C_0}, \qquad k = \frac{KD}{C_0^2 B \phi}$$

Now we look for traveling wave solutions of (20)–(21) of the form (we use the symbol v for the scaled wave speed)

$$c = c(\xi - v\tau), \qquad n = n(\xi - v\tau) \tag{22}$$

Ahead of the wave we assume that $c = n = 0$, that is, the solute and sorbed concentrations are zero; the conditions behind the wave, after its passage, will be determined depending on the velocity v. Thus the boundary conditions are

$$c(+\infty) = n(+\infty) = 0 \tag{23}$$

Substituting (22) into (20) and (21) gives

$$-vc' + cc' - \beta v n' = c'' \tag{24}$$

$$-vn' + k\left(n - \frac{c^2}{1 + c^2}\right) = 0 \tag{25}$$

where a prime denotes d/dz, where $z = \xi - v\tau$. Noting that $cc' = d/dz(c^2/2)$, equation (24) can be integrated with respect to z, and the constant of integration can be determined by the boundary conditions (23). We obtain, after rearrangement,

$$c' = -vc - \beta vn + \frac{c^2}{2} \tag{26}$$

$$n' = kv^{-1}\left(n - \frac{c^2}{1 + c^2}\right) \tag{27}$$

which is a nonlinear dynamical system in two dimensions. We analyze these equations in the phase plane. The nullclines are given by the parabola

$$n = 2(\beta v)^{-1}c(c - 2v) \tag{28}$$

where the vector field is vertical ($c' = 0$), and the equilibrium curve

$$n = \frac{c^2}{1 + c^2} \tag{29}$$

where the vector field is horizontal ($n' = 0$). These curves are shown in Figure 6.7. The point $(0, 0)$, representing the assumed state at plus infinity, is a critical point; there is at least one additional critical point in the first quadrant. The c-coordinates of the critical points are the positive real roots

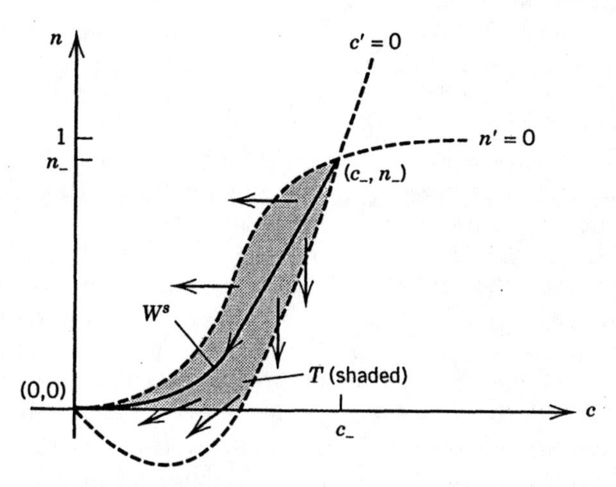

Figure 6.7. Nullclines and direction field associated with the system (26)–(27). W^s is the stable manifold entering the critical point $(0, 0)$.

of the cubic equation

$$c^3 - 2vc^2 + (1 - 2\beta v)c - 2v = 0 \tag{30}$$

which comes from equating (28) and (29). Figure 6.7 shows the case of a single positive root of (30). We shall now argue that for any $v > 0$ there is an orbit connecting (c_-, n_-) to $(0, 0)$, where c_- is the smallest positive root of (30) and $n_- = c_-^2/(1 + c_-^2)$. This heteroclinic orbit will represent a TWS to the system (20)–(21).

Figure 6.7 shows a generic case; the equilibrium curve is fixed, but the parabola depends on the parameters β and v. In any case, the parabola has a negative minimum at $c = v$, and it lies strictly below the equilibrium curve for $0 < c < c_-$. The linearized Jacobi matrix associated with the nonlinear system (26)–(27) is

$$J(c, n) = \begin{pmatrix} c - v & -\beta v \\ -\dfrac{2kcv^{-1}}{1 + c^2} & \dfrac{k}{v} \end{pmatrix}$$

The eigenvalues of $J(0, 0)$ are $-v$ and k/v, and therefore $(0, 0)$ is a saddle point. The eigenvectors are $(1, 0)$ and $(1, -(1 + k/v^2)/\beta)$. Now define the region T by

$$T = \left\{(c, n) | n > 0, c < c_-, n > 2(\beta v)^{-1}c(c - 2v), \quad \text{and} \quad n < \frac{c^2}{1 + c^2}\right\}$$

which is the region in the first quadrant below the equilibrium curve, above the parabola, and to the left of the first positive intersection point of the two curves. This region is shaded in Figure 6.7. Now denote by W^s the one-dimensional stable manifold (the separatrix) entering $(0, 0)$. This manifold must enter $(0, 0)$ from the region T. Now follow this manifold backward in time, as $z \to -\infty$. Because the vector field along the boundaries of T, taken backward in time, point inward, we argue that W^s cannot leave the region T, and consequently it must enter the critical point (c_-, n_-). The argument here can be stated differently: that if we reverse time, the boundary of T consists of ingress points (the vector field is inward), so by the Wazewski retract theorem (see, e.g., Hartman [1964]), the curve W^s must enter (c_-, n_-). We can also note that on the contrary, if W^s did touch the boundary of T, uniqueness would be violated. Therefore, we have shown that there exists an orbit (the separatrix) connecting (c_-, n_-) to $(0, 0)$, and hence we have produced a traveling wave solution to the convection–reaction–diffusion system for any wave speed $v > 0$. Figure 6.8 shows the qualitative behavior of these concentration waves.

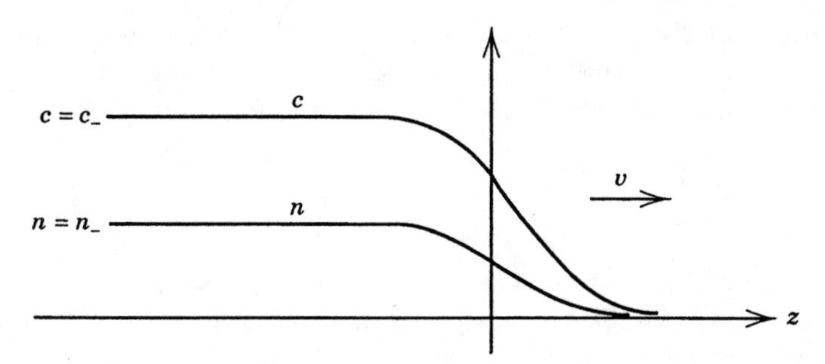

Figure 6.8. Traveling concentration waves.

In the two preceding examples the problem of determining traveling waves reduced to a two-dimensional phase-plane analysis. This is more than can usually be expected in a problem. More often than not, the dynamical system has dimension three (or more), and the analysis is considerably more difficult.

EXERCISES

1. In the rabies epidemic model, discuss the existence of traveling wave solutions in the case that the parameter b, the ratio of the death rate to the infection rate, is larger than unity.

2. Consider the mass balance equation (14) for the groundwater problem.

(a) Derive an equation for the concentration C if the flux is given by $q = -DC_x + \phi VC$ and $N = \alpha C$, where V is the velocity of the bulk movement of the water through the soil and α is a positive constant. Do traveling wave solutions exist in this case?

(b) Along with (14) and (15), assume that the rate of sorption is given by $N_t = \alpha C - \gamma N$, where α and γ are positive constants. Show that traveling waves for the resulting system exist, and describe them.

3. Consider the initial–boundary value problem

$$C_t - KC_{xx} + \beta C_x = 0, \qquad 0 < x < \pi, t > 0$$

$$C(x,0) = C_0(x), \qquad 0 < x < \pi$$

$$C(0,t) = C(\pi,t) = 0, \qquad t > 0$$

(a) Give a physical interpretation of this problem in the setting of groundwater flow.

(b) Because this is a linear problem, the eigenfunction expansion method (or separation of variables) may be expected to yield a solution. Apply this method to determine a solution.

(c) Prove that the solution is unique using the *energy method*. That is, assume that u and v are two solutions and form the difference $w = u - v$. Then prove that $w = 0$ by showing that the *energy*

$$E(t) = \frac{1}{2}\int_0^{\pi} w^2\,dx$$

is identically zero [to this end, show that $E'(t) < 0$].

(d) What is the long-time behavior of the solution $C(x,t)$? Interpret your result physically.

4. *Project*. Consider the reactive–diffusive system

$$u_t - Du_{xx} = u(1 - u - v), \qquad v_t - v_{xx} = av(u - b)$$

where a, b, and D are positive constants.

(a) Interpret this (scaled) model in a predator–prey setting.

(b) In the special case $D = 0$, investigate the existence of traveling waves. [See S. Dunbar, *J. Math. Biol.* **17**, 11–32 (1983).]

5. A simplified scaled model of a detonation process is given by the system

$$u_t + uu_x + \lambda_x = Du_{xx}, \qquad \lambda_t = r(u, \lambda)$$

where u is a scaled temperature and λ is the mass fraction of the product species P in a reversible chemical reaction R $\leftrightarrow$ P with reaction rate r given by

$$r(u, \lambda) = 1 - \lambda - \lambda e^{-1/u}$$

and D is a positive constant. Show that there exist positive traveling wave solutions to this nonlinear convective–reactive–diffusive system if the wave speed c exceeds the value of u at $+\infty$. [See J. D. Logan and S. Dunbar, *IMA J. Appl. Math.* **49**, 103–121 (1992).]

6. Show that traveling waves exist for wave speeds $c > 2$ for the reactive–convective system

$$u_t + uu_x + v_x = (2 - u)(u - 1), \qquad v_t = 1 - v - v\exp(u^{-1})$$

[See J. D. Logan and T. S. Shores, *Math. Models Methods Appl. Sci.* **3**(3), 341–358, (1993).

7. Determine all nonnegative traveling waves for the system

$$u_t + (u + v)_x = u_{xx}, \qquad v_t = \frac{1 + u^2}{(1 + 2u^2)} - v$$

Note: A large number of examples and exercises on traveling waves in biological and chemical systems can be found in Murray [1989].

8. A model for the burning of a solid waste material is given by the system

$$u_t = u_{xx} + kar(u) \qquad a_t = -ar(u)$$

where $u = u(x,t)$ is the temperature, $a = a(x,t)$ is the concentration of the immobile, combustible, unburnt waste material, k is a positive constant, and $r(u)$ is the rate of burning. The problem is to investigate the existence of traveling wave solutions of speed c that satisfy the boundary conditions $(u, a) \to (0, a_+)$ at plus infinity and $(u, a) \to (u_-, 0)$ at minus infinity, where a_+ and u_- are some constants.

(a) Show that if such solutions exist, then necessarily $u_- = ka_+$.

(b) Assume that $r(u) = r_0 u$, $r_0 > 0$, and show that traveling waves exist for wavespeeds c satisfying $c^2 > 4r_0 u_-$. Interpret physically.

(c) Examine the problem when the burning rate is given by $r(u) = 0$ if $u \le u_b$, and $r(u) = r_0(u - u_b)$ if $u > u_b$, where u_b is an ignition temperature where the combustion process switches on. (See Grindrod [1991], pages 64–65.)

6.3 EXISTENCE OF SOLUTIONS

We now take up the question of the existence of a solution to the nonlinear initial value problem

$$u_t - Du_{xx} = f(u), \qquad x \in R, \quad t > 0 \tag{1}$$

$$u(x,0) = u_0(x), \qquad x \in R \tag{2}$$

For the present we consider the scalar case, and conditions on u_0 and the nonlinear reaction term $f(u)$ will be imposed later.

It is easy to write down simple examples of problems for which a solution does not exist for all $t > 0$. The initial value problem

$$u_t - Du_{xx} = u^2, \qquad x \in R, \quad t > 0$$

$$u(x,0) = u_0, \qquad x \in R$$

where u_0 is a positive constant, has a spatially independent solution $u(x,t) = u_0/(1 - u_0 t)$, which blows up in finite time. A less trivial example is given in Section 6.5. Further, if a solution exists, it need not be unique, as the following example shows. Consider the initial–boundary value problem

$$u_t - u_{xx} = 2u^{1/2}, \qquad x \in R, \quad t > 0$$

$$u(x,0) = 0, \qquad x \in R$$

It is easy to check that both $u(x,t) = 0$ and $u(x,t) = t^2$ are solutions.

Therefore, to guarantee existence and uniqueness it is evidently necessary to impose conditions on the initial data and the type of nonlinearity.

Fixed-Point Iteration

Of course, the best strategy for showing that a solution to a given problem exists is actually to exhibit a formula for the solution. For certain linear problems, we can actually proceed in this manner. For example (using Fourier transforms, Duhamel's principle, or Green's functions), one can derive the solution to the linear, nonhomogeneous diffusion problem

$$u_t - Du_{xx} = g(x,t), \qquad x \in R, \quad t > 0 \tag{3}$$

$$u(x,0) = u_0(x), \qquad x \in R \tag{4}$$

For reference (we shall refer to this solution in the sequel), it is given by

$$u(x,t) = \int_R K(x-y,t)u_0(y)\,dy + \int_0^t \int_R K(x-y,t-s)g(y,s)\,dy\,ds \tag{5}$$

where $K(x,t)$ is the diffusion kernel given by

$$K(x,t) = \left(\frac{1}{4\pi Dt}\right)^{1/2} \exp\left(\frac{-x^2}{4Dt}\right) \tag{6}$$

and where u_0 and G are continuous bounded functions. For nonlinear problems, however, it is usually impossible to find such formulas, and alternative methods must be found that prove existence without actually producing a formula. Such existence questions are common in applied analysis, and in this section we discuss a general method that applies to many nonlinear problems; this method, which is called fixed-point iteration, is an impressive illustration of the unifying power of abstraction in analysis. The basic idea is to produce a sequence, through iteration of a certain map, that converges to the solution of the problem, thus showing existence.

The reader may be familiar with this iteration procedure from prior experience in elementary courses. Regardless, we shall now indicate, mainly for the purpose of motivation, the methodology in two different settings, nonlinear algebraic equations and ordinary differential equations, before addressing the existence question for nonlinear PDEs.

Example 1 (*Fixed-Point Iteration*). Consider a nonlinear algebraic equation in the form

$$x = \phi(x) \tag{7}$$

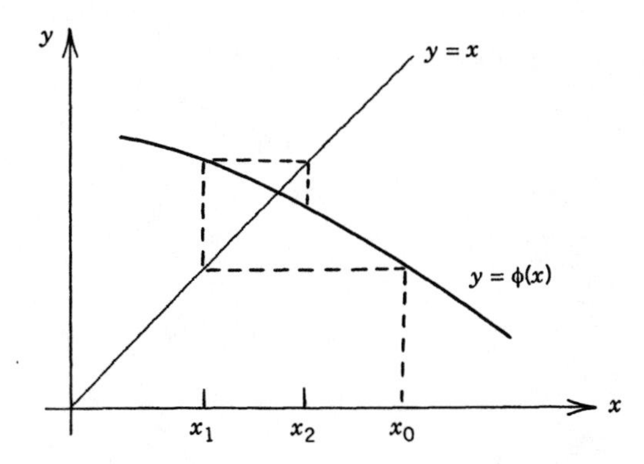

Figure 6.9. Fixed-point iteration in the convergent case.

where ϕ is a given continuous function defined on R. By selecting an initial approximate x_0 and then iterating the map ϕ repeatedly via the formula

$$x_{n+1} = \phi(x_n), \qquad n = 0, 1, 2, \ldots \tag{8}$$

one can produce a sequence of numbers $x_0, x_1, x_2, x_3, \ldots$ which, under certain conditions, will converge to a root of (7). For, suppose that x_n converges to some number x' (i.e., $\lim x_n = x'$). Then, taking the limit of both sides of (8) as $n \to \infty$ and using the continuity of ϕ gives $x' = \lim \phi(x_n) = \phi(\lim x_n) = \phi(x')$, and therefore x' is a root of (7). Thus existence of a root can be proved from a limiting process. Because $\phi(x') = x'$, the function ϕ maps x' to itself, and therefore x' is called a *fixed point* of the mapping ϕ. The process (8) is often called *fixed-point iteration*. The situation is shown geometrically in Figure 6.9; a root x' is the x-coordinate of the intersection of the graphs of $y = x$ and $y = \phi(x)$. The sequence x_n can be determined geometrically as shown, depending on the shape of the graph of ϕ. If the graph is ϕ is too steep at x', as shown in Figure 6.10, the sequence defined by the iterative process (8) will not converge to the root, or fixed point, x'. When will the fixed-point iteration converge? To fix the idea, let ϕ be a continuous function on R and suppose that ϕ satisfies a *Lipschitz condition* of the form

$$|\phi(x) - \phi(z)| \le k|x - z| \qquad \text{for all } x, z \in R \tag{9}$$

where k is a constant with $0 < k < 1$. Notice that the Lipschitz condition puts a bound on all secant lines or chords connecting two points on the graph of ϕ, and therefore it restricts the steepness of the graph. Under condition (9) it is not difficult to observe that for any choice of x_0, the iterative process (8) will converge to the unique root of (7). To prove this fact, we show that x_n is a Cauchy sequence of real numbers, and hence convergent. First we

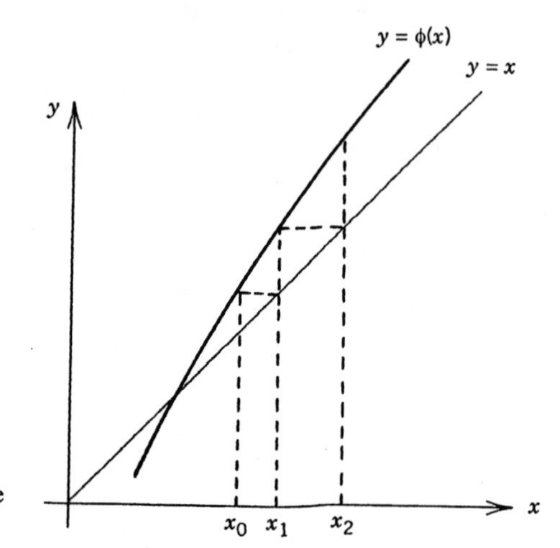

Figure 6.10. Fixed-point iteration in the divergent case.

calculate the distance between consecutive iterates as

$$|x_{n+1} - x_n| = |\phi(x_n) - \phi(x_{n-1})| \leq k|x_n - x_{n-1}|$$

Then, if $|x_1 - x_0| = a$, we have

$$|x_2 - x_1| \leq ka, \quad |x_3 - x_2| \leq k^2 a, \ldots, |x_{n+1} - x_n| \leq k^n a$$

whereupon, for any positive integer p,

$$\begin{aligned} |x_{n+p} - x_n| &\leq |x_{n+p} - x_{n+p-1}| + |x_{n+p-1} - x_{n+p-2}| + \cdots + |x_{n+1} - x_n| \\ &\leq ak^{n+p-1} + ak^{n+p-2} + \cdots + ak^n \\ &\leq ak^n[k^{p-1} + k^{p-2} + \cdots + 1] \\ &< \frac{ak^n}{1-k} \end{aligned}$$

Here, to get the first line we added and subtracted all the terms between x_{n+p} and x_n and then used the triangle inequality; to get the final inequality we used the fact that the sum of the geometric series is $\Sigma k^n = 1/(1-k)$. Thus since $k < 1$, the difference $|x_{n+p} - x_n|$ can be made arbitrarily small provided that n is chosen large enough, for all $p > 0$. In different words, for any $\varepsilon > 0$ there is an integer N such that $|x_{n+p} - x_n| < \varepsilon$ for $n > N$ and any $p > 0$. So, by definition, the sequence x_n is a Cauchy sequence, and therefore it converges to some x'. Our earlier argument showed that $x' = \lim x_n$ is a root of (7). It is straightforward to show that the root x' is unique. For, by way of contradiction, let x' and x'' be two distinct roots of (7); from (9) we have $|x' - x''| = |f(x') - f(x'')| \leq k|x' - x''|$, for $k < 1$, which is a contradiction; consequently, only one root can exist.

To summarize, we have shown that if ϕ is a continuous function on R satisfying the Lipschitz condition (9), then (7) has a unique real root. Thus we have proved the existence of a solution without exhibiting it explicitly; the argument is clearly based on the fact that the set of real numbers R is complete (Cauchy sequences converge) and the assumption that the iteration function ϕ satisfies a Lipschitz condition. The latter condition, equation (9), can be interpreted alternatively by noting that the distance between $\phi(x)$ and $\phi(z)$ is strictly less than the distance between x and z; that is, ϕ is a *contraction mapping*. The ideas of fixed-point iteration, contraction mappings, and completeness can be generalized to a broad setting where they apply to ordinary and partial differential equations, integral and functional equations, and other equations of interest in applied analysis.

Example 2 (*Ordinary Differential Equations*). In ordinary differential equations we are often interested in the existence of a solution to the initial value problem

$$y' = f(t, y), \qquad y(t_0) = y_0 \tag{10}$$

where t_0 and y_0 are given constants; f is assumed to be a continuous function. We can easily rewrite the initial value problem (10) as an integral equation by integrating (10) from t_0 to t to obtain

$$y(t) = y_0 + \int_{t_0}^{t} f(s, y(s))\, ds \tag{11}$$

The right side of (11) can be regarded as a mapping Φ on the set of continuous functions; that is, to each continuous function y there is associated another continuous function $\Phi(y)$ defined at each t by

$$\Phi(y)(t) \equiv y_0 + \int_{t_0}^{t} f(s, y(s))\, ds$$

Therefore, the integral equation (11) may be written in the form

$$y(t) = \Phi(y)(t)$$

which formally has the same structure as the algebraic equation (7) in the preceding example. Therefore, in the same manner we may fashion a sequence of functions $y_0(t), y_1(t), y_2(t), \ldots$ by the iteration formula

$$\begin{aligned} y_{n+1}(t) &= \Phi(y_n)(t) \\ &= y_0 + \int_{t_0}^{t} f(s, y_n(s))\, ds, \qquad n = 0, 1, 2, \ldots \end{aligned} \tag{12}$$

where the first iterate is $y_0(t) = y_0$. The sequence of iterates $y_n(t)$ obtained in this way, in the context of ordinary differential equations, are called *Picard iterates*, and the fixed-point iteration procedure is called *Picard iteration*. One is left with the task of showing that the sequence converges to a solution of the given integral equation, or equivalently, the given initial value problem.

It is not difficult to prove the following theorem (see, e.g., Birkhoff and Rota [1969]):

Theorem 1. Let $f(t, y)$ be a continuous function and assume that f satisfies the Lipschitz condition

$$|f(t,y) - f(t,z)| \le k|y - z| \qquad \text{for all } y, z \in R \quad \text{and} \quad |t - t_0| \le T$$

Then for any y_0 the Picard iterates $y_n(t)$ converge uniformly on $|t - t_0| \le T$, and the limiting function is a unique solution to the initial value problem (10) on the interval $|t - t_0| \le T$.

Reaction–Diffusion Equations

With the ideas in the preceding paragraphs as motivation, we now consider the question of existence of a solution to the nonlinear initial value problem (1)–(2). The last example on ordinary differential equations suggests the strategy of writing the initial value problem as an integral equation. A hint on how to accomplish this is found in the solution (5) to the linear, nonhomogeneous problem (3) and (4). For the moment, assume that f and u_0 are bounded, continuous functions on R. If $u(x, t)$ is a solution of (1)–(2), then

$$u_t(x,t) - Du_{xx}(x,t) = f(u(x,t)) = g(x,t), \qquad u(x,0) = u_0(x) \quad (13)$$

Thus from (5) we expect that

$$u(x,t) = \int_R K(x-y,t)u_0(y)\,dy + \int_0^t \int_R K(x-y,t-s)g(y,s)\,dy\,ds$$

or

$$u(x,t) = \int_R K(x-y,t)u_0(y)\,dy + \int_0^t \int_R K(x-y,t-s)f(u(y,s))\,dy\,ds \tag{14}$$

which is a nonlinear integral equation for $u = u(x, t)$. One can show that

$u = u(x,t)$ is a solution of (1)–(2) if, and only if, $u = u(x,t)$ is a solution of (14). Now we observe that equation (14) has the form

$$u = \Phi(u)$$

where Φ is the mapping defined on the set of bounded continuous functions by

$$\Phi(u)(x,t) \equiv \int_R K(x-y,t)u_0(y)\,dy + \int_0^t \int_R K(x-y,t-s)f(u(y,s))\,dy\,ds$$

Thus one can define a fixed-point, iterative process by $u_{n+1} = \Phi(u_n)$, or

$$u_{n+1}(x,t) = \int_R K(x-y,t)u_0(y)\,dy + \int_0^t \int_R K(x-y,t-s)f(u_n(y,s))\,dy\,ds, \quad n = 1,2,3,\ldots \tag{15}$$

with

$$u_0(x,t) = \int_R K(x-y,t)u_0(y)\,dy \tag{16}$$

The initial approximation $u_0(x,t)$ given by (16) is just the first term on the right side of (14), and it is the solution to the linear, homogeneous diffusion equation with initial condition $u(x,0) = u_0(x)$. There should be no confusion in using the notation $u_0(x,t)$ and $u_0(x)$ for these two different objects; we shall always indicate the arguments so that the context is clear. We are now in position to prove an existence theorem for the initial value problem (1)–(2) under suitable assumptions. This version of the theorem was adapted from Kolmogorov, Petrovsky, and Piscounoff [1937] in their celebrated paper on nonlinear reaction–diffusion equations. In the proof of the following theorem we use the fact that the diffusion kernel K is strictly positive and that

$$\int_R K(x-y,t-s)\,dx = 1 \qquad \text{for all } y \text{ and all } s < t \tag{17}$$

That is, the total area (energy) under the fundamental solution of the linear diffusion equation is unity (see Chapters 1 and 4).

Theorem 2. Consider the initial value problem (1) where $u_0(x)$ is a bounded continuous function on R, and where f is a bounded continuous function on R that satisfies the global Lipschitz condition

$$|f(u) - f(v)| \leq k|u - v| \qquad \text{for all } u, v \in R \tag{18}$$

where k is a positive constant independent of u and v. Then for any $T > 0$, there exists a unique, bounded, solution $u(x, t)$ of (1) for $x \in R$ and $0 \leq t \leq T$.

Proof. We shall show that the sequence $u_n(x, t)$ defined by (15) and (16) converges uniformly on $R \times [0, T]$ to a function that is a solution of (1)–(2). First we note that the Lipschitz condition (18) implies that

$$|f(u_0(x,t)| \leq |f(0)| + k|u_0(x,t)| \leq (1 + k)m,$$
$$m = \max\{f(0), \sup|u_0(x)|\}$$

Then, from (15) and (17) we have

$$|u_1(x,t) - u_0(x,t)| \leq \int_0^t \int_R K(x - y, t - s)|f(u_0(y,s))|\,dy\,ds$$
$$\leq \int_0^t (1 + k)m\,ds = Mt \tag{19}$$

For convenience, let us introduce the notation

$$M_n(t) = \sup\{|u_n(x,s) - u_{n-1}(x,s)| : x \in R, s \leq t\}, \qquad t \leq T$$

Then, from (19),

$$M_1(t) \leq Mt, \qquad 0 \leq t \leq T \tag{20}$$

Now, let us obtain a bound for $|u_{n+1} - u_n|$. From (15) we have

$$|u_{n+1}(x,t) - u_n(x,t)|$$
$$\leq \int_0^t \int_R K(x - y, t - s)|f(u_n(y,s)) - f(u_{n-1}(y,s))|\,dy\,ds$$
$$\leq \int_0^t \int_R K(x - y, t - s)k|u_n(y,s) - u_{n-1}(y,s)|\,dy\,ds$$
$$\leq \int_0^t \int_R K(x - y, t - s)kM_n(s)\,dy\,ds$$
$$= k\int_0^t M_n(s)\,ds$$

Therefore, taking the supremum,

$$M_{n+1}(t) \le k\int_0^t M_n(s)\, ds, \qquad 0 \le t \le T \quad (n = 1,2,3,\ldots) \tag{21}$$

Now, using (20) and (21), we may obtain bounds on each M_n. To this end, we observe that

$$M_2(t) \le k\int_0^t Ms\, ds = \frac{kMt^2}{2}$$

$$M_3(t) \le k\int_0^t \frac{kMs^2}{2}\, ds = \frac{M(kt)^3}{3!}$$

and so on, to obtain finally,

$$M_n(t) \le M(kt)^n/n! \qquad 0 \le t \le T \tag{22}$$

Equation (22) implies that the sequence $u_n(x,t)$ converges uniformly on $R \times [0,T]$ to some continuous, bounded function $u(x,t)$. Then, taking the limit of both sides of (15) and using the uniform convergence to pull the limit under the integral, we observe that the limit function $u(x,t)$ satisfies the integral equation (14) and therefore is a solution to the initial value problem (1).

Uniqueness of the solution $u(x,t)$ can be proved by the standard contradiction argument. Let $v(x,t)$ be another bounded, continuous solution of (14). Then

$$|u(x,t) - v(x,t)| \le \int_0^t \int_R K(x-y, t-s)|f(u(y,s)) - f(v(y,s))|\, dy\, ds$$

$$\le \int_0^t \int_R K(x-y, t-s) k |u(y,s) - v(y,s)|\, dy\, ds$$

Now let $M(t) = \sup|u(x,s) - v(x,s)|$, taken over all $x \in R$ and $s \le t$. It follows that

$$M(t) \le k\int_0^t \int_R K(x-y, t-s) M(s)\, dy\, ds = k\int_0^t M(s)\, ds$$

which is impossible unless $M(t) \equiv 0$. Therefore, $u = v$ and solutions are unique, completing the proof.

The preceding existence and uniqueness theorem has a very strict hypothesis, namely the global Lipschitz condition on the reaction term $f(u)$. The reaction term $f(u) = u(1-u)$ for the Fisher equation, for example, would not satisfy this condition since in this case

$$|f(u) - f(v)| = |u - u^2 - v + v^2| = |1 - u - v|\,|u - v|$$

and the right side cannot be bounded by $k|u - v|$ for all u and v in R. However if u and v are restricted, the factor $|1 - u - v|$ can be bounded and we obtain a *local form* of a Lipschitz condition. Thus we seek to formulate an existence–uniqueness theorem by weakening the hypothesis in Theorem 2 to a local Lipschitz condition, thereby obtaining a theorem with broader applicability. We also take the opportunity to introduce some basic concepts from elementary functional analysis that will permit us to resolve existence–uniqueness questions in a variety of contexts that includes many of the equations occurring in applied mathematics. As mentioned earlier, this unifying approach shows the power of abstraction in mathematics and should convince even the most skeptical of the value of abstract methods.

Definitions and Terminology

We recall from elementary linear algebra that a *real linear space* (or vector space) is a set of objects (numbers, vectors, matrices, functions, or whatever) on which an addition is defined and in which multiplication of the objects by scalars (real numbers) is defined. That is, if X denotes the set of objects and u and v are in X, then $u + v$ is defined and belongs to X, and au is defined and belongs to X for any u in X and any real number a. The addition must satisfy the usual rules of addition: commutivity, associativity, existence of a zero element, and existence of an inverse element $(-u)$ corresponding to each u in X. More specifically:

(i) $u + v = v + u$, $u + (v + w) = (u + v) + w$.
(ii) There is an element 0 in X such that $u + 0 = u$ for all u in X.
(iii) For each u in X there is an inverse (denoted by $-u$) in X for which $u + (-u) = 0$.

Similarly, the scalar multiplication must satisfy a minimal set of rules:

(iv) $1u = u$ for all u in X.
(v) $(ab)u = a(bu)$, $(a + b)u = au + bu$, $a(u + v) = au + av$, for all u and v in X and all a and b in R.

Any set of objects on which addition and scalar multiplication is defined and the rules (i)–(v) hold is called a real *linear space*. Thus a linear space is a set on which there is well-defined algebraic structure.

Given a linear space X with its algebraic structure, one may also impose geometry, that is, a measure of the size of a given element of the space. If to each u in a linear space X there is associated a nonnegative number, denoted by $\|u\|$, satisfying the three properties

$$\|u\| = 0 \text{ if, and only if, } u = 0; \qquad \|au\| = |a|\,\|u\| \quad \text{for } a \in R \text{ and } u \in X;$$
$$\|u + v\| \le \|u\| + \|v\| \quad \text{for } u, v \in R \tag{23}$$

then X is called a *normed linear space*, and the measure of size $\|\cdot\|$ is called a *norm* on X. In a natural way, a *distance function* (i.e., the distance between two elements u and v) may be defined in a normed linear space by the formula

$$\operatorname{dist}(u, v) = \|u - v\|$$

Common examples of normed linear spaces are:

(a) $X =$ the real numbers R, with $\|x\| = |x|$; the distance between x and y is $|x - y|$.
(b) $X = R^n = \{(x_1, x_2, \dots, x_n) | x_i \in R\} =$ the set of all ordered n-tuples of real numbers; one norm is the *Euclidean* norm $\|(x_1, \dots, x_n)\| = (x_1^2 + \cdots + x_n^2)^{1/2}$; two other norms are $\|(x_1, \dots, x_n)\| = |x_1| + \cdots + |x_n|$ and $\|(x_1, \dots, x_n)\| = \max\{|x_1|, \dots, |x_n|\}$. Thus it is possible to define different norms on a given linear space.
(c) $X =$ the set of real n by n matrices, with $\|M\| = \Sigma|m_{ij}|$, where $M = (m_{ij})$.

Many of the normed linear spaces of interest in applied mathematics are spaces of functions. For example:

(d) $X =$ the set of continuous functions defined on a closed interval $[a, b]$. This linear space is usually denoted by $C[a, b]$. It can be made into a normed linear space by defining $\|u\| = \max|u(x)|$, where the maximum is taken over $a \le x \le b$. Another norm on $C[a, b]$ is $\|u\| = \int_a^b |u(x)|\, dx$.
(e) $X =$ the set of bounded, uniformly continuous functions on R. This linear space is often denoted by BC. A norm on BC is $\|u\| = \sup|u(x)|$, $x \in R$.

The reader should verify that some of the norms defined in the preceding examples do indeed satisfy the conditions (23).

It is a fundamental property of the real number of system that Cauchy sequences converge. That is, if x_n is a sequence of real numbers having the property that for any $\varepsilon > 0$ there exists a positive integer N such that $|x_{n+p} - x_n| < \varepsilon$ for all $n > N$ and all $p > 0$, the sequence x_n must converge to a real number. This fact follows from the completeness axiom of the real number system, which states that every bounded set of real numbers must have a least upper bound (supremum) and a greatest lower bound (infimum). We can generalize this notion to an arbitrary normed linear space X in the following manner. Let u_n be a sequence of elements in a normed linear space X. We say that u_n *converges* to an element u if for any $\varepsilon > 0$ there is a positive integer N such that $\|u_n - u\| < \varepsilon$ for all $n > N$; this type of convergence is also called *norm convergence*, or convergence in the norm topology. We say that u_n is a *Cauchy sequence* in X if, and only if, for any $\varepsilon > 0$ there is a positive integer N such that $\|u_{n+p} - u_n\| < \varepsilon$ for all $n > N$ and $p > 0$. A normed linear space X is said to be *complete* if it has the property that every Cauchy sequence in X converges to an element of X. A complete normed linear space is called a *Banach space*.

Example 3. Consider the normed linear space $C[0,1]$ consisting of continuous functions on $[0,1]$ with the norm $\|u\| = \max|u(x)|$. It is not difficult to prove that $C[0,1]$ with this norm is a Banach space. On the other hand, if the norm is defined by $\|u\| = \int_0^1 |u(x)|\,dx$, then $C[0,1]$ with this norm is not complete (the reader is asked to demonstrate this fact in an exercise at the end of the section); however, $C[0,1]$ with this norm may be completed in much the same way that the rational numbers are completed to obtain the real numbers, namely, by appending to $C[0,1]$ the limits of all Cauchy sequences of functions in $C[0,1]$. The resulting space is denoted by $L^1[0,1]$ and is called the space of Lebesgue integrable functions on $[0,1]$. We remark that limits of Cauchy sequences in $C[0,1]$ may not be Riemann integrable, and therefore the integral in the definition of the norm must be generalized to the Lebesgue integral. In a similar manner, the $L^p[a,b]$ spaces are Banach spaces that arise by completing the continuous functions in the norm

$$\|u\|_p = \left[\int_a^b |u(x)|^p\,dx\right]^{1/p}$$

Now we state the fundamental result which will be the vehicle to an existence theorem for nonlinear partial differential equations. Let X be a Banach space and suppose that $\phi\colon X \to X$ is a mapping on X. If ϕ has the property that $\|\phi(u) - \phi(v)\| \le \alpha\|u - v\|$ for all u and v in X, for some positive constant $\alpha < 1$, then ϕ is called a *contraction mapping*. Then we have:

Theorem 3 (*Banach Fixed-Point Theorem*). If $\phi\colon X \to X$ is a contraction mapping on a Banach space X, then ϕ has precisely one fixed point [i.e., there exists a unique $u \in X$ such that $\phi(u) = u$].

Proof. Earlier in this section we proved this theorem for a contraction mapping on the reals. The general proof for a Banach space is exactly the same argument with the absolute values replaced by norms.

Remark. It often occurs in applications that ϕ is not a contraction on the entire Banach space, but rather only on a closed subset in the space. (A set G in a Banach space is *closed* if any convergent sequence in G has its limit in G.) It is straightforward to prove that the fixed-point theorem remains valid on closed subsets of a Banach space.

Using the machinery of a Banach space and the fixed-point theorem, we now formulate and prove an existence theorem for the semilinear problem (1)–(2), where only a local Lipschitz condition is required. First we establish some notation. Consider a function $u = u(x,t)$. For each *fixed* time t we can regard u as a function of x (on R) representing a slice of the surface $u(x,t)$ at time t, or equivalently, a time snapshot of the wave. The underlying Banach space B in the following formulation will specify the type of functions that these time snapshots or wave profiles can be. Therefore, let B be the Banach space of all bounded, continuous functions $u(x,t)$ on R (t fixed), and let $\|u(\cdot,t)\|_B$ denote the norm of a function $u(x,t)$ in B. Specifically, we take

$$\|u(\cdot,t)\|_B = \sup_{x\in R} |u(x,t)| \qquad \text{for } t \text{ fixed} \tag{24}$$

This norm is called the *sup norm* (this norm is usually denoted with the subscript ∞ rather than B; however, we shall maintain this generic notation because the following results can be extended to other Banach spaces with other norms). Now, let T be positive and let $C([0,T]; B)$ be the set of all continuous functions defined on the interval $0 \le t \le T$ that have values in the Banach space B. That is, to each $t \in [0,T]$ we associate a bounded, continuous function $u(x,t)$ on $x \in R$ (t fixed) which is an element of the Banach space B. The set $C([0,T]; B)$, whose elements will be denoted by either u or $u(x,t)$, is a Banach space in its own right with norm

$$\|u\| = \sup_{t\in[0,T]} \|u(\cdot,t)\|_B \tag{25}$$

One must take care not to confuse the various objects that we have defined; there are two Banach spaces here, B and $C([0,T], B)$. The objects in B are denoted by either $u(x,t)$ or $u(\cdot,t)$ and are considered to be a functions of x, with t fixed; the norm in B is denoted with the subscript B, and we prefer to use the norm notation $\|u(\cdot,t)\|_B$ with the generic *dot* argument because the spatial variable has been "sup-ed out" in taking the norm. Regarded as an object in $C([0,T], B)$, $u = u(x,t)$ is a function of both x and t, such that for each t on $[0,T]$ the association of $u(x,t)$ in B is continuous; in this space there is no subscript on the norm symbol. Another way of thinking about the objects in $C([0,T], B)$ is as continuous curves in the Banach space B; that is, to each t in the parameter interval $[0,T]$ we associate an element $u(\cdot,t)$ of B. Interpreted still differently, to each t we associate a wave profile in B, and the totality of all the wave profiles forms the surface $u = u(x,t)$. We also introduce the convolution operation

$$(K * u)(x, t) = \int_R K(x - y, t) u(y, t)\, dy$$

where $K(x, t)$ is the diffusion kernel. $K * u$ is called the convolution of K with u, and $(K * u)(\cdot, t)$ also belongs to B. Perhaps the greatest difficulty in proving the existence theorem is sorting out the notation.

Theorem 4 (*Local Existence*). Consider the initial value problem (1)–(2) where $u_0 \in B$ and f satisfies the conditions

(i) $f \in C^1(R)$.

(ii) $f(0) = 0$, and for each fixed t in $[0, T]$, $f(u(x, t)) \in B$ for each $u(x, t) \in B$.

(iii) For any $M > 0$ there exists a constant k, depending only on M, such that

$$\| f(u(\cdot, t)) - f(v(\cdot, t)) \|_B \le k \| u(\cdot, t) - v(\cdot, t) \|_B$$

for all $t \in [0, T]$ and all $u(x, t)$ and $v(x, t)$ in B with $\|u(\cdot, t)\|_B \le M$ and $\|v(\cdot, t)\|_B \le M$.

Then there exists $t_0 > 0$, where t_0 depends only on f and $\|u_0(\cdot)\|_B$, such that the initial value problem (1)–(2) has a unique solution $u = u(x, t)$ in $C([0, t_0]; B)$, and $\|u\| \le 2\|u_0(\cdot)\|_B$.

Proof. We shall define a closed subset G of the Banach space $C([0, t_0]; B)$ and show that the mapping

$$\phi(u)(x, t) = \int_R K(x - y, t) u_0(y)\, dy + \int_0^t \int_R K(x - y, t - s) f(u(y, s))\, dy\, ds \tag{26}$$

is a contraction mapping on G. Then we may apply the Remark following the Banach fixed-point theorem to produce a solution to $u = \phi(u)$, which by our previous remarks is a solution to the initial value problem. To this end, define

$$G = \{u \in C([0, T]; B): \| u(\cdot, t) - (K * u_0)(\cdot, t) \|_B \le \| u_0(\cdot) \|_B, \text{ for } 0 \le t \le t_0\}$$

where $t_0 = 1/2k$. The set G is closed and nonempty (e.g., zero is in G). Also, the defining property of G, the triangle inequality, and the fact that

$$\|(K * u)(\cdot, t)\|_B \le \| u(\cdot, t) \|_B \tag{27}$$

imply that $\|u(\cdot,t)\|_B \le 2\|u_0(\cdot)\|_B$, whereupon taking the supremum on t gives

$$\|u\| \le 2\|u_0(\cdot)\|_B \tag{28}$$

This proves the last statement in the theorem. Now, we have from (ii), for any $0 \le t \le t_0$,

$$\begin{aligned}\|f(u(\cdot,t)) - f(v(\cdot,t))\|_B &\le k\|u(\cdot,t) - v(\cdot,t)\|_B \\ &\le k\|u - v\|\end{aligned} \tag{29}$$

Because k depends only on the sup norm of u and v, and of course on f, it is clear that t_0 depends only on f and the sup norm of u_0, by virtue of the inequality (28).

The fact that ϕ maps G into G [i.e., $\phi(u) \in G$ if $u \in G$] follows from the sequence of inequalities

$$\begin{aligned}\|\phi(u)(\cdot,t) - (K * u_0)(\cdot,t)\|_B &= \left\|\int_0^t (K * f(u))(\cdot,t-s))\,ds\right\|_B \\ &= \int_0^t \|(K * f(u))(\cdot,t-s))\|_B\,ds \\ &\le \int_0^t \|f(u(\cdot,s))\|_B\,ds \\ &\le \int_0^t 2k\|u_0(\cdot)\|_B\,ds \\ &= 2kt\|u_0(\cdot)\|_B \\ &< \|u_0(\cdot)\|_B\end{aligned}$$

Now we prove that ϕ is a contraction. We have

$$\begin{aligned}&\|\phi(u)(\cdot,t) - \phi(v)(\cdot,t)\|_B \\ &\quad\le \sup_{x\in R}\int_0^t\int_R K(x-y,t-s)|f(u(y,s)) - f(v(y,s))|\,dy\,ds \\ &\quad\le \int_0^t \|f(u(\cdot,s)) - f(v(\cdot,s))\|_B\,ds \\ &\quad\le \int_0^t k\|u - v\|\,ds \\ &\quad\le kt_0\|u - v\| = \frac{\|u - v\|}{2}\end{aligned}$$

Taking the supremum over $t \in [0, T]$ then gives

$$\|\phi(u) - \phi(v)\| \leq \tfrac{1}{2}\|u - v\|$$

which completes the proof that ϕ is contraction on the closed subset G of the Banach space $C([0, T]; B)$. By the Remark following the fixed-point theorem it follows that there is a unique fixed point in G [i.e., a unique solution to the initial value problem (1)–(2) in G].

It remains to show that there are no solutions outside the set G. This fact results from the following general argument. If u and v are two solutions lying in $C([0, T]; B)$ that satisfy the same initial condition, then

$$u(x,t) - v(x,t) = \int_0^t \int_R K(x-y, t-s)[f(u(y,s)) - f(v(y,s))]\, dy\, ds$$

from which it easily follows that

$$\|u(\cdot, t) - v(\cdot, t)\|_B \leq k\int_0^t \|u(\cdot, s) - v(\cdot, s)\|_B\, ds$$

Multiplying this inequality by $\exp(-kt)$ gives the result that

$$\frac{d}{dt}\{e^{-kt}\|u(\cdot, t) - v(\cdot, t)\|_B\} \leq 0$$

which in turn implies that $\|u(\cdot, t) - v(\cdot, t)\|_B = 0$, or $u = v$. Consequently, the theorem is proved.

The theorem is only a local existence result, guaranteeing a solution for $0 \leq t \leq t_0$, for some t_0. Under certain conditions we may extend the solution to any finite time T. We have the following result.

Theorem 5 (*Global Existence*). Under the same hypotheses of the last theorem, assume in addition that the solution is *a priori* bounded in the sup norm on $0 \leq t \leq T$; that is, assume there is a constant $C > 0$ depending only on $\sup_{x \in R}|u_0(x)|$ such that if u is any solution of (1)–(2) in $0 \leq t \leq T$, then $\sup_{x \in R}|u(x, t)| \leq C$. Then the solution to (1)–(2) exists for all t in $[0, T]$, and $u(x, t) \in B$. Here T may be infinity, giving global existence.

Proof. The local theorem guarantees a solution u on $[0, t_0]$. Then we can apply the local theorem again with initial condition $u(x, t_0)$ to get a solution on $[t_0, 2t_0]$. Continuing in this manner we can obtain, after a finite number of steps, a solution on $[0, T]$.

Example. Consider the initial value problem for *Fisher's equation*:

$$u_t - Du_{xx} = u(1-u), \qquad x \in R, \quad t > 0$$
$$u(x,0) = u_0(x), \qquad x \in R$$

Here $f(u) = u(1-u)$. Clearly, $f(0) = 0$, f is smooth, and

$$\sup|f(u) - f(v)| = \sup(|1-u-v|\,|u-v|) \le k \sup|u-v|$$

where $k = 1 + \sup|u| + \sup|v|$, and the sup is taken over $x \in R$. Therefore, the local Lipschitz condition holds. Further, if $u \in B$, so is $f(u)$. The local existence theorem guarantees a solution on $[0, t_0]$, for some positive t_0. To prove existence for all $t > 0$ we would need to have an *a priori* bound on solutions. Such a bound can be obtained from the comparison theorems in the next section.

The proofs above go through easily for *systems* of reaction–diffusion equations of the form

$$\mathbf{u}_t - \mathbf{D}\mathbf{u}_{xx} = \mathbf{f}(\mathbf{u}), \qquad x \in R, \quad t > 0$$
$$\mathbf{u}(x,0) = \mathbf{u}_0(x), \qquad x \in R$$

where $\mathbf{u}$ and $\mathbf{f}$ are n-vectors, and $\mathbf{D} = \operatorname{diag}(d_1, \ldots, d_n)$ is a diagonal matrix with positive diagonal entries (see Smoller [1983, Chap. 14, Sec. A]). Now the Banach space B is a space of vector-valued functions on R with values in R^n, and each component is bounded, continuous function. The equivalent integral equation is

$$\mathbf{u}(x,t) = \int_R \mathbf{K}(x-y,t)\mathbf{u}_0(y)\,dy + \int_0^t \int_R \mathbf{K}(x-y,t-s)\mathbf{f}(\mathbf{u}(y,s))\,dy\,ds$$

where $\mathbf{K}(x,t)$ is a diagonal matrix with diagonal elements

$$k_i(x,t) = (4\pi d_i t)^{-1/2} \exp\left(-\frac{x^2}{4d_i t}\right), \qquad i = 1, \ldots, n$$

EXERCISES

1. Consider the linear initial value problem

$$u_t - Du_{xx} = au, \quad x \in R, \quad t > 0; \qquad u(x,0) = u_0(x), \quad x \in R$$

where $u_0 \in B$, where a and D are positive constants, and where B is the space of all continuous, bounded functions on R with the sup norm. Prove that a solution exists for all $t \geq 0$.

2. Consider the initial value problem

$$u_t - Du_{xx} = u^p, \quad x \in R, t > 0; \qquad u(x,0) = u_0(x), \quad x \in R$$

where $p > 1$, and u_0 is a bounded continuous function on R.
   (a) Use the local existence theorem to prove that a unique solution exists for $0 \leq t \leq t_0$.
   (b) Show that a global solution does not, in general, exist. *Hint*: Take $u_0(x) = u_0 =$ constant.
3. Prove that the linear space $C[0,1]$ with norm $\|u\| = \int_0^1 |u(x)|\, dx$ is not a Banach space by considering the (Cauchy) sequence of continuous functions

$$u_n(x) = \begin{cases} 0 & \text{if } 0 \leq x \leq \dfrac{1}{2} \\ 1 & \text{if } \dfrac{1}{2} + \dfrac{1}{n} \leq x \leq 1 \\ \text{linear} & \text{if } \dfrac{1}{2} \leq x \leq \dfrac{1}{2} + \dfrac{1}{n} \end{cases}$$

4. Consider the convection–diffusion problem

$$u_t - Du_{xx} = F(u)_x, \qquad x \in R, \quad t > 0 \qquad (D > 0)$$
$$u(x,0) = u_0(x), \qquad x \in R$$

   (a) Show that the integral representation of a solution to this initial value problem is given by

$$u(x,t) = \int_R K(x-y,t)u_0(y)\, dy$$
$$+ \int_0^t \int_R K_x(x-y,t-s)F(u(y,s))\, dy\, ds$$

   where K is the diffusion kernel.
   (b) Let $F(u) = u^2/2$. Formulate and prove a local existence theorem for the initial value problem.

6.4 MAXIMUM PRINCIPLES AND COMPARISON THEOREMS

In Section 6.3 we dealt with the question of existence of solutions to the initial value problem for certain nonlinear reaction–diffusion equations. Now we wish to formulate two other theoretical concepts usually associated with diffusion problems and parabolic equations: maximum principles and comparison theorems.

A maximum principle is a statement about where a solution to a given problem takes on its maximum value. There are correspondingly minimum principles. Generally, these problems can be initial value problems or boundary value problems, or mixed problems. For example, diffusion processes, by their very physical nature, tend to smear out the density function u, and this precludes "clumping" of the solution in the interior of the given spacetime domain; therefore, diffusion problems often have solutions whose maximum occurs on the boundary of the domain. Comparison theorems, on the other hand, are statements that compare the solutions to two similar problems, say differing only in their boundary or initial values. If one of the problems can be solved, a comparison theorem can give bounds on the solution of the problem that perhaps cannot be solved. This information can be used, for example, to obtain the asymptotic behavior of the solution as t gets large or produce a priori bounds which guarantee the existence of global solutions (see Section 6.3).

Maximum Principles

As stated above, a maximum principle is a theorem indicating where a solution to a given PDE must assume its maximum value. A simple example involving the linear diffusion equation, or heat equation, will well illustrate the concept.

Example 1. Consider a solution $u = u(x,t)$ to the equation

$$Hu \equiv u_t - ku_{xx} = f(x,t) \tag{1}$$

on the bounded spacetime domain D shown in Fig. 6.11. We assume that D is open (i.e., D does not include any of its boundary), and its boundary consists of a segment S (excluding the endpoints) at time $t = T > 0$ and a lower boundary B which does include the endpoints of S. We assume that D is also convex (contains the line segment connecting any two points in the domain) and lies strictly in the upper half plane $t > 0$. Further, assume that $u \in C(\overline{D})$ and $u \in C^2(D)$, where $\overline{D} = D \cup S \cup B$ is the closure of D, that is, D along with its boundary. We note that portions of $\overline{D}$ can lie on the initial time line $t = 0$; for example, D could be a rectangular region with a segment of B lying along $t = 0$. Now, if the source term f in (1) is strictly negative, or precisely,

$$Hu = f < 0, \qquad (x,t) \in D \tag{2}$$

it is easy to show that a local maximum of u cannot occur in D or on S. By

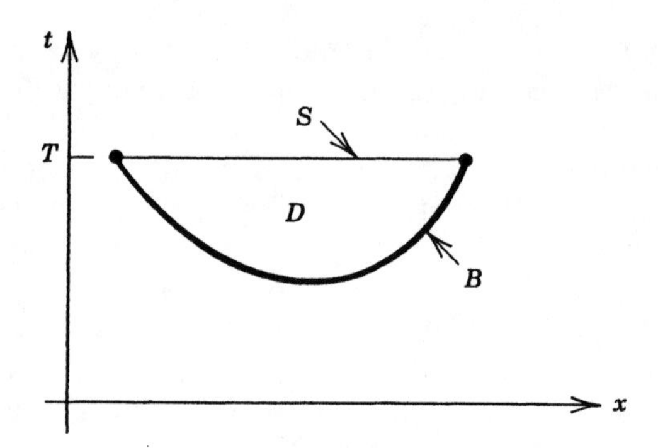

Figure 6.11. Open domain D.

way of contradiction, assume there is such a point (x_0, t_0) in $D \cup S$. Then, from calculus, we must have

$$u_x(x_0, t_0) = 0, \qquad U_{xx}(x_0, t_0) \le 0, \qquad u_t(x_0, t_0) \ge 0$$

But then $Hu \ge 0$ at (x_0, t_0), contradicting (2). Consequently, if the source term is strictly negative, the maximum value of any solution to (1) must occur on B, the lower portion of the boundary.

We want to extend the notions in Example 1 to other cases with more general differential operators or perhaps allowing equality in (2). Therefore, consider the operator P defined by

$$Pu = a(x,t)u_{xx} + b(x,t)u_x - u_t \tag{3}$$

where a and b are continuous functions on $\overline{D}$. We say that P is *uniformly parabolic* in D if $a(x,t) \ge m > 0$ in D, for some constant m. First, we remark that the result of Example 1 can be extended at once to the operator P. If $Pu > 0$, a local maximum of u cannot occur in D or on S since, at such a point, $u_{xx} \le 0$, $u_x = 0$, and $u_t \ge 0$, which violate $Pu > 0$.

We now state and prove a general theorem, called the *weak maximum principle*, which essentially states that functions u satisfying $Pu + c(x,t)u \ge 0$ on D, where $c \le 0$ on D, must assume their maximum on B. The result is called *weak* since it does not preclude the maximum also occurring at an interior point in D. There is also a *strong maximum principle*, which we shall formulate later, but not prove, that does preclude a maximum occurring in D unless u is constant up to that point of time.

Theorem 1 (*Weak Maximum Principle*). Let $u \in C(\overline{D}) \cap C^2(D)$ be a solution of the equation

$$Pu + cu = f(x,t), \qquad (x,t) \text{ in } D \tag{4}$$

where c and f are $C(D)$, $c(x,t) \le 0$, $f(x,t) \ge 0$ in D, and P is uniformly parabolic in D. Then u assumes its positive maximum on B, provided that it exists.

Proof. The idea of the proof is to assume the contrary and construct a function that violates the usual calculus conditions for a maximum. Most proofs of maximum principles use this strategy in one way or another, but some proofs require clever constructions.

Suppose that u has a positive maximum in $\overline{D}$. To show that this maximum must occur somewhere on B we first define the auxiliary function

$$v(x,t) = u(x,t) - \varepsilon t$$

If u has a positive maximum in $\overline{D}$, so does v for ε sufficiently small. By way of contradiction, assume that the positive maximum of v occurs at a point (x_0, t_0) in $D \cup S$. Then

$$v_x(x_0,t_0) = 0, \qquad v_{xx}(x_0,t_0) \le 0$$

Now calculate v_t. We have

$$\begin{aligned} v_t &= u_t - \varepsilon = au_{xx} + bu_x + cu - f - \varepsilon \\ &= av_{xx} + bv_x + c(v + \varepsilon t) - f - \varepsilon \\ &\le cv + \varepsilon tc - \varepsilon \qquad \text{at } (x_0,t_0) \\ &\le -\varepsilon \end{aligned}$$

Now choose a number $d > 0$ such that $v_t(x_0,t) \le -\varepsilon/2$ for $t_0 - d \le t \le t_0$. Then integration gives

$$v(x_0,t_0) - v(x_0,t_0 - d) \le \int_{t_0-d}^{t_0} -\frac{\varepsilon}{2}\,dt = -\frac{\varepsilon d}{2} < 0$$

This contradicts the assumption that v has a positive maximum at (x_0, t_0). So the positive maximum of v must occur on B. Then we can make the following estimates:

$$\begin{aligned} \max_{\overline{D}} u &= \max_{\overline{D}} (v + \varepsilon t) \le \max_{\overline{D}} v + \varepsilon T \\ &= \max_{B} v + \varepsilon T = \max_{B} (u - \varepsilon t) + \varepsilon T \le \max_{B} u + \varepsilon T \end{aligned}$$

Consequently, since ε is arbitrary, $\max_{\overline{D}} u \le \max_B u$, and therefore the maximum must occur someplace on B, completing the proof.

If $c = 0$ and $f = 0$, we have the linear homogeneous equation $Pu = 0$. If u is a solution to this equation, so is u + constant, for any constant. Thus for this equation we may always assume that u has a positive maximum, and this hypothesis in the weak maximum principle may be deleted. We therefore have the following corollary.

Corollary 1. Let $u \in C(\overline{D}) \cap C^2(D)$ be a solution of

$$Pu = 0 \qquad \text{on } D$$

where P is uniformly parabolic on D. Then u assumes its maximum and minimum on B.

We may now take the analysis one step further and prove the following theorems with no assumption on the sign of the function $c = c(x, t)$. This theorem will yield simple proofs of uniqueness theorems, which will be requested in the Exercises.

Theorem 2. Let $u \in C(\overline{D}) \cap C^2(D)$ be a solution of

$$Pu + cu = f(x,t) \qquad \text{on } D$$

$$u = 0 \qquad \text{on } B$$

where $f \geq 0$ on D. Then $u = 0$ in D.

Proof. First consider the case $c \leq 0$ in D. By way of contradiction assume that u is strictly positive at some point of D. Then u has a positive maximum in D, and by the weak maximum principle u must have a positive maximum on B, contradicting the hypothesis that $u = 0$ on B. If $u < 0$ at some point in D, then in the same way it must have a negative minimum on B, again a contradiction. So in the case $c \leq 0$ we must have $u = 0$ in D.

Now assume that $c \leq 0$ is not satisfied. In this case choose a constant $\lambda \geq c$ in D and set $u = we^{\lambda t}$. Then w satisfies the equation

$$Pw + (c - \lambda)w = fe^{-\lambda t} \qquad \text{in } D$$

$$w = 0 \qquad \text{on } B$$

But now $c - \lambda \leq 0$, and therefore the weak maximum principle may be applied to get $w = 0$ in D. Hence we must have $u = 0$ in D, which completes the argument.

As we mentioned earlier, there is a strong maximum principle that states the consequences of having a maximum inside D. We state, without proof, a simplified version of this theorem that applies to the simply connected, convex region D shown in Figure 6.11. An easily accessible, general version is stated and proved in Smoller [1983], Protter and Weinberger [1967], or Friedman [1964].

Theorem 3 (*Strong Maximum Principle*). Let $u \in C(\overline{D}) \cap C^2(D)$ be a solution of the equation

$$Pu + c(x,t)u = f(x,t) \qquad \text{in } D$$

where P is uniformly parabolic in D, where $c \le 0$ and $f \ge 0$ (respectively, $f \le 0$) in D, and c and f are continuous in D. Let m be a nonnegative maximum (respectively, nonpositive minimum) of u in D, and suppose that $u(x_0, t_0) = m$ at some point of D. Then $u = m$ at all points (x, t) in D with $t < t_0$.

Remark. The strong maximum principle is also valid on unbounded domains (that are open and connected), provided that the functions a, b, c, and f are continuous and bounded on D, and the terms *max* and *min* in the statement of the theorem above are replaced by *sup* and *inf*.

Example 2. In general, as the following example shows, maximum principles do not hold for *systems* of diffusion equations. Consider the system

$$u_{xx} - u_t \ge 0, \qquad v_{xx} - 9u_x - v_t \ge 0$$

on the unit square $0 \le x \le 1$, $0 \le t \le 1$. A solution is

$$u(x,t) = -\exp(x+t), \qquad v(x,t) = t - 4(x - 0.5)^2$$

Here u and v are negative on the bottom and side boundaries, yet $v > 0$ along $x = \frac{1}{2}$.

Comparison Theorems

In the preceding discussion we alluded to the fact that a comparison theorem is a result that compares solutions to similar problems. We are now in a position to prove such a theorem for certain nonlinear partial differential equations. For example, consider the semilinear equation

$$u_t - a(x,t)u_{xx} - b(x,t)u_x = f(x,t,u) \tag{5}$$

on the convex domain D shown in Figure 6.11. We assume that a and b are continuous functions on $\overline{D}$ and $a(x,t) \ge \mu > 0$. The reaction term f is assumed to be continuously differentiable on $D \times R$. Now let $v = v(x,t)$ and $u = u(x,t)$ be two solutions of (5) in D, and suppose that $u \le v$ on the lower boundary B of D. Then it follows that $u \le v$ on the entire domain D. To prove this fact, let $w = u - v$. Then $w \le 0$ on B and

$$aw_{xx} + bw_x - w_t = f(x,t,v) - f(x,t,u) \qquad \text{on } D$$

Now, by the mean value theorem,

$$f(x,t,u) - f(x,t,v) = f_u(x,t,u^*)w,$$
$$u^* = \theta u + (1-\theta)v, \quad 0 < \theta < 1$$

Consequently,

$$aw_{xx} + bw_x + f_u w - w_t = 0 \qquad \text{on } D$$
$$w \le 0 \qquad \text{on } B$$

We now have a form of the problem for w in which we may apply the weak maximum principle. Suppose first that $f_u \le 0$ on D. Directly from the weak maximum principle we conclude that $w \le 0$ in D (otherwise, if w were positive at some point of D, a positive maximum would have to occur on B, a contradiction). If the coefficient f_u does not satisfy the condition $f_u \le 0$ on D, choose a number λ such that $\lambda \ge f_u$ in D and let $w = W\exp(\lambda t)$. Then W satisfies the problem

$$aW_{xx} + bW_x + (f_u - \lambda)W - W_t = 0 \qquad \text{on } D$$
$$W \le 0 \qquad \text{on } B$$

Because $f_u - \lambda \le 0$ we may now apply the weak maximum principle to conclude that $W \le 0$ on D and thus ascertain that $w \le 0$ on D in this case as well. Here is a formal statement of what we proved.

Theorem 4. Let $u, v \in C(\overline{D}) \cap C^2(D)$ be two solutions of the semilinear equation (5) under the stated assumptions. If $u \le v$ on B, then $u \le v$ in D.

It is not difficult to generalize and extend this result to general nonlinear equations. To this end we require one bit of terminology. Let us assume that $F(x,t,u,p,r)$ is a given continuously differentiable function of its five arguments. We say that F is *elliptic* with respect to a function $u = u(x,t)$ at a point (x,t) if $F_r(x,t,u(x,t),u_x(x,t),u_{xx}(x,t)) > 0$.

Theorem 5. Let D be the domain in Figure 6.11, and consider the equation

$$L[u] = F(x,t,u,u_x,u_{xx}) - u_t = f(x,t) \qquad \text{in } D \tag{6}$$

where $F = F(x,t,u,p,r)$ is a given continuously differentiable function of its five arguments, and f is continuous on $\overline{D}$. Suppose that u, w, and W are continuous in $\overline{D}$ and C^2 on D and that u is a solution of (6). Further, assume that F is elliptic on D with respect to the functions $\theta w + (1-\theta)u$ and $\theta W + (1-\theta)u$, where $\theta \in [0,1]$. If

$$L[W] \le f(x,t) \le L[w] \qquad \text{in } D$$

and

$$w \le u \le W \quad \text{on } B$$

then

$$w \le u \le W \quad \text{in } D \tag{7}$$

Proof. First, let $v = w - u$. Then $v \le 0$ on B and we must show that $v \le 0$ in D. We have

$$f(x,t) = F(x,t,u,u_x,u_{xx}) - u_t \le F(x,t,u,w,w_x,w_{xx}) - w_t$$

or

$$F(x,t,w,w_x,w_{xx}) - F(x,t,u,u_x,u_{xx}) - v_t \ge 0 \quad \text{on } D$$

By Taylor's theorem we have

$$F(x,t,w,w_x,w_{xx}) - F(x,t,u,u_x,u_{xx}) = F_u v + F_p v_x + F_r v_{xx}$$

where the partial derivatives F_u, F_p, and F_r are evaluated at $(x, t, \theta w + (1-\theta)u, \theta w_x + (1-\theta)u_x, \theta w_{xx} + (1-\theta)u_{xx})$ and $0 < \theta < 1$ Therefore, we have

$$F_r v_{xx} + F_p v_x + F_u v - v_t \ge 0 \quad \text{on } D$$

This equation is now in the form, with the coefficients being continuous functions of x and t on D, where we can apply the weak maximum principle. As in the proof of the Theorem 4, the principle can be applied regardless of the sign of F_u. All we require is that F_r, evaluated at the point indicated above, is positive and bounded away from zero on D. Thus the weak maximum principle implies that $v \le 0$ in D, giving the leftmost inequality in (7). We leave the right inequality as an exercise (see Exercise 2).

EXERCISES

1. Consider the nonlinear diffusion problem

$$\begin{aligned} (K(u)u_x)_x - u_t &= 0, \quad 0 < x < 1, \ 0 < t < T \\ u(x,0) &= 1 + x(1-x), \quad 0 < x < 1; \\ u(0,t) &= u(1,t) = 1 \quad \text{for } 0 < t < T \end{aligned}$$

where the diffusion coefficient $K = K(u)$ is a continuously differentiable function and $K(u) > 0$. Show that if u is any solution to this problem, then $1 \le u(x,t) \le \frac{5}{4}$.

2. Prove the rightmost inequality in (7) of the comparison theorem, Theorem 5.
3. Let D be the infinite domain $R \times (0, T)$, and let u be continuous on $\overline{D}$ and C^2 on D, and assume that

$$u_t - u_{xx} = 0 \qquad \text{in } D$$

Let $M = \sup_D u$. Prove that if M is finite, $\sup_{x \in R} u(x, 0) = M$. *Suggestion:* Proceed by contradiction and then apply the weak maximum principle on a sufficiently large bounded domain to the function $v = u - \varepsilon(2t + x^2)$.
4. Consider the initial–boundary value problem

$$u_{xx} + 2u - tu_t = 0, \qquad 0 < x < \pi, \quad t > 0$$

$$u(x, 0) = 0, \quad 0 < x < \pi; \qquad u(0, t) = u(\pi, t) = 0, \quad t > 0$$

Using the method of separation of variables, show that this problem has nonzero solutions. Does the maximum principle apply? Discuss.
5. Consider the equation

$$x^2 u_{xx} - u_t = 0, \qquad -1 < x < 1, \quad 0 < t < \tfrac{1}{2}$$

Does the maximum principle apply? *Hint:* Consider $u = -x(x + 2t)$.
6. (a) Let D be the infinite strip $-\infty < x < \infty$, $0 < t < T$, and consider the differential operator P defined by

$$Pu = a(x, t)u_{xx} + b(x, t)u_x + c(x, t)u - u_t$$

where $a(x, t) \geq \mu > 0$ in D, $c(x, t) \leq 0$ in D, and a, b, and c are continuous functions on D. Prove that if $Pu > 0$ in D, or if $Pu \geq 0$ and $c < 0$ in D, then u cannot have a positive maximum in D.

(b) Formulate and prove a corresponding theorem regarding a negative minimum.
7. Prove the following one-dimensional version of a comparison theorem. Let f and g be continuous functions on $[a, b]$ with continuous second derivatives on (a, b). Assume that $f(a) \geq g(a)$, $f(b) \geq g(b)$, and $g''(x) \geq f''(x)$ for x in (a, b). Prove that $f \geq g$ on (a, b).

6.5 ENERGY ESTIMATES AND ASYMPTOTIC BEHAVIOR

In this section we introduce an important technique that is applicable to both parabolic and hyperbolic equations; this technique and its variations, which are collectively called energy methods, allow us to obtain bounds on certain quantities associated with the solution, prove uniqueness, show that solutions blow up, and obtain other important information about the behavior of solutions. In Chapter 8 we show that similar techniques are applicable to elliptic problems. One vehicle for establishing energy estimates is integration by parts, one of the basic techniques in partial differential equations; another is a fundamental set of inequalities that is developed in the sequel.

Let $u = u(x, t)$ be the solution of a given evolution problem on the domain $0 \le x \le 1$, $t \ge 0$. The quantity $E(t)$, defined by

$$E(t) \equiv \int_0^1 u^2(x,t)\, dx$$

is called the *energy* at time t; it is the total area under the wave profile squared. One of the important problems for solutions of evolution equations is to obtain bounds on $E(t)$. For example, if we could obtain an estimate of the form $E(t) \le C/t$, where C is a positive constant, we could conclude that the energy decays like $1/t$, and in fact, goes to zero as t gets large. In a generalized sense, this would mean that the solution u itself would have to go to zero asymptotically. Another quantity of interest is the integral of the square of the gradient, that is,

$$Q(t) \equiv \int_0^1 u_x^2(x,t)\, dx$$

If one can show, for example, that $Q(t)$ tends to zero as t goes to infinity, then u_x has to go to zero in a generalized sense and thus u must tend to a constant. Results like this are important, for example, in showing that certain reaction–diffusion equations occurring in the biological sciences cannot give rise to spatial patterns involving density variations.

Calculus Inequalities

To obtain energy and gradient estimates, we shall require some basic inequalities. One cannot underestimate the role that inequalities play in the theory of PDEs. Here we introduce only a few of the basic inequalities; a detailed development of other important inequalities, for example the Sobolev inequalities, can be found in the references.

The first result we present is Young's inequality, which permits a product to be bounded by a sum; it is a generalization of the inequality between the arithmetic and geometric mean.

Young's Inequality. Let f and g be positive quantities, and let p and q be positive real numbers satisfying the condition $1/p + 1/q = 1$. Then

$$fg \le \frac{f^p}{p} + \frac{g^q}{q}$$

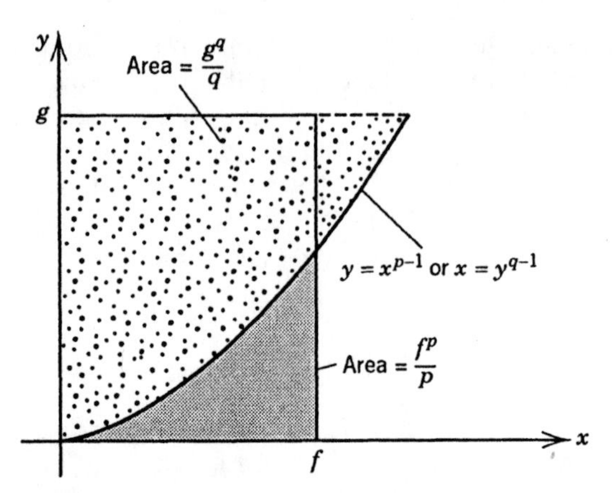

Figure 6.12. Geometric description of Young's inequality.

Example 1. If we take $f = \int_a^b u(x)^2\,dx$ and $g = \int_a^b v(x)^2\,dx$, and $p = q = 2$, then Young's inequality gives

$$\int_a^b u(x)^2\,dx \cdot \int_a^b v(x)^2\,dx \le \tfrac{1}{2}\left[\int_a^b u(x)^2\,dx\right]^2 + \tfrac{1}{2}\left[\int_a^b v(x)^2\,dx\right]^2$$

A geometric proof of Young's inequality follows easily from the graph in Figure 6.12. Here fg is the area of the rectangle, while f^p/p is the area under the curve $y = x^{p-1}$ from 0 to f (shown shaded). Because $q - 1 = 1/(p-1)$, the area under the curve $x = y^{q-1}$ from 0 to g, which is g^q/q, is the dotted area. Geometrically, the sum of the shaded area and the dotted area exceeds the area of the rectangle.

Another important inequality is the Hölder inequality, which is a generalization of the Cauchy–Schwarz inequality encountered in elementary courses; we leave its proof (which follows from Young's inequality) to the reader.

Hölder's Inequality. If p and q are positive real numbers satisfying $1/p + 1/q = 1$, then

$$\int_a^b |u(x)v(x)|\,dx \le \left[\int_a^b |u(x)|^p\,dx\right]^{1/p}\left[\int_a^b |v(x)|^q\,dx\right]^{1/q}$$

provided that the two integrals on the right exist.

Finally, we need the following results, known as Poincaré inequalities. These inequalities relate integrals of functions to integrals of their derivatives when certain boundary conditions hold true. We state the inequalities on the interval $[0, 1]$, but they hold true on any interval with appropriate adjustment of the constants.

Poincaré Inequalities. Let $u = u(x)$ be a twice continuously differentiable function on $[0, 1]$.

(i) If $u(0) = u(1) = 0$, then $\int_0^1 u'(x)^2\,dx \geq \pi^2 \int_0^1 u(x)^2\,dx$.

(ii) If $u'(0) = u'(1) = 0$, then $\int_0^1 u''(x)^2\,dx \geq \pi^2 \int_0^1 u'(x)^2\,dx$.

We shall prove (i) and leave (ii) as an exercise. First, consider the eigenvalue problem

$$u'' + \lambda u = 0 \quad (0 < x < 1), \qquad u(0) = u(1) = 0$$

which has eigenvalues $\lambda_n = n^2\pi^2$ and corresponding orthonormal eigenfunctions $u_n(x) = 2^{1/2} \sin n\pi x$, $n = 1, 2, \ldots$. It is well known that a function u can be expanded in terms of these eigenfunctions, or

$$u(x) = \sum a_n u_n(x)$$

where the a_n are the Fourier coefficients given by

$$a_n = (1/\sqrt{2}) \textstyle\int_0^1 u(x) \sin n\pi x\,dx$$

Therefore, using integration by parts and then the orthogonality property of the $u_n(x)$, that is, $\int_0^1 u_n(x) u_m(x)\,dx = 0$ if $m \neq n$; $= 1$ if $m = n$, we have

$$\begin{aligned}
\int_0^1 u'(x)^2\,dx &= -\int_0^1 u''(x) u(x)\,dx \\
&= -\int_0^1 \left(\sum a_n u_n''\right)\left(\sum a_n u_n\right) dx \\
&= \int_0^1 \left(\sum a_n \lambda_n u_n\right)\left(\sum a_n u_n\right) dx \\
&= \sum a_n^2 \lambda_n \geq \pi^2 \sum a_n^2
\end{aligned}$$

But

$$\int_0^1 u(x)^2\,dx = \int_0^1 \left(\sum a_n u_n\right)\left(\sum a_n u_n\right) dx = \sum a_n^2$$

again using the orthogonality. This completes the argument.

The Poincaré inequalities can be generalized to higher dimensions where they relate the norms (the L_2-norm) of functions to the norms of their gradients (see Smoller [1983], pp. 112ff.).

Energy Estimates

We are now in position to obtain estimates of $E(t)$ and $Q(t)$ for various initial–boundary value problems.

Example 2. To illustrate the various techniques, we consider the semilinear reaction–diffusion equation

$$u_t - u_{xx} = f(u), \qquad 0 < x < 1, \quad t > 0 \tag{1}$$

$$u(x,0) = u_0(x), \qquad 0 < x < 1 \tag{2}$$

$$u_x(0,t) = u_x(1,t) = 0, \qquad t > 0 \tag{3}$$

where the reaction term f is assumed to be continuously differentiable and $\sup_{u \in R} |f'(u)| = M < \infty$. Let

$$Q(t) = \int_0^1 u_x^2 \, dx$$

Then, differentiating Q gives

$$\begin{aligned} Q'(t) &= 2\int_0^1 u_x u_{xt} \, dx = 2u_x u_t|_0^1 - 2\int_0^1 u_{xx} u_t \, dx \\ &= -2\int_0^1 u_{xx}(u_{xx} + f(u)) \, dx \\ &= -2\int_0^1 u_{xx}^2 \, dx + 2\int_0^1 u_x^2 f'(u) \, dx \end{aligned}$$

where we have used integration by parts and applied (1) and (2). We use Poincaré's inequality on the first term on the right side of the last equation, and we bound the second integral by

$$\int_0^1 u_x^2 f'(u) \, dx \le \int_0^1 u_x^2 |f'(u)| \, dx \le M \int_0^1 u_x^2 \, dx$$

Therefore,

$$Q'(t) \le -2\pi^2 \int_0^1 u_x^2 \, dx + 2M \int_0^1 u_x^2 \, dx = 2(M - \pi^2)Q(t) = \mu Q(t)$$

where $\mu \equiv 2(M - \pi^2)$. We have obtained a differential inequality for the quantity $Q(t)$ of the form

$$Q'(t) - \mu Q(t) \le 0$$

Multiplying by the integrating factor $e^{-\mu t}$ gives

$$(Qe^{-\mu t})' \le 0$$

and then integrating from 0 to t yields

$$Q(t) \le Q(0)e^{\mu t}$$

We finally have obtained a bound on $Q(t)$. If $\mu < 0$ (i.e., $M < \pi^2$), then $Q(t)$ decays exponentially as t tends to infinity. So if the nonlinear problem (1)–(3) has a solution $u = u(x,t)$, and if f is a smooth function with the property that $\sup|f'(u)| = M < \pi^2$, then

$$\lim_{t\to\infty} \int_0^1 u_x^2(x,t)\,dx = 0$$

Physically, for example, this result precludes spatial pattern formation (e.g., spatial striations) in reactive–diffusive systems governed by (1)–(3).

We now consider an example, a reaction–convection–diffusion equation, where inequalities are used in a clever way to show that if the initial wave profile is square integrable, then the wave profile for any $t > 0$ is square integrable. Another way of stating this is to say that the L_2-norm of the solution remains bounded, or stays under control, as t gets large.

Example 3. Consider the initial boundary value problem

$$u_t - Du_{xx} + uu_x = f(u), \qquad 0 < x < 1, \quad t > 0 \tag{4}$$

$$u(x,0) = u_0(x), \qquad 0 < x < 1 \tag{5}$$

$$u(0,t) = u(1,t) = 0, \qquad t > 0 \tag{6}$$

We assume that $u_0(0) = u_0(1) = 0$ and u_0 is continuous and nonnegative on $[0,1]$; we assume that the reaction term f is a bounded continuous function on R, and $M = \sup|f|$. D is a positive diffusion constant. Now let us show that $E(t) = \int_0^1 u(x,t)^2\,dx$ stays bounded for all t and, in fact, determine an upper estimate. We have, using (4) and (6),

$$
\begin{aligned}
E'(t) &= 2\int_0^1 uu_t\,dx \\
&= 2\int_0^1 u(Du_{xx} - uu_x + f(u))\,dx \\
&= 2D\int_0^1 uu_{xx}\,dx - 2\int_0^1 \left(\frac{u^3}{3}\right)_x dx + 2\int_0^1 uf(u)\,dx
\end{aligned}
$$

The second integral is zero by the boundary conditions on u at $x = 0$ and $x = 1$; the first integral can be integrated by parts to obtain

$$
\int_0^1 uu_{xx}\,dx = uu_x|_0^1 - \int_0^1 u_x^2\,dx
$$

The third integral can be bounded by the following sequence of inequalities:

$$
\begin{aligned}
\int_0^1 uf(u)\,dx &= \left(\int_0^1 u^2\,dx\right)^{1/2}\left[\int_0^1 f(u)^2\,dx\right]^{1/2} && \text{(Hölder's inequality)} \\
&\le M\left(\int_0^1 u^2\,dx\right)^{1/2} \\
&\le \frac{M}{\pi}\left(\int_0^1 u_x^2\,dx\right)^{1/2} && \text{(Poincaré's inequality)} \\
&= \frac{MD^{-1/2}}{\pi}\left(D\int_0^1 u_x^2\,dx\right)^{1/2} \\
&\le \frac{1}{2}\left(\frac{MD^{-1/2}}{\pi}\right)^2 + \frac{D}{2}\int_0^1 u_x^2\,dx && \text{(Young's inequality)}
\end{aligned}
$$

Putting all the inequalities together yields

$$
\begin{aligned}
E'(t) &\le -2D\int_0^1 u_x^2\,dx + \frac{M^2D^{-1}}{\pi^2} + D\int_0^1 u_x^2\,dx \\
&= -D\int_0^1 u_x^2\,dx + \frac{M^2D^{-1}}{\pi^2} \\
&\le -D\pi^2\int_0^1 u^2\,dx + \frac{M^2D^{-1}}{\pi^2}
\end{aligned}
$$

where, in the last step, Poincaré's inequality is applied again. Therefore, we have derived the differential inequality

$$E'(t) + D\pi^2 E(t) \leq \frac{M^2}{D\pi^2}$$

Multiplying by $\exp(D\pi^2 t)$ and then integrating from 0 to t yields

$$E(t) \leq E(0)\exp(-\pi^2 Dt) + \frac{M^2}{D^2\pi^4}\left[1 - \exp(-\pi^2 Dt)\right]$$

Therefore, the energy remains bounded for all $t > 0$, and an upper bound on the energy has been obtained.

Another important problem in nonlinear partial differential equations is the problem of determining if solutions blow up in finite time. Again, an energy-type argument is relevant.

Example 4 (Blowup). Consider the reactive–diffusive system

$$u_t = u_{xx} + u^3, \qquad 0 < x < \pi, \quad t > 0 \tag{7}$$

$$u(0,t) = u(\pi,t) = 0, \qquad t > 0 \tag{8}$$

$$u(x,0) = u_0(x), \qquad 0 < x < \pi \tag{9}$$

where u_0 is assumed to be continuous and nonnegative on $[0, \pi]$. We shall demonstrate that if

$$\int_0^\pi u_0(x)\sin x\,dx > 2 \tag{10}$$

then the solution blows up in finite time. To this end we observe that the maximum principle implies that $u \geq 0$, so long as the solution exists, and we define

$$s(t) \equiv \int_0^\pi u(x,t)\sin x\,dx$$

Then, using integration by parts,

$$s'(t) = \int_0^\pi u_t \sin x\,dx = \int_0^\pi \left(u_{xx}\sin x + u^3 \sin x\right)dx = -s(t) + \int_0^\pi u^3 \sin x\,dx$$

Now apply Hölder's inequality with $p = 3$ and $q = \frac{3}{2}$ to get

$$s(t) = \int_0^\pi u \sin x \, dx = \int_0^\pi \sin^{2/3} x u \sin^{1/3} x \, dx$$

$$\leq \left(\int_0^\pi (\sin^{2/3} x)^{3/2} \, dx \right)^{2/3} \left(\int_0^\pi (u \sin^{1/3} x)^3 \, dx \right)^{1/3}$$

$$\leq 2^{2/3} \left(\int_0^\pi u^3 \sin x \, dx \right)^{1/3}$$

Therefore,

$$s(t)^3 \leq 4 \int_0^\pi u^3 \sin x \, dx$$

and hence

$$s'(t) \geq -s(t) + \frac{s(t)^3}{4}, \quad t > 0; \qquad s(0) > 2 \tag{11}$$

We shall now show that the inequality (11) implies that $s(t) \to +\infty$ at a finite t, and thus the solution blows up in finite time. The expression (11) reminds us of a Bernoulli equation in ordinary differential equations, except that it is an inequality. But the same technique applies. Namely, let $v = 1/s^2$. Then (11) becomes a linear inequality in $v(t)$ given by

$$v'(t) \leq 2v(t) - \tfrac{1}{2}$$

Multiplying through by the integrating factor e^{-2t} and then integrating from 0 to t yields

$$v(t) \leq \frac{1 - e^{-2t}}{4} + v(0)e^{2t}$$

Consequently,

$$s(t) \geq \left[\left(e^{2t} \left(\frac{1}{s(0)^2} - \frac{1}{4} \right) + \frac{1}{2} \right] \right]^{-2} \tag{12}$$

But since $s(0) > 2$, the right side of (12) goes to infinity at a finite value of t, showing that $s(t)$ blows up at finite time.

Example 5 (Uniqueness). In this example we use energy estimates to show that an initial–boundary value problem associated with Burgers' equation can have at most one solution. Consider the problem

$$u_t + uu_x = Du_{xx}, \qquad 0 < x < 1, \quad t > 0 \tag{13}$$

$$u(x,0) = u_0(x), \qquad 0 < x < 1 \tag{14}$$

$$u(0,t) = u(1,t) = 0, \qquad 0 < x < 1 \tag{15}$$

In the usual manner assume that both u and v are solutions to (13)–(15); then $w = u - v$ must satisfy the equation

$$w_t + \left(\frac{aw}{2}\right)_x = Dw_{xx}, \qquad a = a(u,v) \equiv u + v \tag{16}$$

with $w = 0$ along $x = 0$, $x = 1$, and $t = 0$. Now define $E(t) = \frac{1}{2}\int_0^1 w(x,t)^2\,dx$. Then

$$E'(t) = \int_0^1 ww_t\,dx = \int_0^1 Dww_{xx}\,dx - \tfrac{1}{2}\int_0^1 (aw)_x w\,dx \tag{17}$$

The first term on the right can be integrated by parts to obtain

$$\int_0^1 Dww_{xx}\,dx = -D\int_0^1 w_x^2\,dx$$

The second integral on the right of (17) can be written

$$\int_0^1 (aw)_x w\,dx = \int_0^1 a_x w^2\,dx + \int_0^1 aww_x\,dx = \int_0^1 a_x w^2\,dx - \int_0^1 (aw)_x w\,dx$$

where integration by parts has been applied yet again. Thus

$$\int_0^1 (aw)_x w\,dx = \tfrac{1}{2}\int_0^1 a_x w^2\,dx$$

and (17) becomes

$$\begin{aligned} E'(t) &= -D\int_0^1 w_x^2\,dx - \tfrac{1}{4}\int_0^1 a_x w^2\,dx \\ &\le -\frac{1}{4}\int_0^1 a_x w^2\,dx \le \frac{\max|a_x|}{4}\int_0^1 w^2\,dx \end{aligned}$$

Therefore, we have shown that $E'(t) \le cE(t)$ for all $t > 0$, where c is a positive constant. As in previous examples we multiply by e^{-ct} and integrate from 0 to t to obtain the inequality $E(t) \le E(0)e^{ct}$. But $E(0) = 0$ and therefore $E(t) = 0$ for all $t > 0$, showing that w is identically zero. Thus

$u = v$ and uniqueness is established. Note that uniqueness could also be proved by using the Cole–Hopf transformation to transform Burgers' equation to the diffusion equation.

Invariant Sets

Another theoretical method that predicts information about the longtime behavior of solutions of certain classes of reaction–diffusion equations is the method of invariant sets. To introduce the concept we first consider the plane autonomous system of ordinary differential equations

$$\frac{du}{dt} = f(u,v) \qquad \frac{dv}{dt} = g(u,v) \tag{18}$$

A domain $\Sigma \subset R^2$ enclosed by a smooth, simple closed curve $\partial\Sigma$ is an invariant set for (18) if any solution $(u(t), v(t))$ with $(u(0), v(0))$ in Σ remains in Σ for all $t > 0$. It is clear that if the vector field $\mathbf{f} = (f, g)$ along $\partial\Sigma$ points into the region Σ, which is implied by the condition $\mathbf{f} \cdot \mathbf{n} < 0$ on $\partial\Sigma$, where $\mathbf{n}$ is the outward unit normal, then Σ is an invariant set. Having information about the existence of bounded, invariant sets allows us to extract information about the longtime behavior of solutions of (18); for example, the Poincaré–Bendixson theorem states that if $(u(t), v(t))$ remains bounded as t becomes infinite, then the orbit tends to a critical point, is a periodic solution, or approaches a periodic solution.

This concept extends in a natural way to systems of reaction–diffusion equations. The idea is to find a closed bounded region in the space of dependent variables that traps the solution for all $t > 0$, or at least up until the time the solution ceases to exist. If such a region can be found, then one automatically obtains a priori bounds on the solution (recall that a priori bounds are often required to obtain global existence of solutions). Even if the region is unbounded, useful information can often be extracted.

To fix the concept we consider a system of two reaction–diffusion equations of the form

$$u_t = d_1 u_{xx} + f(u,v) \qquad v_t = d_2 v_{xx} + g(u,v), \qquad x \in I, \quad t > 0 \tag{19}$$

where $\mathbf{u} = (u, v)$ is the unknown solution vector, $\mathbf{f} = (f, g)$ is the vector field of nonlinear reaction terms, d_1 and d_2 are nonnegative diffusion constants, and I is an interval in R, possibly all of R. The functions f and g are assumed to be continuous, and initial conditions are given by

$$u(x,0) = u_0(x), \qquad v(x,0) = v_0(x), \qquad x \in I \tag{20}$$

If I is not all of R, then we assume Dirichlet or Neumann boundary conditions at the ends of the interval.

Definition 1. Let Σ be a closed set in R^2. If $\mathbf{u}(x,t)$ is a solution to (19)–(20) for $0 \le t < \delta \le \infty$, with given boundary conditions, and the initial values and boundary values are in Σ, and $\mathbf{u}(x,t)$ is in Σ for all $x \in I$ and $0 < t < \delta$, then Σ is called an *invariant set* for the solution $\mathbf{u}(x,t)$.

The following theorem asserts that if the vector field $\mathbf{f}$ points inward along the boundary of a rectangle, then the rectangle must be an invariant set.

Theorem 1. Let $\Sigma = [a,b] \times [c,d]$ be a rectangle in uv-space, and let Σ^0 denote its interior and $\partial\Sigma$ denote the boundary, with $\mathbf{n}$ the outward unit normal. If

$$\mathbf{f}(\mathbf{u}) \cdot \mathbf{n} < 0 \qquad \text{on} \quad \partial\Sigma \tag{21}$$

then Σ is an invariant set for (19)–(20).

Proof. We proceed by contradiction and assume that Σ is not an invariant set. Then, without loss of generality, we may suppose that $u(x_0,t_0) = b$ for some (x_0,t_0) with $u(x,t) < b$ for all $x \in I$, $0 < t < t_0$, yet

$$u_t(x_0,t_0) \ge 0 \tag{22}$$

Then the function $u(x,t_0)$, regarded as a function of x, must have a maximum at $x = x_0$. (For example, let $h(x) = u(x,t_0)$, so that $h(x_0) = b$. If $h'(x_0) > 0$ then $h(x_1) > b$ for some $x_1 > x_0$ sufficiently close to x_0. Then $u(x_1,t_0) > b$, and because u is continuous, $u(x,t) > b$ in some neighborhood of (x_1,t_0). In particular, $u(x,t) > b$ for some $x \in I$ and some $t < t_0$, which is contrary to our assumption. Similarly, we cannot have $h'(x_0) < 0$.) Therefore $u_{xx}(x_0,t_0) \le 0$. Moreover, at (x_0,t_0), we have

$$u_t = d_1 u_{xx} + f(u,v) \le f(u,v) = \mathbf{f}(\mathbf{u}) \cdot \mathbf{n} < 0 \tag{23}$$

the latter statement following from the fact the $\mathbf{n} = (1,0)$ on the boundary $u = b$. But (23) contradicts (22), and therefore Σ must be an invariant set, completing the proof.

Example 6. The FitzHugh–Nagumo equations are

$$u_t = u_{xx} + u(1-u)(u-a) - v \qquad v_t = \alpha v_{xx} + \sigma u - \gamma v$$

where $0 < a < 1$, $\alpha \ge 0$, and $\sigma, \gamma > 0$; they represent a simplified set of reaction–diffusion equations that models conduction along nerve fibers. Here,

$$f(u,v) = u(1-u)(u-a) - v \qquad g(u,v) = \sigma u - \gamma v$$

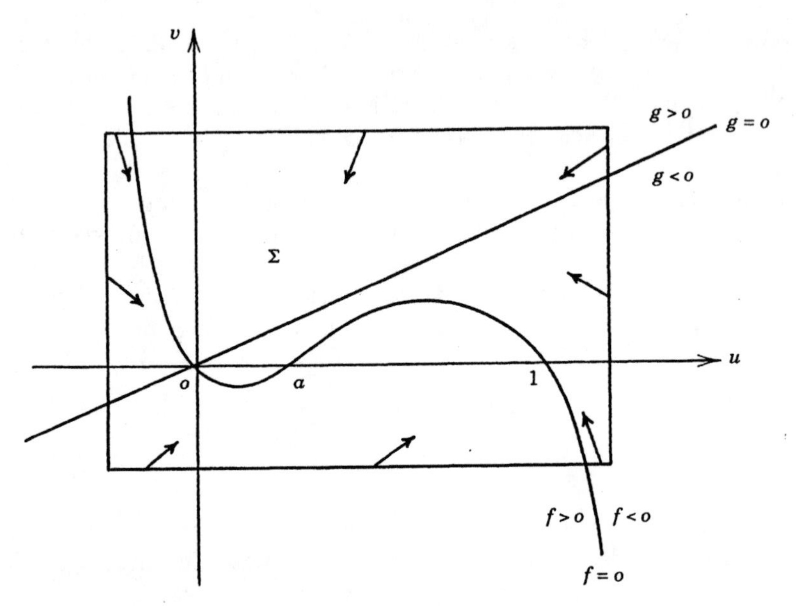

Figure 6.13. Invariant rectangle for the FitzHugh–Nagumo equations.

are the reaction terms. Let Σ be the rectangle shown in Figure 6.13. It is easy to observe that the vector field $\mathbf{f} = (f, g)$ points inward along the boundaries of the rectangle; thus (21) holds and Σ is an invariant set.

REMARK 1. Theorem 1 can be generalized immediately to an n-dimensional system

$$\mathbf{u}_t = D(\mathbf{u})\mathbf{u}_{xx} + C(\mathbf{u})\mathbf{u}_x + \mathbf{f}(\mathbf{u}, t), \quad x \in I, \quad t > 0 \tag{24}$$

where D and C are n-by-n diagonal matrices with $D \geq 0$ (i.e., D is positive semi-definite), $\mathbf{f}$ is continuously differentiable, and $\mathbf{u} = (u_1, \ldots, u_n)$ is the unknown solution vector. Then any region (a box in R^n) of the form

$$\Sigma = [a_1, b_1] \times \cdots \times [a_n, b_n]$$

is an invariant set for (24), with appropriate initial and boundary conditions, provided that

$$\mathbf{f}(\mathbf{u}, t) \cdot \mathbf{n} < 0 \quad \text{for all} \quad \mathbf{u} \in \partial\Sigma, \quad t > 0$$

where $\mathbf{n}$ is the outward unit normal to Σ. The proof is left as an exercise.

The result mentioned in Remark 1 can be generalized under certain conditions to permit invariant sets to have the form of a convex subset of R^n.

We shall now formulate this general result and refer the reader to Smoller [1983] for a proof; the proof is not difficult.

Consider the reactive–diffusive system (24) where $D \geq 0$ and C are continuous matrices, not necessarily diagonal, and $\mathbf{f}$ is continuously differentiable. Let G_i, $i = 1, \ldots, p$, be continuously differentiable, real-valued functions on R^n and define the regions $E_i = \{\mathbf{u} \in R^n | G_i(\mathbf{u}) \leq 0\}$ for $i = 1, \ldots, p$. We assume that $\mathbf{grad}\, G_i$ is never zero. Recall that the vector $\mathbf{grad}\, G_i$ is normal to the locus $G_i(\mathbf{u}) = 0$, which is a level curve of the surface $z = G_i(\mathbf{u})$ in R^{n+1}. Now define

$$\Sigma = \bigcap_{i=1}^{p} E_i$$

so that Σ is the intersection of half-spaces. Clearly Σ is closed. To show that Σ is an invariant set for (24) we must have three conditions hold true at every boundary point of Σ. First, the gradient of the G_i that defines the particular section of the boundary under consideration must be a left eigenvector of the diffusion matrix D and the convection matrix C. Second, the vector field $\mathbf{f}$ along the boundary must point into the region. Third, on every section of the boundary the function G_i defining that portion of the boundary must be a convex function; this implies, of course, that the region Σ will be convex. To formulate these conditions mathematically, we say that a function $\mathbf{G}$ is *quasi-convex* at a point $\mathbf{v}$ in R^n if $\mathbf{grad}\, G(\mathbf{v}) \cdot \mathbf{h} = 0$ implies that $\mathbf{h}^T Q(\mathbf{v})\mathbf{h} \geq 0$, where Q is the matrix of second partial derivatives of G, i.e., $Q = (\partial^2 G/\partial u_i \partial u_j)$. Then one can prove the following: Σ is an invariant set for (24) provided the following three conditions hold: for all $t > 0$ and for every $\mathbf{u}^* \in \partial\Sigma$ (so that $G_i(\mathbf{u}^*) = 0$ for some i)

(a) $\mathbf{grad}\, G_i(\mathbf{u}^*)$ is a left eigenvector of $D(\mathbf{u}^*)$ and of $C(\mathbf{u}^*)$
(b) $\mathbf{grad}\, G_i(\mathbf{u}^*) \cdot \mathbf{f}(\mathbf{u}^*, t) < 0$
(c) If $\mathbf{grad}\, G_i(\mathbf{u}^*) D(\mathbf{u}^*) = \mu\, \mathbf{grad}\, G_i(\mathbf{u}^*)$ and $\mu \neq 0$, then G_i is quasi-convex at $\mathbf{u}^*$.

EXERCISES

1. Prove the Poincaré inequality (ii).

2. Consider the initial value problem for the diffusion equation

$$u_t - Du_{xx} = 0, \qquad x \in R, \quad t > 0$$
$$u(x,0) = u_0(x), \qquad x \in R,$$

where u_0 is bounded and continuous, and $\int_R u_0^2(x)\, dx$ is finite. Prove there exists a constant C such that $|u(x,t)| \leq C/t^{1/4}$ for all $t > 0$.

3. Consider the problem

$$u_t - u_{xx} = f(u), \qquad 0 < x < 1, \quad t > 0$$
$$u(x,0) = u_0(x), \qquad 0 < x < 1$$
$$u(0,t) = u(1,t) = 0, \qquad t > 0$$

where f is continuously differentiable and $\sup|f'(u)| < \pi^2$. Prove that if a unique solution $u = u(x,t)$ exists, then

$$\lim_{t\to\infty} \int_0^1 u^2(x,t)\,dx = 0$$

4. Consider the reactive–diffusive system

$$u_t - D_1 u_{xx} = f(u,v), \qquad v_t - D_2 v_{xx} = g(u,v), \qquad 0 < x < 1, \quad t > 0$$
$$u(x,0) = u_0(x), \qquad v(x,0) = v_0(x), \quad 0 < x < 1$$
$$u_x = v_x \quad \text{at } x = 0 \quad \text{and} \quad x = 1 \quad \text{for all } t > 0$$

where f and g are continuously differentiable and f_u, f_v, g_u, and g_v are bounded. Let

$$D = \min(D_1, D_2), \qquad M = \sup\left\{\left(f_u^2 + f_v^2 + g_u^2 + g_v^2\right)^{1/2} : u, v \in R\right\}$$

Show that if $4M - 2\pi^2 D < 0$, then $\lim_{t\to\infty} q(t) = 0$, where

$$q(t) = \tfrac{1}{2}\textstyle\int_0^1 \left(u_x^2 + v_x^2\right) dx$$

5. (a) Show that the boundary value problem

$$-u'' = \delta e^u, \quad 0 < x < 1; \qquad u(0) = u(1) = 0$$

has no solution for $\delta > \pi^2/e$. *Suggestion:* Show that $\int_0^1 \phi_1(\delta e^u - \lambda_1 u)\,dx = 0$, where λ_1 is the smallest eigenvalue of the linear problem $-\phi'' = \lambda\phi$, $\phi(0) = \phi(1) = 0$, and ϕ_1 is a corresponding eigenfunction.

(b) *Gelfand Problem.* Show that the initial–boundary value problem

$$u_t - u_{xx} = \delta e^u, \qquad 0 < x < 1, \quad t > 0$$
$$u(x,0) = 0, \quad 0 < x < 1; \qquad u(0,t) = u(1,t) = 0, \quad t > 0$$

does not have a global solution if $\delta > \pi^2/e$. *Suggestion:* Show that

$$E'(t) + \lambda_1 E(t) = \delta \int_0^1 \phi_1 e^u \, dx \geq \delta e^{E(t)}$$

where $E(t) = \int_0^1 \phi_1(x) u(x,t) \, dx$.

6. Consider the initial value problem for the Schrödinger equation

$$u_t = i u_{xx}, \quad x \in R, \quad t > 0; \qquad u(x,0) = u_0(x), \quad x \in R$$

and assume that a solution $u = u(x,t)$ exists for all $t > 0$ and $x \in R$ with u and all of its derivatives vanishing at $|x| = \infty$. Prove that $E(t) = \int_R u\bar{u} \, dx$ is constant for $t > 0$.

7. Consider the system

$$\mathbf{u}_t = A\mathbf{u}_x, \qquad 0 < x < 1, \quad t > 0$$

$$\mathbf{u}(0,t) = \mathbf{u}(1,t) = \mathbf{0}, \quad t > 0; \qquad \mathbf{u}(x,0) = \mathbf{u}_0(x), \quad 0 < x < 1$$

where $\mathbf{u} = \mathbf{u}(x,t)$ is an n-vector and A is a constant matrix with $A^T = A$ (T denotes transpose). Prove that $E'(t) = 0$ where $E(t) = \int_0^1 \mathbf{u}^T \mathbf{u} \, dx$.

8. Consider the problem

$$u_t - u_{xx} = f(u), \qquad 0 < x < \pi, \quad t > 0$$

$$u(0,t) = u(\pi,t) = 0, \quad t > 0; \qquad u(x,0) = u_0(x), \quad 0 < x < \pi$$

where u_0 is nonnegative and continuous, and

$$f(u) > 0, \qquad f'(u) > 0, \qquad f''(u) > 0 \quad \text{for} \quad u > 0$$

and $\int^\infty du/f(u)$ is finite. If $\int_0^\pi u_0(x) \, dx$ is sufficiently large, prove that the solution blows up in finite time. *Hint:* Consider $s(t) = \int_0^\pi u(x,t) \sin x \, dx$ and use the fact that $\int_0^\pi f(u) \sin x \, dx > f(\int_0^\pi u \sin x \, dx)$ to get $s'(t) \geq -\lambda s(t) + f(s(t))$ for some $\lambda > 0$.

9. Assume that $u \in C^\infty(R)$ and that $u(x+1) = u(x)$ for all x, and denote $\|u\|^2 = \int_0^1 |u(x)|^2 \, dx$. Prove for all $c > 0$ and all integers $j, k \geq 1$ that

$$\|u^{(j)}\|^2 \leq c\|u^{(j+k)}\|^2 + c^{-j/k}\|u\|^2$$

This is an example of a *Sobolev-type* inequality.

10. *Project*. The Navier–Stokes equations, which govern a Newtonian viscous fluid, are given by

$$\mathbf{u}_t - \nu \, \Delta \mathbf{u} + (\mathbf{u} \cdot \mathbf{grad})\mathbf{u} + \mathbf{grad} \, p = \mathbf{f}(\mathbf{x}) \qquad \operatorname{div} \mathbf{u} = 0$$

where $\mathbf{x} = (x, y, z)$, and where $\mathbf{u} = \mathbf{u}(\mathbf{x}, t)$ is the velocity vector, $p = p(\mathbf{x}, t)$ the pressure, $\mathbf{f}$ the body force, $\nu > 0$ the viscosity, and Δ is the three-dimensional Laplacian. Let Ω be an open, bounded region in space with smooth boundary $\partial\Omega$, and assume that the Navier–Stokes equations hold in Ω with $\mathbf{u} = \mathbf{0}$ on $\partial\Omega$ for all $t > 0$, and $\mathbf{u}(\mathbf{x}, 0) = \mathbf{u}_0(\mathbf{x})$ on Ω. Prove that

$$\|\mathbf{u}(t)\|^2 \leq \|\mathbf{u}(0)\|^2 e^{-a\nu t} + \frac{C}{\nu}(1 - e^{-a\nu t}) \qquad (C, a > 0)$$

for all $t > 0$, where $\|\mathbf{u}(t)\|$ denotes the L_2-norm

$$\|\mathbf{u}(t)\| = \left[\int_\Omega \mathbf{u}(\mathbf{x}, t)^T \mathbf{u}(\mathbf{x}, t)\, d\mathbf{x}\right]^{1/2} \qquad d\mathbf{x} = dx\, dy\, dz$$

Thus the energy remains bounded for all time. *Hint:* Integrate over Ω, then use integration by parts followed by three-dimensional versions of Young's and Poincaré's inequalities. Exercise 1 of Section 8.1 may be helpful.

11. The flow of an ideal, incompressible fluid of constant density in an open bounded domain Ω in R^3 with a smooth boundary is governed by the momentum law and the incompressibility condition:

$$\mathbf{u}_t + (\mathbf{u} \cdot \mathbf{grad})\mathbf{u} + \mathbf{grad}\, p = \mathbf{0} \qquad \operatorname{div} \mathbf{u} = 0, \qquad x \in \Omega, \quad t > 0$$

where $\mathbf{u}$ is the velocity and p is the pressure. Assume that the flow is parallel to the boundary, i.e., $\mathbf{u} \cdot \mathbf{n} = 0$ on $\partial\Omega$, and initially $\mathbf{u}(\mathbf{x}, 0) = \mathbf{u}_0(\mathbf{x})$, $\mathbf{x} \in \Omega$. Prove that

$$\|\mathbf{u}(t)\| = \|\mathbf{u}_0\|, \qquad t > 0$$

(the notation is the same as in Exercise 10).

12. Consider the initial boundary value problem

$$u_t = u_{xx} + u(1 - v), \qquad v_t = v_{xx} + av(u - 1), \qquad 0 < x < 1, \quad t > 0$$
$$u(x,0) = u_0(x), \qquad v(x,0) = v_0(x), \qquad 0 < x < 1$$
$$u_x = v_x = 0 \quad \text{at} \quad x = 0, 1$$

where $a > 0$. Prove that $\lim u_x = \lim v_x = 0$ as $t \to \infty$. *Suggestion:* Define $s(x, t) = a(u - \ln u) + v - \ln v$ and $S(t) = \int_0^1 s(x, t)\, dx$. Show that

$$s_t - s_{xx} = -au_x^2/u^2 - v_x^2/v^2$$

and thus $dS/dt \leq 0$. Use the fact that $s \geq a + 1$ to show $S(t)$ tends to a finite limit.

13. Consider the reactive-diffusive system

$$u_t = u_{xx} + (1 - u^2 - v^2)u \qquad v_t = v_{xx} + (1 - u^2 - v^2)v$$

Show that the unit circle in uv-space is an invariant set. *Hint:* Consider the set $\Sigma_\varepsilon = \{(u, v) | u^2 + v^2 \leq 1 + \varepsilon\}$ for $\varepsilon > 0$.

14. Prove Remark 1.

15. In combustion theory the solid fuel model is

$$y_t = Dy_{xx} - yr(T) \qquad T_t = kT_{xx} + qyr(T)$$

where y is the mass fraction of the reactant, T is the temperature, and D, k, and q are positive constants. The reaction rate is $r(T) = \exp(-E/RT)$, where E and R are positive constants. Show that $\Sigma = \{(y, T) | 0 \leq y \leq 1,\ T \geq a > 0\}$ is an invariant set. *Hint:* First consider y in the range $-\varepsilon \leq y \leq 1$.

16. Consider the reaction–diffusion problem

$$u_t = u_{xx} - u + u^p, \qquad 0 < x < \pi, \quad t > 0$$

$$u(x,0) = u_0(x) > 0, \qquad 0 < x < \pi; \qquad u(0,t) = u(\pi,t) = 0, \quad t > 0$$

where $p > 1$. If

$$\int_0^\pi u_0(x) \sin x \, dx > 2^{p/(p-1)}$$

prove that the solution blows up in finite time.

REFERENCES

The recent literature on reaction–diffusion equations is extensive. A starting place is Murray [1989], Smoller [1983], Fife [1979], Britton [1986], and Grindrod [1991]. All have detailed bibliographies that can serve as a guide to further reading. The applications in these references focus on problems in biological settings; for applications of reaction–diffusion equations in chemically reacting systems, especially combustion phenomena, see the reference section of Chapter 7. A large number of conferences in recent times have been devoted to reaction–diffusion equations, and many proceedings have been published on the topic.

Birkhoff, G., and G.-C. Rota, 1965. *Ordinary Differential Equations*, Blaisdell, Waltham, Mass.

Britton, N. F., 1986. *Reaction-Diffusion Equations and Their Applications to Biology*, Academic Press, London.

Fife, P. C., 1979. *Mathematical Aspects of Reacting and Diffusing Systems*, Springer-Verlag, New York.

Friedman, A., 1964. *Partial Differential Equations of Parabolic Type*, Prentice Hall, Englewood Cliffs, N.J.

Grindrod, P., 1991. *Patterns and Waves*, Oxford Univ. Pr., Oxford.

Guenther, R. B., and J. W. Lee, 1988. *Partial Differential Equations of Mathematical Physics and Integral Equations*, Prentice Hall, Englewood Cliffs, N.J.

Hartman, P., 1964. *Ordinary Differential Equations*, Wiley, New York.

Kolmogorov, A., I. Petrovsky, and N. Piscounoff, 1937. Study of the diffusion equation with growth of the quantity of matter and its application to a biology problem, in *Dynamics of Curved Fronts*, edited by P. Pelce, Academic Press, San Diego, 1988. Originally published in *Bull. Univ. Etat Moscou Ser. Int. A* **1** (1937).

Murray, J. D., 1989. *Mathematical Biology*, Springer-Verlag, New York.

Protter, M. H., and H. Weinberger, 1967. *Maximum Principles in Differential Equations*, Prentice Hall, Englewood Cliffs, N.J.

Smoller, J., 1983. *Shock Waves and Reaction–Diffusion Equations*, Springer-Verlag, New York.

See the article by S. Cohn and J. D. Logan, *Math. Models and Methods Appl. Sci.* (1995), for a fuller discussion of the groundwater model equations (20)–(21) of Section 6.2.

7

MODELS FOR CHEMICALLY REACTING FLUIDS

In this chapter we take up a significant application of nonlinear PDEs, namely that of combustion and detonation phenomena. The model equations for such phenomena are the equations of chemically reacting fluid flow, that is, the equations of gas dynamics (the mass, momentum, and energy balance laws) coupled with chemical species equations that govern the evolution of the chemical reactions involved; the energy equation contains a source term representing the thermal energy liberated (or absorbed) by the various reactions. The development of the underlying theory of combustion has led to specialized mathematical methods that extend far beyond the bounds of combustion, and the methods and techniques are applicable to problems in the biosciences and the entire realm of reaction–diffusion equations. Thus the subject of combustion has spawned a variety of interesting ODEs and PDEs, and the methods of analysis of these equations are general enough to be relevant in many other areas of science and technology.

From a historical perspective, the study of combustion and detonation phenomena arose out of experiments on flames propagating in tubes of combustible gases that were conducted by scientists in the late nineteenth century. We give a simplistic view of the results of these experiments. The scientists distinguished two types of flame fronts, one a relatively slow front whose speed was on the order of meters per second, and a second whose speed was a thousand times faster. The former, at small Mach numbers, was called combustion (or deflagration); the latter, at high Mach number, was called detonation. Combustion is characterized by an exothermic surface reaction and is supported by heat transfer from the hot, gaseous products; mass diffusion and viscous effects play important roles. On the other hand, detonations are exothermic reactions accompanied by shock waves; the chemical reaction is initiated by the heating that accompanies shock compression—heat transfer and viscosity play almost no role and the process is dominated by energy and momentum transfer. Simplistically, combustion is a reaction–diffusion process, and detonation is a hyperbolic wave process.

The first theoretical model of these process was proposed around the turn of the century by Chapman and Jouget. Their theory, now called CJ theory, employs an idealized one-dimensional model in which it is assumed that the chemical energy is released instantaneously in a narrow flame front (modeled mathematically by a shock) propagating into the combustible gas. It defines a unique steady-state velocity of the front and permits the calculation of the state (pressure, temperature, etc.) just behind the flame front in terms of the quiescient state ahead of the front. The CJ equations are just algebraic formulas that express the fact that mass, momentum, and energy must be balanced across the discontinuous flame front. A more detailed gas dynamic model that resolves the chemistry over a finite region was developed independently during the 1940s by von Neumann, Zeldovich, and Doering. This model, the ZND model, forms the basis of the modern theory, and it is based on the nonlinear PDEs of gas dynamics and equations governing the evolution of the chemical species.

In Section 1.1 we derive from first principles the basic conservation laws governing the motion of a multicomponent fluid, and we introduce the fundamental constitutive relations that supplement the balance laws to form a complete set of equations for reactive fluid flow. At the end of the section we obtain a set of working equations that describe the behavior of a two-component reacting fluid (a reactant and a product), and it is special cases of these equations that we examine in the sequel. In Section 1.2 we investigate the Semenov model for homogeneous explosions, and we introduce the detonation model. In Section 1.3 we take up some basic concepts from detonation theory, including CJ theory and steady ZND waves. Finally, in Section 1.4 we study the equations of combustion phenomena with particular emphasis on obtaining a mathematically tractable system by specializing the general equations in special limits, for example, low Mach number or high activation energy.

We emphasize that the treatment herein is only introductory, and the focus is on developing the model equations in an elementary setting. The material should motivate some of the material on elliptic equations in Chapter 8 as well as illustrating some of the hyperbolic and reactive–diffusive systems studied in previous chapters.

7.1 EQUATIONS OF REACTIVE FLOW

Classical gas dynamics focuses on the motion of a fluid consisting of a single inert component. In the study of reacting fluid systems, however, the description of the flow involves many different fluid components coexisting simultaneously. These components are often designated as chemical species since their concentrations are changing via one or more chemical reactions.

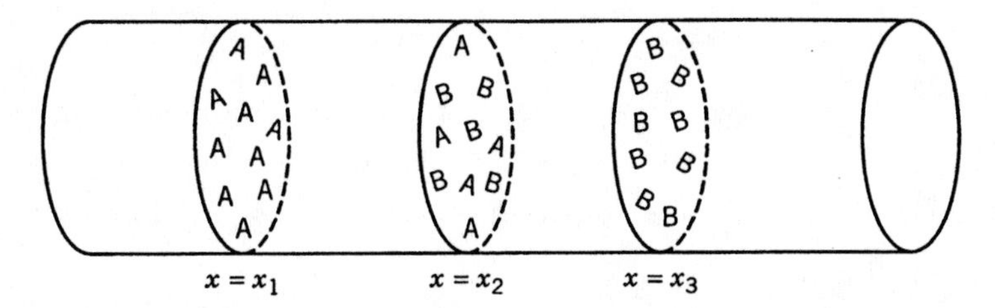

Figure 7.1. Chemical composition observed at three different locations in a reacting flow.

Consequently, reactive gas dynamics is classical gas dynamics of multicomponent fluids coupled with chemistry. In brief, reactive gas dynamics is governed by the same equations as gas dynamics—conservation of mass, momentum, and energy—with the important difference that the energy equation contains source terms because of the heat given off, or taken up, by the chemical reactions. Our goal in this section is to obtain a relatively simple set of working equations for general reactive flow that can be specified to particular problems and that can be used as the basis for the development of mathematical machinery to analyze those problems. We formulate the equations in the simplest possible context, that is, in a one-dimensional medium using a fixed control volume, much the same as was used in Chapter 5 for gas dynamics. The article by Truesdell [1965] and the book by Williams [1985] are excellent sources for the equations of motion of multicomponent fluids and transport properties.

We start with a simplified, qualitative description of reactive flow where there are just two components, a reactant A and a product B undergoing an irreversible chemical reaction $A \to B$, proceeding at some given rate. We refer to Figure 7.1, where the fluid (gas) is shown flowing through a tube. Three different observers, located at the fixed laboratory positions x_1, x_2, and x_3, can record the concentrations of each chemical species. As the reactant moves through the tube from left to right it begins to react, and the observer at x_1 sees only the reactant A present; observer x_2 sees both A and B present, and observer x_3 sees only the product B present, and the reaction has gone to completion. The goal is to determine the concentrations of these chemical species at any location and at any time, as they are coupled to the gas dynamic flow with pressure, temperature, and velocity all changing. We emphasize that at any instant, both the reactant and product species may be present in a given cross section; both coexist simultaneously, each as a continuum.

The Conservation Laws

In general, we consider the motion of a fluid in one dimension which consists of N different species, each coexisting simultaneously as a continuum. We

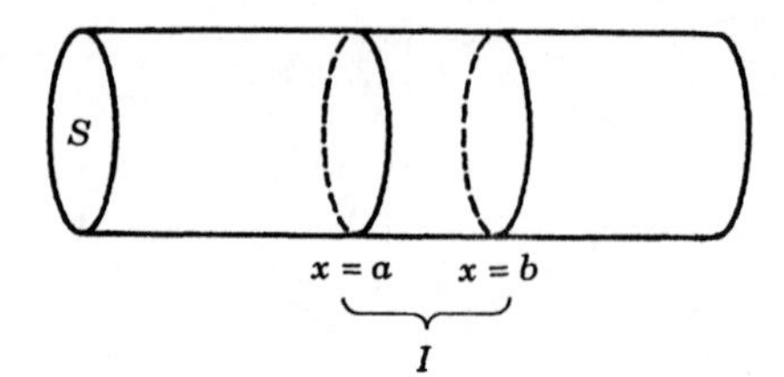

Figure 7.2. Cylindrical tube of cross-sectional area S.

index the N continua by the letter A; that is $A = 1, \ldots, N$. The fluid is imagined to be flowing through a tube of cross-sectional area S. The one-dimensional assumption implies that all physical parameters and states do not vary in any cross section, and the variations are only in the x direction along the tube. We now fix an arbitrary section $I = [a, b]$ of the tube and write down the conservation and balance laws that apply (see Figure 7.2).

Mass conservation requires that the mass of each species be balanced in the section I. This means that the time rate of change of the total amount of species A in the section I must equal the rate that A that flows in at $x = a$ minus the rate that A flows out at $x = b$, plus the rate that A is produced, or depleted, by chemical reaction. More precisely:

Mass Balance. Let ρ_A be the density of species A and let u_A be its velocity, both as functions of x and t, and let r_A denote the mass production of species A. Then we assume for any section I that

$$\frac{d}{dt}\int_a^b \rho_A \, dx = \rho_A(a,t)u_A(a,t) - \rho_A(b,t)u_A(b,t) + \int_a^b r_A \, dx, \qquad A = 1, \ldots, N \tag{1}$$

Equation (1) is the integral form of the mass balance law for each of the N species; we remark that the constant cross-sectional area S has been canceled from each term in (1). The production rates r_A have dimensions mass/(volume · time) and they may depend on various flow and thermodynamic variables. We specifically *postulate* that

$$\Sigma r_A = 0 \tag{2}$$

since, as one species appears, others disappear because of the chemical reactions; thus the sum of the rates must be zero. We shall dispense with the indices on the summand sign if the context is clear.

In the usual manner, equations (1) may be reformulated as PDEs if the functions ρ_A and u_A are C^1. In this case we obtain the *species equations*

$$(\rho_A)_t + (\rho_A u_A)_x = r_A, \qquad A = 1, \ldots, N \tag{3}$$

Equation (3) is the continuity equation for the Ath species with a source term.

We now catalog some formal definitions that will be needed in the sequel.

Definition. The *total density* $\rho = \rho(x,t)$ is defined by

$$\rho = \Sigma \rho_A$$

The *mass fraction* $y_A = y_A(x,t)$ of species A is defined by

$$y_A = \frac{\rho_A}{\rho}$$

The *mass averaged velocity* $u = u(x,t)$ is

$$u = \Sigma y_A u_A$$

and the *diffusion velocity* $U_A = U_A(x,t)$ of species A is

$$U_A = u_A - u$$

Therefore, we have defined a weighted average velocity u for the flow, weighted by the densities of the species. The diffusion velocity U_A is the actual velocity u_A relative to the average velocity, and later we shall assume that the diffusion velocities are small. It easily follows from the definitions that

$$\Sigma y_A U_A = 0, \qquad \Sigma y_A = 0 \tag{4}$$

These two relations will be used frequently in the sequel. From the above definitions and relations, the following proposition also holds true. We leave the proof as a straightforward exercise.

Proposition 1. The mass balance equations (3) and the assumption (2) imply that

$$\rho_t + (\rho u)_x = 0 \tag{5}$$

$$\rho \frac{Dy_A}{Dt} + (\rho_A y_A)_x = r_A \tag{6}$$

where $A = 1, \ldots, N$ and $D/Dt = \partial/\partial t + u\partial/\partial x$.

We notice that (5) is the usual continuity equation of gas dynamics with the total density ρ and the average velocity u. Equations (6) are referred to as the species equations, which govern the concentrations of the individual chemical species. The material derivative D/Dt in (6) is the time derivative being convected with the average velocity of the multicomponent fluid. It is

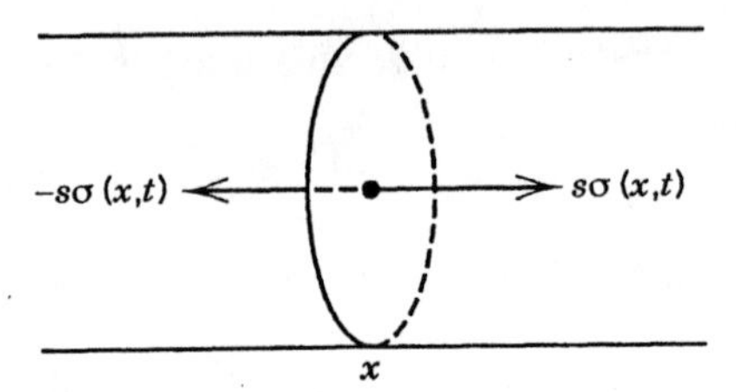

Figure 7.3

clear that (5) may also be written

$$\frac{D\rho}{Dt} + \rho u_x = 0 \tag{7}$$

Equations (5) and (6) are two of the basic governing equations of reactive fluid flow.

Now we demand that momentum also balance. In words, for each species, the time rate of change of total momentum inside I must equal the rate that momentum flows into I at $x = a$, less the rate that momentum flows out of I at $x = b$, plus the forces (due to stresses, e.g., pressure) acting on the region I at $x = a$ and $x = b$. We recall that the stress is the force per unit area acting acting across a cross section of fluid; the stress, by convention, is the force per unit area on the fluid to the left of the cross section caused by the fluid to the right of the cross section (see Figure 7.3). We assume here that there are no body forces (e.g., gravity). Formally, then, we may state:

Momentum Balance. Let $\sigma_A = \sigma_A(x, t)$ be the stress component on the Ath species. Then for any interval $[a, b]$, momentum balance requires

$$\frac{d}{dt}\int_a^b \rho_A u_A \, dx = \rho_A(a,t)u_A(a,t)^2 - \rho_A(b,t)u_A(b,t)^2 - \sigma_A(a,t) + \sigma_A(b,t) \tag{8}$$

for A $= 1, \ldots, N$, where the constant cross-sectional area S has been canceled from each term.

Equation (8) is the integral form of conservation of momentum for the Ath species. If a body force f_A (in force per unit mass per unit volume) acts on the Ath species, a term $\int_a^b \rho_A f_A \, dx$ must be added to the right side of (8). Also, (8) assumes that no momentum is produced by the chemical reactions as one chemical species changes to another. These additional effects could be included, but the resulting complications would take us away from our objective of a simple set of working equations. Now, in the usual way, it is possible, under the assumption of smoothness, to write (8) in the form of a partial differential equation. Carrying out this reduction and then summing over the indices A = $1, \ldots, N$ yields

$$(\rho u)_t = -(\Sigma \rho_A u_A^2)_x + (\Sigma \sigma_A)_x \tag{9}$$

By straightforward algebra it also follows that

$$\Sigma \rho_A u_A^2 = \rho u^2 + \Sigma \rho_A U_A^2 \tag{10}$$

Consequently, we have obtained the following result:

Proposition 2. The principle of momentum balance (8) implies, assuming that the state variables are smooth, that

$$(\rho u)_t + (\rho u^2 - \sigma)_x = 0 \tag{11}$$

where

$$\sigma = \Sigma(\sigma_A - \rho_A U_A^2) \tag{12}$$

Equation (11) is the third of our basic equations that govern reactive flow, expressing conservation of momentum. We observe that it is the same form as the gas dynamic equation for inert flow [compare equation (16) of Section 5.1], except that (11) contains averaged quantities ρ and u, and the stress σ contains terms involving the diffusion velocities.

A discussion of energy balance is more intricate than that of mass and momentum. It is suggested that the reader review the derivation of the energy equation in gas dynamics in Section 5.1. The fundamental principle is the first law of thermodynamics, which relates changes in the energy of a system to the work done by the various forces on the system and to the heat energy flowing into the system. The heat flow is characterized by N different heat flux quantities $q_A(x, t)$ which measure the heat energy, per unit time per unit area, that flows across a cross section at x into the Ath continuum, with the convention that q_A is positive when the net flow is in the positive x direction. For the system we take the interval $I = [a, b]$. Then, for the Ath continuum, the amount of heat per unit time flowing into I is given by $(q_A(a, t) - q_A(b, t))S$. We also note that the rate that work is done by the stress σ_A is $-Su_A(a, t)\sigma_A(a, t)$ at $x = a$ and $Su_A(b, t)\sigma_A(b, t)$ at $x = b$. Finally, the total energy is composed of two parts, the kinetic energy and the internal energy, and we recall that to obtain energy flux terms we must multiply energy by velocity. So we have the following principle, which states the rate of change of the total energy in I must equal the net energy flux plus the rate that the forces do work, plus the rate that heat energy flows into I.

Energy Balance. Let $q_A = q_A(x, t)$ be the heat flux and let $e_A = e_A(x, t)$ be the specific internal energy (energy per mass) of the Ath continuum. The for any interval $I = [a, b]$ the energy balance principle requires that

$$\begin{aligned}\frac{d}{dt}\int_a^b \rho_A\left(e_A + \frac{u_A^2}{2}\right)dx = {} & \rho_A(a,t)\left[e_A(a,t) + \frac{u_A(a,t)^2}{2}\right]u_A(a,t)\\ & - \rho_A(b,t)\left[e_A(b,t) + \frac{u_A(b,t)^2}{2}\right]u_A(b,t)\\ & - \sigma_A(a,t)u_A(a,t) + \sigma_A(b,t)u_A(b,t)\\ & + q_A(a,t) - q_A(b,t)\end{aligned} \tag{13}$$

for $A = 1, \ldots, N$. The constant area S has been canceled from each term.

Equations (13) are energy balance statements in integral form for each of the N species. Our goal is to obtain a single energy balance relation in terms of the averaged quantities. As has been our previous strategy, we eliminate actual velocities u_A in favor of diffusion velocities U_A and the overall average velocity u. To this end we sum the expressions (13) over $A = 1, \ldots, N$ and remove the integrals (using smoothness and the arbitrariness of the interval I of integration) to obtain

$$\left[\Sigma\left(\rho_A e_A + \frac{\rho_A u_A^2}{2}\right)\right]_t = -\left[\Sigma\left(\rho_A e_A + \frac{\rho_A u_A^2}{2}\right)u_A - \Sigma\sigma_A u_A + \Sigma q_A\right]_x \tag{14}$$

Now simplify each term, in turn, using the relation $u_A = U_A + u$. First,

$$\Sigma\left(\rho_A e_A + \frac{\rho_A u_A^2}{2}\right) = \rho e + \frac{\rho u^2}{2} \tag{15}$$

where e is defined by

$$e = \Sigma\left(y_A e_A + \frac{y_A U_A^2}{2}\right) \tag{16}$$

Next,

$$\Sigma\sigma_A u_A = \Sigma\sigma_A U_A + u\left(\sigma + \Sigma r_A U_A^2\right) \tag{17}$$

Finally,

$$\Sigma\rho_A\left(e_A + \frac{u_A^2}{2}\right)u_A = \left(\rho e + \frac{\rho u^2}{2}\right) + \Sigma\rho_A e_A U_A + \frac{\Sigma\rho_A U_A^3}{2} + u\Sigma\rho_A U_A^2 \tag{18}$$

Using (15)–(18) in (14) gives the following result, which we state as a proposition.

Proposition 3. Assuming smoothness of all the functions, the energy balance principle implies that

$$\left(\rho e + \frac{\rho u^2}{2}\right)_t = -\left[\rho u\left(e + \frac{u^2}{2}\right) - \sigma u\right]_x - q_x \tag{19}$$

where

$$q = \Sigma\left[q_A - \sigma_A U_A + \rho_A\left(e_A + \frac{U_A^2}{2}\right)U_A\right] \tag{20}$$

Equation (19) has the same form as the energy equation for a single-component fluid, except for the fact that the heat flux q, the specific internal energy e, and the stress σ contain terms with diffusion velocities. Note that the chemical energy is contained in e.

In summary, we have obtained a basic set of equations governing reactive flow consisting of (5), (6), (11), and (19). These represent $N + 3$ equations for the $2N + 5$ unknowns $y_1, \ldots, y_N, U_1, \ldots, U_N, u, \rho, e, \sigma$, and q. We also have the relation $\Sigma y_A = 1$. To make further progress additional assumptions must be made; these assumptions come in the form of material properties, or constitutive relations, that define the makeup of the fluid.

Constitutive Assumptions

We now pose specific assumptions regarding the form of e, σ, q, and the diffusion velocities.

The specific internal energy e will be specified by assuming that each species or fluid component is an ideal gas with the same molecular weight and that the specific heats of each species are constant and equal. Therefore, we assume that each continum satisfies the equation of state

$$p_A = R^0 \frac{\rho_A T}{w_A}$$

where w_A is the molecular weight of species A, R^0 is the *universal gas constant*, and T is the common *temperature* of all the species. We define the *total pressure* by (the sum of the partial pressures)

$$p = \Sigma p_A$$

It follows that

$$p = R^0 T \Sigma\left(\frac{\rho_A}{w_A}\right) = R^0 \rho T \Sigma\left(\frac{y_A}{w_A}\right) = \frac{R^0}{W} \rho T \tag{21}$$

where, as we stated, $w_A = W$ for $A = 1, \ldots, N$. The internal energy e_A of the Ath species is related to the *enthalpy* h_A of the Ath species by

$$e_A = h_A - \frac{p_A}{\rho_A}$$

This equation is the actual definition of the enthalpy for any substance. For ideal gases, the enthalpy has the form

$$h_A = h_A^0 + c_{pA}(T - T_0)$$

where h_A^0 is a constant and c_{pA} is the constant *specific heat at constant pressure* of species A. The h_A^0 are the *heats of formation* at temperature T_0. As we remarked, our assumption is $c_{pA} = c_p$ for $A = 1, \ldots, N$, where c_p is a constant. Therefore, (16) becomes, with these assumptions,

$$e = \Sigma y_A h_A^0 + c_p(T - T_0) + \frac{\Sigma y_A U_A^2}{2} - \frac{p}{\rho}$$

Now comes an important assumption about the diffusion velocities, namely that they are small so that quadratic and cubic terms in U_A can be neglected. This means, of course, that the actual velocities of the species are close to the average velocity u. Thus we shall retain first order terms in U_A in the various equations, but we shall neglect higher order terms such as U_A^2, U_A^3, and also $y_A(U_A)_x$. Consequently, we make the first constitutive assumption.

Constitutive Assumption 1 (Energy). The specific internal energy e has the form

$$e = \Sigma y_A h_A^0 + c_p(T - T_0) - \frac{p}{\rho} \tag{22}$$

One way to view a multicomponent fluid in many real situations is that of a dilute solution where one of the components is inert and serves as a solution bath in which the remaining reacting components have small concentrations. In air mixtures nitrogen might play this role. In this case the reacting components would take on the bulk properties of the bath, for example its specific heat, viscosity, or conductivity, and we might expect that the component velocities u_A would differ little from u, the bulk velocity of the inert. This context, of a dilute mixture, is one important way to understand the assumptions in the preceding paragraphs, where the physical properties are defined in terms of a single component.

Neglecting the quadratic diffusion velocity term, the stress σ defined by (12) becomes $\sigma = \Sigma\sigma_A$. We assume that the stress σ_A has the form

$$\sigma_A = -p_A + \frac{4k_A}{3}(u_A)_x$$

composed of a pressure term and a shear viscosity term. The constant k_A is the *viscosity* of the Ath species given in dimensions of mass per length per unit time. It is difficult to motivate the viscosity term in one-dimensional flow, and the real justification of this term must come from the form of the stress tensor in higher dimensions. In any case, our assumptions imply that

$$\sigma = -p + \tfrac{4}{3}u_x(\Sigma k_A) + \tfrac{4}{3}\Sigma k_A y_A (U_A)_x$$

Neglecting the last term and defining $k = k_A / y_A$ gives the next constitutive assumption, which has the exact form as the stress assumption in ordinary gas dynamics, provided that a viscous term is included.

Constitutive Assumption 2 (Stress). We assume that

$$\sigma = -p + \frac{4k}{3}u_x \tag{23}$$

Finally we come to the heat flux q. We assume that Fourier's law holds for the Ath component, namely that the heat flux is proportional to the temperature gradient:

$$q_A = -\lambda_A T_x$$

where λ_A is the *thermal conductivity* of the Ath component, assumed to be constant. Defining $\lambda = \Sigma\lambda_A$ gives $\Sigma q_A = -\lambda T_x$ and (20) becomes, upon ignoring the cubic term in U_A,

$$q = -\lambda T_x - \Sigma\sigma_A U_A + \Sigma\rho_A e_A U_A \tag{24}$$

The second term is

$$\Sigma\sigma_A U_A = \Sigma\left(-p_A + \frac{4k_A}{3}(u_A)_x\right)U_A$$

which becomes, upon neglecting the viscosity term,

$$\Sigma\sigma_A U_A = -\Sigma p_A U_A \tag{25}$$

The final term in (24) is

$$\Sigma\rho_A e_A U_A = \Sigma\rho_A\left(h_A - \frac{p_A}{\rho_A}\right)U_A \tag{26}$$

Substituting (25) and (26) into (24) yields the next constitutive assumption.

Constitutive Assumption 3 (Heat Flux). We assume that the heat flux is given by

$$q = -\lambda T_x + \Sigma\rho_A h_A U_A \tag{27}$$

We point out that (27) recognizes a contribution to the heat flux from the diffusion velocities. Using the relation for the h_A given by

$$h_A = h_A^0 + c_p(T - T_0)$$

we note that the constitutive equation for q can be written as

$$q = -\lambda T_x + \rho\Sigma y_A h_A^0 U_A \tag{28}$$

Our final constitutive assumptions concern the form of the diffusion velocities. Here we are really making an assumption about how mass diffuses, or how one substance diffuses into another. The simplest assumption is that a substance satisfies Fick's law, where the mass flux of a substance is proportional to its concentration gradient, that is,

$$\rho_A U_A = -D_A(y_A)_x \tag{29}$$

for each A, where D_A is the *mass diffusion coefficient*. This means that the diffusion of A does not depend on the concentrations of the other species or their spatial concentration gradients. A more general assumption would be

$$\rho_A U_A = -\Sigma_{AB}(y_B)_x$$

where the sum is over B = 1, . . . , N, and the matrix D_{AB} represents a set of diffusion coefficients defining how A diffuses into B, for all species B. Thus, in (29), the matrix D_{AB} is a diagonal matrix. For just two species, say A and B, assuming (29) would mean

$$\rho_A U_A = -D_A(y_A)_x, \qquad \rho_B U_B = -D_B(y_B)_x \tag{30}$$

Using the fact that $y_A + y_B = 1$, it is straightforward to show that $D_A = D_B$. Therefore, when we deal with just two species, which will be the common model in the sequel, we shall assume that

$$\rho_A U_A = -D(y_A)_x, \qquad A = 1,2 \tag{31}$$

where $D > 0$ is the mass diffusion constant.

Reactive Flow Equations

We now summarize our work so far and then put all the information together to obtain a final set of evolution equations for reacting fluids. We have conservation of mass,

$$\frac{D\rho}{Dt} + \rho u_x = 0 \tag{32}$$

the species equations,

$$\rho \frac{Dy_A}{Dt} + (\rho_A U_A)_x = r_A \qquad (A = 1, \ldots, N) \tag{33}$$

momentum balance,

$$(\rho u)_t = -(\rho u^2 - \sigma)_x \tag{34}$$

energy balance,

$$\left(\rho e + \frac{\rho u^2}{2}\right)_t = -\left[\rho u\left(e + \frac{u^2}{2}\right) - \sigma u\right]_x - q_x \tag{35}$$

with the following constitutive assumptions:

$$e = \Sigma h_A^0 y_A + c_p(T - T_0) - \frac{p}{\rho} \qquad \text{(internal energy)} \tag{36}$$

$$\sigma = -p + \frac{4k}{3} u_x \qquad \text{(stress)} \tag{37}$$

$$q = -\lambda T_x + \rho \Sigma h_A^0 y_A U_A \qquad \text{(heat flux)} \tag{38}$$

$$\rho_A U_A = -D_A(y_A)_x \qquad \text{(diffusion velocities)} \tag{39}$$

$$\Sigma y_A = 1 \tag{40}$$

and the ideal gas law $p = (R^0/W)\rho T = R\rho T$. Now it is easy to see that (36)–(39) can be used to reduce the problem to $N + 4$ equations in the $N + 4$ unknowns ρ, u, T, p, and $y_1, \ldots, y_N$. These $N + 4$ equations will be (32)–(35) along with (40). To determine the actual equations is a little more difficult and involves some algebra, especially to reduce the energy equation (35) to the desired form. Easily, the species equations (33) become, using (39),

$$\rho\frac{Dy_A}{Dt} - (D_A(y_A)_x)_x = r_A \qquad (A = 1,\ldots,N) \tag{41}$$

and with only a little effort the momentum equation becomes, using (37) and (32),

$$\rho\frac{Du}{Dt} + p_x = \frac{4}{3}ku_{xx} \tag{42}$$

We leave it as an exercise (that everyone should do once in a lifetime) to show that the energy equation (35) reduces to

$$c_p\rho\frac{DT}{Dt} - \frac{Dp}{Dt} = -\Sigma h_A^0 r_A + \frac{4}{3}ku_x^2 + \lambda T_{xx} \tag{43}$$

Consequently, we finally have a set of equations that govern a multicomponent reacting fluid: equations (32), (41), (42), (43), (40), and the ideal gas equation $p = R\rho T$.

In the case that there are only two species A and B, where $A \to B$ and the reaction is exothermic, we showed that $D_A = D$ for $A = 1, 2$; and that case we also have

$$-\Sigma h_A^0 r_A = -h_1^0 r_1 - h_2^0 r_2 = (h_1^0 - h_2^0)r = Qr$$

where $r = -r_1 = r_2$ (the rate of production of B must equal the rate of depletion of A), and

$$Q = h_1^0 - h_2^0 > 0$$

Here Q is the positive *heat of reaction*, or the heat energy given off in the exothermic reaction (the energy of formation of the reactant A is greater that the energy of formation of the product B). Therefore, if we denote $y = y_1$, the mass fraction of the reactant A, the governing equations are

$$\frac{D\rho}{Dt} + \rho u_x = 0 \tag{44}$$

$$\rho\frac{Dy}{Dt} - Dy_{xx} = -r \qquad \text{one-dimensional reactive flow} \tag{45}$$

$$\rho\frac{Du}{Dt} + p_x = \frac{4}{3}ku_{xx} \qquad \text{equations for } A \to B \text{ chemistry} \tag{46}$$

$$\rho c_p\frac{DT}{Dt} - \frac{Dp}{Dt} = Qr + \frac{4}{3}ku_x^2 + \lambda T_{xx} \tag{47}$$

$$p = R\rho T \tag{48}$$

The only thing missing from the last set of equations is the form of the chemical reaction rate r. A standard assumption is embodied in the *law of mass action*, which requires that the rate r be proportional to the concentration of the reactant species (i.e., ρy) entering the reaction. The proportionality constant, called the rate constant (a misnomer), is generally a function of temperature. Therefore, we assume that $r = K(T)\rho y$. The form of the rate constant $K(T)$ can be obtained from kinetic theory in some cases, and it relates to the number of effective collisions between the molecules A that produce a reaction. The rate constant for elementary reactions is almost always expressed as

$$K(T) = k_0 T^\beta e^{-E/RT}$$

where E is the *activation energy* (the excess energy required to produce the reaction), k_0 is a constant factor, and β is small nonnegative power. The case $\beta = 0$ corresponds to classical Arrhenius theory. Therefore, we assume the *Arrhenius law*

$$r = k_0 \rho y e^{-E/RT} \tag{49}$$

for the form of the chemical reaction rate. (For a complete discussion of rate laws in combustion theory, especially, see Williams [1985], and in general, for example, see Nauman [1987].)

Now that we have obtained a full set of equations, we are in a position to solve some specific problems. In the next section we examine some particular models where the equations of reacting fluids apply.

EXERCISE

1. Verify (43).

7.2 REACTIVE FLOW MODELS

In Section 7.1 we formulated a general set of nonlinear PDEs that model a reacting fluid having two components, a reactant A and a product B. What kinds of problems are we interested in solving? First, the general set of equations cannot be solved analytically. It is possible to develop computer algorithms that give numerical solutions to these equations for diverse initial and boundary conditions; in fact, there is a substantial, ongoing effort in this regard. However, to make analytical progress on the equations, we must resort to simplified models where, in certain limits, some of the terms in the equations are small and may safely be neglected. This strategy is the theme of this entire chapter. Setting up and solving simplified equations give insight into the behavior of the general system and also indicates the breadth of applicability of the general equations.

The simplest thing that can be done, and this is true for any system of PDEs, is to examine a problem that is spatially homogeneous, where there is no dependence on the spatial variable x. For the reactive fluid equations this problem is called the Semenov problem. It answers the temporal question of how combustion occurs in a homogeneous system when initial data are imposed.

For reference, we write down the governing PDEs obtained in Section 7.1.

$$\frac{D\rho}{Dt} + \rho u_x = 0 \tag{1}$$

$$\rho \frac{Dy}{Dt} - Dy_{xx} = -r \tag{2}$$

$$\rho \frac{Du}{Dt} + p_x = \frac{4}{3} k u_{xx} \tag{3}$$

$$\rho c_p \frac{DT}{Dt} - \frac{Dp}{Dt} = Qr + \frac{4}{3} k u_x^2 + \lambda T_{xx} \tag{4}$$

$$p = R\rho T \tag{5}$$

where $r = r(\rho, y, T)$ is the chemical reaction rate for A $\rightarrow$ B.

Before proceeding, however, we make one remark about scaling, or reformulating the problem in dimensionless form. It is common practice in applied problems to select dimensionless variables and rewrite the problem in terms of those scaled quantities. In fact, this practice is essential in problems where perturbation methods are to be applied. However, we postpone this general normalization procedure on the reactive flow equations (1)–(5) until Section 7.4, where we shall actually be interested in the order of magnitude of the various terms in the equations.

Semenov Problem

We imagine a long tube containing a combustible reactant A at constant temperature $T = T_0$ and composition $y = y_0$. We assume that there is no fluid motion ($u = 0$) for all times, and the density is constant, $\rho = \rho_0$. We ask how the system evolves in time assuming that there is no dependence on the spatial coordinate x. Thus all the remaining state variables, T, p, and y, depend only on the time t, and the governing equations (1)–(5) reduce immediately to two ODEs,

$$\rho_0 y' = -r(\rho_0, y, T) \tag{6}$$

$$c_p \rho_0 T' - p' = Qr(\rho_0, y, T) \tag{7}$$

and the equation of state

$$p = R\rho_0 T \tag{8}$$

Equations (6) and (7) are the species and energy equations [(2) and (4)]; the continuity equation (1) and the momentum equation (3) are satisfied identically. Here, of course, a prime denotes d/dt. The reaction rate r has not been specified yet, but later we shall assume the Arrhenius form. Now the pressure p in (7) can be eliminated using (8), and we obtain a system of two equations for the temperature T and composition y, which can be written in the form

$$y' = \rho_0^{-1} r(\rho_0, y, T) \tag{9}$$

$$T' = \frac{\gamma Q}{c_p \rho_0} r(\rho_0, y, T) \tag{10}$$

where we have defined γ by $\gamma^{-1} = 1 - R/c_p$. The initial conditions that go with (9) and (10) are

$$T(0) = T_0, \qquad y(0) = y_0 \tag{11}$$

The two equations (9) and (10) can be easily reduced to a single first order differential equation for the temperature T. Multiplying (9) by $\gamma Q/c_p$ and adding the result to (10) gives

$$\frac{\gamma Q}{c_p} y' + T' = 0$$

which is instantly integrable to

$$\frac{\gamma Q}{c_p} y + T = \text{constant} = \frac{\gamma Q}{c_p} y_0 + T_0 \tag{12}$$

where the constant of integration has been determined using the initial condition (11). Consequently, y can be eliminated in (10) and we obtain

$$T' = \frac{\gamma Q}{c_p \rho_0} r\left(\rho_0, \frac{T_b - T}{\gamma Q/c_p}, T\right), \qquad T(0) = T_0 \tag{13}$$

where T_b is the constant defined by

$$T_b \equiv \frac{\gamma Q}{c_p} y_0 + T_0 \tag{14}$$

Therefore, for a given reaction rate r, we have formulated an initial value problem for the temperature T in a homogeneous, reacting medium, initially at temperature T_0.

When Arrhenius kinetics is imposed, the reaction rate r takes the form

$$r = K\rho y \exp\left(\frac{E}{R^0 T}\right) \tag{15}$$

Substituting this expression into (13) gives, after simplification,

$$T' = K(T_b - T)\exp\left(\frac{E}{R^0 T}\right) \qquad T(0) = T_0 \tag{16}$$

In this ordinary differential equation the variables separate and the initial value problem can be solved exactly in terms of a special function $\mathrm{Ei}(z)$, called the exponential integral, defined by

$$\mathrm{Ei}(z) = -\int_z^\infty u^{-1} e^{-u}\, du$$

This exact solution is given by (Exercise 3)

$$\begin{aligned} Kt = \mathrm{Ei}\left(\frac{E}{R^0 T}\right) - \mathrm{Ei}\left(\frac{E}{R^0 T_0}\right) \\ + \exp\left(\frac{E}{R^0 T_b}\right)\left[\mathrm{Ei}\left(\frac{E}{R^0 T_0} - \frac{E}{R^0 T_b}\right) - \mathrm{Ei}\left(\frac{E}{R^0 T} - \frac{E}{R^0 T_b}\right)\right] \end{aligned} \tag{17}$$

This formula gives T implicitly in terms of t, and, despite it being exact, it is reasonably opaque. One recourse would be to graph (17) using a software package (e.g., MATHEMATICA) that can handle special functions such as $\mathrm{Ei}(z)$. Another, probably simpler strategy would be to forego the exact solution and solve the initial value problem (16) directly using a numerical differential equation solver. We adopt this second procedure. But first, let us reformulate the problem in dimensionless form. Scaling temperature by T_0 and time by $1/K$ (as we shall observe later, this time scale is not the best choice if a perturbation solution is sought in the case of large activation energy), we introduce the dimensionless variables

$$U = \frac{T}{T_0}, \qquad \tau = Kt$$

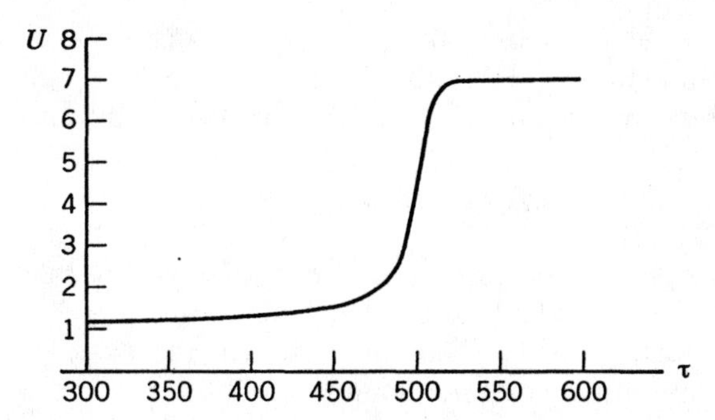

Figure 7.4. Numerical solution of (18) using MATHEMATICA for $\theta = 10$ and $b = 7$.

Then, in terms of τ and U the initial value problem (16) becomes

$$\frac{dU}{d\tau} = (b - U)e^{-\theta/U}, \qquad U(0) = 1 \tag{18}$$

where

$$b \equiv \frac{T_b}{T_0}, \qquad \theta \equiv \frac{E}{R^0 T_0}$$

A typical value for the ratio b is 7 (see, e.g., Zel'dovich et al. [1985]), and θ is usually large, perhaps even the range 30 to 100. For our numerical calculation, however, we take $\theta = 10$. The result of the calculation (using the package NDSolve in MATHEMATICA) is shown in Fig. 7.4. Interestingly enough, the temperature experiences almost no change for a long induction period; then, near $t \cong 500$, there is a rapid transient (an explosion) and the reaction goes to completion almost immediately, leveling off at the temperature T_b. The composition y, which can be calculated from (12), has a similar form as the temperature curve, but reversed; y remains close to its initial value during the long induction period and then experiences a rapid decrease in the explosion regime as the reactant is consumed by the chemical reaction. Further numerical experiments show that as θ is increased, the induction period becomes extremely long.

Can we predict where the rapid transient occurs, that is, the length of the induction period? The answer to this question takes us into the subject of high-activation-energy asymptotics, which we now introduce. High-activation-energy asymptotics is a perturbation method that uses to advantage the fact that the activation energy E is usually large in combustion problems. Our excursion will be fairly brief since a detailed analysis would require singular perturbation techniques and take us too far afield.

As we remarked, the dimensionless activation energy θ introduced in the prior discussion is generally a large parameter. Therefore, its reciprocal

$\varepsilon = 1/\theta$ is a small parameter, and problems containing a small parameter suggest perturbation methods where the solution is sought as a series expansion in the small parameter. However, if such an expansion, say

$$U(\tau) = U_0(\tau) + \varepsilon U_1(\tau) + \varepsilon^2 U_2(\tau) + \cdots$$

is substituted into (17), coefficients of powers of ε cannot be balanced since the right side is exponentially small. To remedy this problem we require a different time scale, and therefore we return to the unscaled dimensioned problem (16) and introduce dimensionless variables

$$U = \frac{T}{T_0}, \qquad \tau = \frac{t}{t_0} \tag{19}$$

where the time scale t_0 will be chosen later. In the variables (19) the initial value problem (16) becomes

$$\frac{dU}{d\tau} = Kt_0(b - U)e^{-\theta/U}, \qquad U(0 = 1) \tag{20}$$

We now pull a factor $\exp(-\theta)$ out of the exponential term and write the differential equation in (20) as

$$\frac{dU}{d\tau} = Kt_0 e^{-\theta}(b - U)e^{\theta - \theta/U} \tag{21}$$

or, in terms of ε,

$$\frac{dU}{d\tau} = Kt_0 e^{-1/\varepsilon}(b - U)e^{(1-1/U)/\varepsilon} \tag{22}$$

Now we assume a perturbation expansion of the form

$$U(\tau) = 1 + \varepsilon\phi(\tau) + O(\varepsilon^2) \tag{23}$$

which represents a small deviation from the initial, scaled temperature $U = 1$. Substituting (23) into (22) gives

$$\varepsilon\phi' + O(\varepsilon^2) = t_0 K e^{-1/\varepsilon}(b - 1 - \varepsilon\phi + O(\varepsilon^2))e^{\phi}(1 + O(\varepsilon)) \tag{24}$$

where we have used the expansion

$$e^{(1-1/U)/\varepsilon} = e^{(1-1/(1+\varepsilon\phi+\cdots))/\varepsilon} = e^{\phi}(1 + O(\varepsilon)) \tag{25}$$

Close examination of (24) suggests that the coefficient $t_0 K e^{-1/\varepsilon}$ should be order $O(\varepsilon)$ so that the chemical reaction can affect the solution at leading order. This means that the time scale t_0 should be chosen by

$$t_0 = K^{-1}\varepsilon e^{1/\varepsilon} \tag{26}$$

Consequently, at order ε in (24) we have the equation

$$\phi' = (b - 1)e^{\phi} \quad \text{with} \quad \phi(0) = 0 \tag{27}$$

Equation (27) can be solved by separation of variables to get the first correction term ϕ as

$$\phi(\tau) = -\ln[1 - (b - 1)\tau] \tag{28}$$

Therefore, in the scaled coordinates (19) we have an approximate solution

$$U(\tau) = 1 - \theta^{-1}\ln[1 - (b - 1)\tau] + O(\theta^{-2}) \tag{29}$$

It is clear that this expansion becomes invalid near $\tau = 1/(b - 1)$ where the logarithmic correction term becomes singular. In real time t, this singularity occurs at

$$t_{bl} = \frac{[K(b - 1)]^{-1}e^{\theta}}{\theta} \tag{30}$$

which is large; this time is called the blowup time, and it is the time at which thermal runaway, or the explosion, occurs.

The perturbation approximation (29) is valid only for $\tau < 1/(b - 1)$, and it is graphed in Figure 7.5. One should now determine how this segment of the solution connects to the constant value $U = b$, when the reaction is

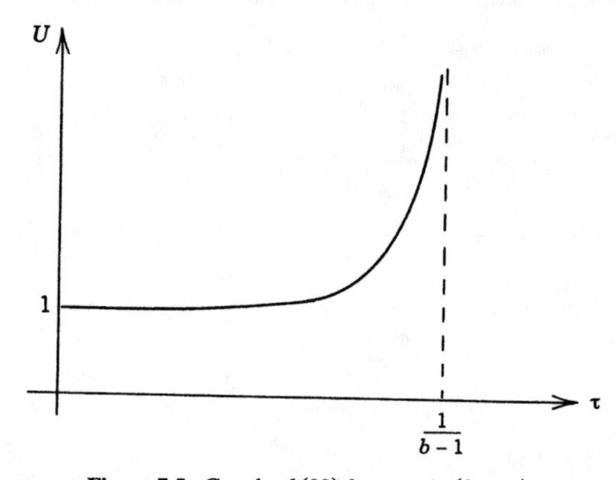

Figure 7.5. Graph of (29) for $\tau < 1/(b - 1)$.

completed. A very narrow transition region at $\tau = 1/(b-1)$ which asymptotically connects these two portions of the solution can be determined by singular perturbation methods (the method of matched asymptotic expansions); we refer the interested reader to Kapila [1983] for a complete discussion.

Detonations

One usually thinks of a detonation, or an explosion, as a process where some combustible material (say, TNT) reacts almost instantaneously and releases large amounts of heat energy. In the preceding paragraphs a simple model of this process, where the explosion occurred uniformly throughout the material, was discussed. In a real explosion, however, there is spatial structure to the process, and if one wishes to understand the details of detonation, as one must in certain high-tech operations, a better model is required. One common model, which is the subject of the next section, is to regard a detonation as a shock wave propagating into a combustible material with velocity $D(t)$ (not to be confused with the mass diffusion coefficient D); the compression that accompanies the shock raises the temperature of the material, initiating the chemical reaction, which is assumed to occur in a narrow region, called the reaction zone, just behind the shock. Figure 7.6 indicates this idea schematically, showing a typical resolved pressure profile at a fixed moment of time throughout the medium, which is assumed to represent a one-dimensional reactive material. The detonation process is extremely rapid; for example, the detonation velocity $D(t)$ is on the order of thousands of meters per second, and the pressures involved are on the order of thousands of atmospheres. In the reaction zone behind the shock the physical parameters (pressure, density, temperature, particle velocity, and

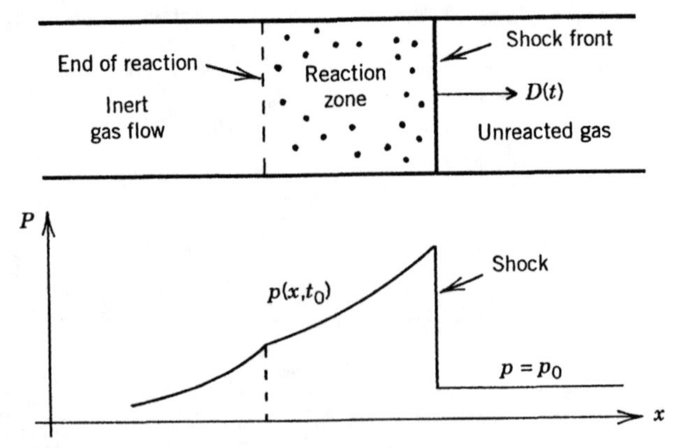

Figure 7.6. Detonation in a tube with a resolved reaction zone showing a typical pressure profile at a fixed time t_0.

composition) are governed by the basic reactive flow equations obtained in Section 7.1 and repeated as equations (1)–(5) at the beginning of this section. But on this rapid time scale, diffusion does not have time to occur, and the viscous forces that transfer momentum are also slow. Thus, in a detonation process we may take $D = 0$ (mass diffusion coefficient), $\lambda = 0$, and $k = 0$ in equations (1)–(5), obtaining

$$\rho_t + u\rho_x + \rho u_x = 0 \tag{31}$$

$$\rho(y_t + uy_x) = -r \tag{32}$$

$$\rho(u_t + uu_x) + p_x = 0 \tag{33}$$

$$c_p\rho(T_t + uT_x) - (p_t + up_x) = Qr \tag{34}$$

where $p = R\rho T$ and $r = r(y, T, \rho)$ is the reaction rate. [Again, one should not confuse the mass diffusion coefficient D with the detonation velocity $D(t)$; in the sequel and in the next section the mass diffusion coefficient is taken to be be identically zero, and therefore $D = D(t)$ will denote the detonation velocity.] To state it differently, we have taken the stress σ, the flux q, and the diffusion velocities U_A to be given by the constitutive relations

$$\sigma = -p, \qquad q = 0, \qquad U_A = 0$$

Thus detonations are rapid exothermic processes dominated by convection and energy and momentum transfer; the governing equations are hyperbolic, and they strongly resemble the equations of gas dynamics with a source term in the energy equation (Qr), due to thermal energy released by the chemical reaction, and with an additional equation, the species equation, that governs the composition y.

What kinds of problems are studied in detonation theory? First, it should be clear that we cannot solve (31)–(34); only in some rare cases, with special reaction rates, have similarity solutions been found, and ultimately this has required numerical resolution of the ordinary differential equations resulting from the similarity transformation (see Chapter 4 for a discussion of this method and J. D. Logan and J. J. Perez, SIAM J. Appl. Math **39**, 512–527 (1980) for the application to detonation). The problem that is of general interest is the so-called *piston problem*, which can be stated as follows. At time $t = 0$ a piston (modeling, for example, a detonator) located at $x = 0$ moves forward into a reactive material, generating a shock wave with speed $D(t)$. The shock initiates the chemical reaction which occurs in a region behind the shock; the end of the reaction zone, where the chemistry has gone to completion, is marked by a curve behind an inert, nonreactive flow ($r = 0$) connects to the piston path. This state of affairs is depicted in the spacetime diagram in Figure 7.7. Mathematically, one must solve the governing equations (31)–(34) in the reaction zone subject to boundary conditions (jump conditions

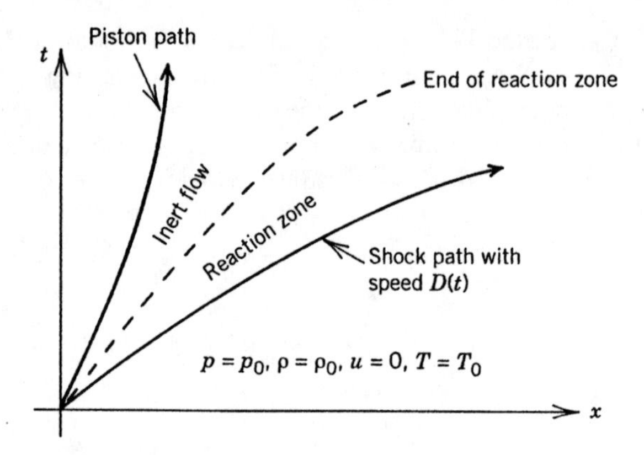

Figure 7.7. Spacetime diagram of a typical piston problem.

across the shock) along the shock path, and then connect the flow to the solution in the nonreactive region of (31)–(34) with $r = 0$. Analytically, of course, the solution of the piston problem is impossible in most cases, and there has been a longtime effort to obtain numerical solutions and approximate solutions in special cases.

One possibility is to search for traveling wave solutions where the shock velocity $D(t)$ is constant and the flow is steady state. The resulting solution is called the ZND solution, and the details of the calculation will be presented in the next section.

EXERCISES

1. Reformulate the governing detonation equations (31)–(34) in shock-fixed coordinates defined by $\tau = t$ and $\eta = \int_0^t D(s)\,ds - x$. Here τ measures time and η is the distance to a point in the flow measured by an observer riding on the shock with velocity $D(t)$.
2. Reformulate the governing detonation equations (31)–(34) in shock-time coordinates defined by $\tau = t - \int_0^x D^{-1}(s)\,ds$ and $\xi = x$. Here τ measures time after arrival of the shock.
3. Derive formula (17).
4. Carry out some numerical experiments using an ordinary differential equations solver to calculate the numerical solution to the initial value problem (18) for various values of θ. Use $b = 7$ and $\theta = 1, 5, 10, 15$, and 20. What happens for large θ?
5. A continuously stirred tank reactor has temperature $T = T(t)$ governed by

$$\frac{dT}{dt} = qe^{-\theta/T} - k(T - T_f)$$

where q, k, θ, and T_f are positive constants representing the rate of heat generation, the heat loss, the activation energy, and the ambient temperature, respectively. Let $\mathscr{D} = q/k$, which is the Damköhler number, and assume that $\theta > 4T_f$. The reactor is in an equilibrium state T_e if $T = T_e$ is a constant solution of the governing equation. Sketch a graph of the equilibrium states T_e versus the Damköhler number $\mathscr{D}$. Determine which equilibrium states on this locus are stable and which are unstable, and describe qualitatively the temperature behavior of the reactor if the Damköhler is slowly increased from a small positive value to a large value, at each instant allowing the reactor to come to equilibrium.

6. In Exercise 5 take $k = 0$ (no heat loss) and assume that the initial condition is $T(0) = T_0$. Solve the resulting initial value problem in terms of the exponential function $\text{Ei}(y)$ to obtain $qt = \theta[f(T_0) - f(T)]$, where $f(z) = \text{Ei}(\theta/z) - z\theta^{-1}e^{\theta/z}$. Use the asymptotic formula

$$\text{Ei}(y) \sim e^y\{1/y + 1/y^2 + \cdots\}$$

for large y to obtain the implicit, asymptotic form of the solution

$$\delta t \sim 1 - \left(\frac{T}{T_0}\right)^2 e^{\theta/T - \theta/T_0} \qquad \text{as } \theta \to \infty$$

where δ is a known constant. Finally, assume that T has an expansion

$$T(t) = T_0 + T_0^2\phi(t)\theta^{-1} + \cdots \qquad \text{as } \theta \to \infty$$

and show that $\phi(t) = -\ln(1 - \delta t)$, thus obtaining a first order approximation for the temperature in the induction zone for large values of θ. When will the temperature blow up?

7. Consider the boundary value problem for $T = T(x)$:

$$T'' = -qe^{-\theta/T}, \qquad -1 < x < 1$$

$$T(-1) = T(1) = T_f$$

where θ is large and $|(T - T_f)/T_f| \ll 1$. Assume a solution of the form

$$T(x) = T_f + T_f^2\phi(x)\theta^{-1} + \cdots$$

and show that the first order correction term $\phi = \phi(x)$ satisfies the problem

$$\phi'' = -\lambda e^{\phi}, \qquad \phi(-1) = \phi(1) = 0$$

where λ is a known positive constant, assumed to be order 1. Show that

$$\phi(x) = \phi_m + 2 \ln \operatorname{sech}\left(\sqrt{\frac{\lambda}{2}}\, e^{\phi_m/2} x\right)$$

where $\phi_m = \phi(0)$. Finally, derive the relationship

$$\sqrt{\frac{\lambda}{2}} = e^{-\phi_m/2} \cosh^{-1}(e^{\phi_m/2})$$

between λ and ϕ_m and show there is a critical value λ_c above which the boundary value problem for ϕ has no solution. *Hint*: Write $e^{-\theta/T} = e^{-\theta/T_f} e^{\theta(T-T_f)/T_f^2}$.

7.3 DETONATION WAVES

In Section 7.2 we showed how a model for detonations could be obtained from the general equations of reactive flow, assuming that viscous and diffusion effects could be neglected. The governing equations are

$$\rho_t + u\rho_x + \rho u_x = 0 \tag{1}$$

$$\rho(y_t + uy_x) = -r \tag{2}$$

$$\rho(u_t + uu_x) + p_x = 0 \tag{3}$$

$$c_p\rho(T_t + uT_x) - (p_t + up_x) = Qr \tag{4}$$

which represent, respectively, mass conservation , chemical species balance, momentum balance, and energy balance. We shall refer to these equations as the *strong detonation model equations.* Here, to review, ρ is the density, u the particle velocity, p the pressure, and T the temperature. The given function $r = r(y, T, \rho)$ is the rate of an irreversible, exothermic, one-step chemical reaction A → B, and y measures the mass fraction of the reactant A. The constants c_p and Q are the specific heat of the fluid and the heat of reaction, respectively.

Equations (1)–(4) govern the reactive flow in regions where the flow is smooth. If a shock is present, certain jump conditions must hold true across the discontinuity as we noted in Chapters 3 and 5. To obtain these conditions easily, we recall the result that if an equation is in conservation form

$$g_t + f_x = 0$$

then the corresponding jump condition is given by

$$-D[g] + [f] = 0$$

where D is the shock velocity and $[g] = g_b - g_a$ is the jump in the quantity g across the shock; here g_a denotes the value of g just ahead of the shock, and g_b denotes the value of g just behind the shock (a similar formula holds for the jump in f). Therefore, if we write each equation in (1)–(4) in conservation form, the corresponding jump conditions can be written down immediately. But equations (1)–(4) were obtained from conservation laws in the first place (in Section 7.1), so we just have to look up those equations in Section 7.1. They are

$$\rho_t + (\rho u)_x = 0 \tag{5}$$

$$(\rho u)_t + (\rho u^2 + p)_x = 0 \tag{6}$$

$$\left(\rho e + \frac{\rho u^2}{2}\right)_t + \left[\rho u\left(e + \frac{u^2}{2}\right) + pu\right]_x = 0 \tag{7}$$

where

$$e = \Sigma y_A h_A^0 + c_p(T - T_0) - \frac{p}{\rho} = \frac{p}{(\gamma - 1)\rho} + Qy + \text{const} \tag{8}$$

Equations (5), (6), and (7) are conservation of mass, momentum, and energy, respectively, and (8) is the equation of state; we also have the ideal gas law $p = R\rho T$ and the relation

$$\gamma = \frac{c_p}{c_p - R}$$

The reader can easily verify that (6), (7), and (8) give (1), (3), and (4) with the given assumptions regarding the equations of state. We shall handle the jump in the reaction coordinate y in a slightly different manner, and thus we do not require a conservation form for equation (2).

From equations (6), (7), and (8) we may therefore infer the following proposition:

Proposition 1. Across a shock with speed D,

$$-D[\rho] + [\rho u] = 0 \tag{9}$$

$$-D[\rho u] + [\rho u^2 + p] = 0 \tag{10}$$

$$-D\left[\rho e + \frac{\rho u^2}{2}\right] + \left[\rho u\left(e + \frac{u^2}{2}\right) + pu\right] = 0 \tag{11}$$

The jump conditions (9)–(11) are called the *Rankine–Hugoniot conditions*, and in this context, the shock velocity D is called the *detonation velocity*, and it need not be constant. To summarize, therefore, equations (1)–(4) govern the one-dimensional flow in regions where th flow is smooth, and the Rankine–Hugoniot conditions hold across simple discontinuities or shocks.

In detonation theory it is common practice to work with the specific volume v rather than the density ρ (by definition $v = 1/\rho$), especially in the jump conditions. We shall also assume that a shock wave, if present, is moving into a region where the particle velocity u ahead of the wave is zero. This simplifies the jump conditions considerably, and one can easily show the following result.

Proposition 2. If $u_a = 0$, the Rankine–Hugoniot relations (9)–(11) can be written

$$\frac{v_b}{v_a} = 1 - \frac{u_b}{D} \tag{12}$$

$$p_b - p_a = \frac{Du_b}{v_a} \tag{13}$$

$$e_b - e_a = \tfrac{1}{2}(p_a + p_b)(v_a - v_b) \tag{14}$$

where the subscripts a and b denote values ahead and behind the shock, respectively. It also follows that

$$p_b - p_a = -\left(\frac{D}{v_a}\right)^2 (v_b - v_a) \tag{15}$$

It is of course assumed that the state ahead of the shock is known, and therefore equations (12)–(14) give three equations for the four states v_b, u_b, p_b, and e_b behind the shock, and the unknown detonation velocity (shock speed) D. Equation (15), although important in the subsequent discussion, is not independent. Equation (5) allows us to eliminate the energy e from (14) to obtain

$$(\gamma - 1)^{-1}(p_b v_b - p_a v_a) + Q(y_b - y_a) = \tfrac{1}{2}(p_a + p_b)(v_a - v_b) \tag{16}$$

which can be simplified to

$$\left(\frac{p_b}{p_a} + m\right)\left(\frac{v_b}{v_a} - m\right) = 1 - m^2 + \frac{2mQ(y_a - y_b)}{v_a p_a},$$

$$m = (\gamma - 1)/(\gamma + 1) \tag{17}$$

Thus (12), (13), and (17) give three algebraic equations for v_b, u_b, p_b, and D, provided that the jump in the reaction coordinate $[y]$ is known.

Now let us focus on the reaction coordinate. First, we shall assume that $y_a = 1$, that is, the material ahead of any shock consists entirely of reactant; no chemical reaction has yet occurred. We shall say that a shock is *reactive* if all of the chemical reaction occurs in the shock (then $y_b = 0$), and we say that a shock is *nonreactive*, or *inert*, if no chemical reaction occurs in the shock (then $y_b = 1$). We will not consider the case where partial reaction occurs in the shock. These two cases define the two problems that we wish to examine in this section. In the first case, where the shock is reactive, we will obtain the Chapman–Jouget (CJ) model, where a detonation is considered to be a shock wave propagating into a chemically reactive material in a quiescent state; all of the chemistry is assumed to take place in the shock front itself, and behind the shock is an inert gas dynamic flow. In the second case, where no reaction occurs in the shock, we obtain the Zeldovich–von Neumann–Doering (ZND) model. In the ZND model a shock moves into the quiescent, reactive material; the compression accompanying the shock raises the temperature of the material, initiating the chemical reaction. No chemistry occurs in the shock, but instead it is resolved in a reaction zone behind the shock according to the governing equations (1)–(4).

CJ Model

The CJ model is the simplest model of a detonation process. It was proposed circa 1900 independently by Chapman and Jouget. Their theory employs an idealized one-dimensional model in which it is assumed that the chemical energy is released instantaneously in a discontinuous shock front propagating into a combustible material. The theory defines a unique, steady-state velocity of the shock front and permits the calculation of the state behind this front. Mathematically, the CJ theory is elementary, for it involves only algebraic relationships expressed by the Rankine–Hugoniot equations. There is no attempt to resolve the chemistry using the equations of reactive flow.

To fix the idea we consider a long cylindrical tube filled with a combustible material; it is characterized by its pressure p_a, density ρ_a, and internal energy e_a. At time $t = 0$ the left end of the tube is ignited in such a way as to introduce a sharply defined planar flame front that is propagated to the right. For the present we ignore how such an actual ignition mechanism could be constructed; it could be, for example, that a piston strikes the material with enough force to generate the flame front. For $t > 0$ it is observed experimentally that after a brief time interval when the flame front velocity is time dependent, a steady-state situation is reached where the velocity of the front becomes constant, or nearly so. The CJ model deals with this steady regime. The sharply defined flame front is modeled as a propagating discontinuity or a constant-velocity shock with speed D. The simplicity of the model lies in the assumption that the chemical reaction, which in real

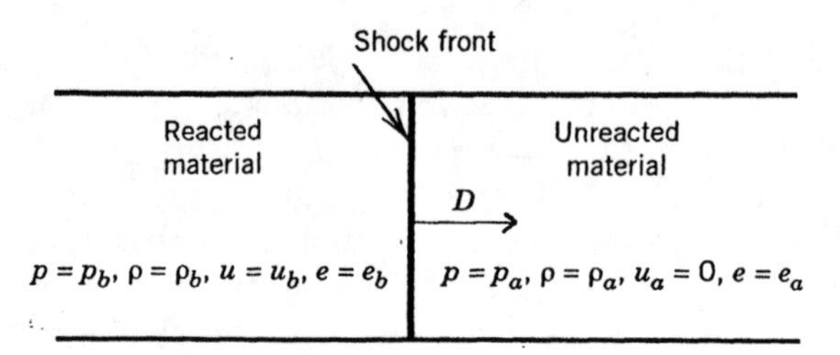

Figure 7.8. CJ model for a detonation.

phenomena may be quite complicated, takes place instantaneously in the shock front. Hence the material emerging from behind the shock is composed only products of the reaction and is characterized by a constant pressure p_b, density ρ_b, velocity u_b, and internal energy e_b. We are going to consider the case where the shock is compressive, that is, $p_b > p_a$. Figure 7.8 diagrams the physical situation.

The test of any physical model and its mathematical formulation is in its ability to predict the results of experiments. The CJ model, although ignoring the transients and the details of the chemistry, has met with high success in its predictive capability, especially for gases; the theory is one of the basic tools in detonation science. We remark that the diameter of the tube must be large enough so that dissipative effects from the sides of the tube do not affect the planarity of the front; for small diameters the flame front is distinctly nonplanar, and below a critical diameter, called the failure diameter, it is often not possible to propagate a front at all.

The equations governing the CJ model are the Rankine–Hugoniot equations, which express conservation of mass, momentum, and energy across a shock front. These equations are (12), (13), and (17), the latter with $y_a = 1$ and $y_b = 0$ (since all the chemistry is assumed to occur in the shock). These are three equations in four unknowns: u_b, p_b, v_b, and D, representing the states behind the shock and the shock velocity. It is convenient to regard the solution as a one-parameter family depending on D; we shall show that each choice of D above some minimal value leads to a state (not necessarily unique) p_a, u_a, and v_a behind the compressive shock. For a special value of D, we shall obtain a unique state called the CJ state. The situation is best illustrated graphically rather than analytically. On a pv = diagram (Figure 7.9) we graph equation

$$\left(\frac{p}{p_a} + m\right)\left(\frac{v}{v_a} - m\right) = 1 - m^2 + \frac{2mQ}{p_a v_a} \tag{18}$$

which is a hyperbola called the reacted Hugoniot curve. The pv-state behind shock, (v_b, p_b), must lie on this Hugoniot curve, by (17). Further, consider

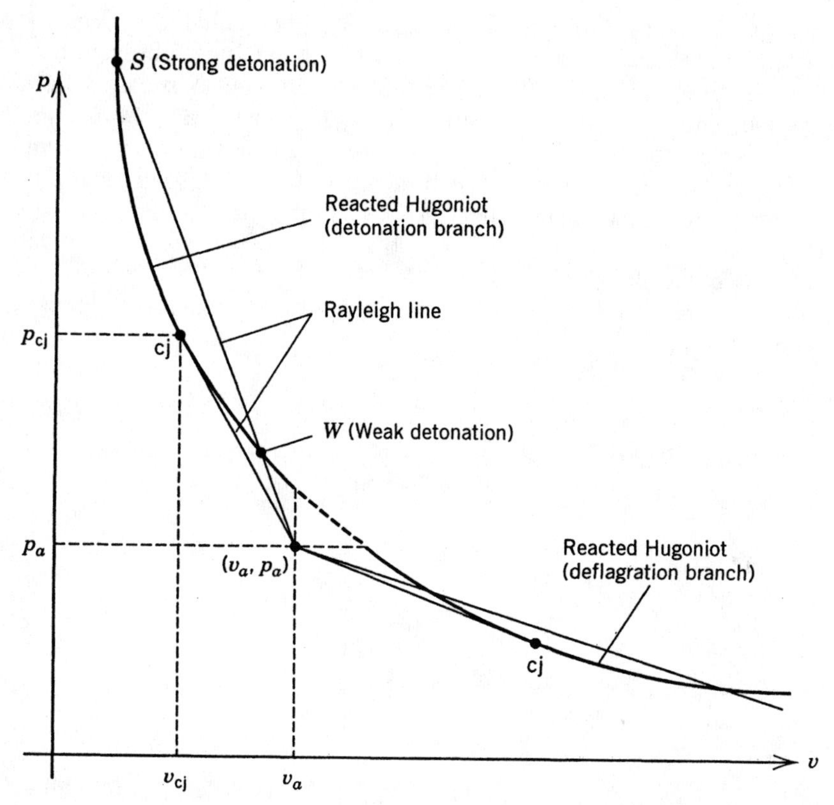

Figure 7.9. A pv-Diagram showing the detonation (upper) and deflagration (lower) branches of the reacted Hugoniot curve.

the straight line

$$p - p_a = -\left(\frac{D}{v_a}\right)^2 (v - v_a) \tag{19}$$

which passes through the state (v_a, p_a). This line, called the Rayleigh line, has negative slope and the state (v_b, p_b) must, by (15), lie on this line as well. Consequently, the pv-state behind the shock must be at the intersection of the Hugoniot and the Rayleigh line. Because the Hugoniot curve is concave up, the Rayleigh line will intersect the Hugoniot twice, once, or not at all, depending on the value of D that determines the slope of the Rayleigh line. Several such Rayleigh lines are shown in Figure 7.9. The upper branch of the Hugoniot, where the shock is compressive ($p_b > p_a$), is called the *detonation branch*, and the lower branch, where $p_b < p_a$, is called the *deflagration branch*. Here we shall only be concerned with the detonation branch. If there are two intersection points, the upper intersection point S is called a *strong detonation*, and the lower intersection point W is called a *weak detonation*. The case where the Rayleigh line is tangent to the Hugoniot defines a unique state (v_{cj}, p_{cj}) behind the shock, called the CJ state, and this minimal value of D, denoted by D_{cj}, is called the CJ detonation velocity. For $D < D_{cj}$, no intersections occur and we do not obtain a solution for the Hugoniot equations behind the shock.

To determine D_{cj} we equate the derivatives dp/dv taken from equations (18) and (19). From (18),

$$\frac{dp}{dv} = \frac{p + mp_a}{v - mv_a}$$

and from (19),

$$\frac{dp}{dv} = -\left(\frac{D}{v_a}\right)^2$$

Equating and using the jump conditions (12) and (13) gives, after simplification and using the subscript cj to denote the state behind the shock in this special case, the additional condition

$$u_{cj} = \frac{D_{cj}(1 - m)}{2} - \frac{v_a p_a(+m)}{2D_{cj}} \tag{20}$$

Equation (20), appended to (12), (13), and (17), with $y_a = 1$ and $y_b = 0$, gives four equations that uniquely determine a state u_{cj}, v_{cj}, p_{cj} behind the shock, as well as a detonation velocity D_{cj}. This special detonation, where the Rayleigh line is tangent to the Hugoniot, is called a *CJ detonation*, and it is the one often observed in experiment (strong detonations can be observed when a piston is supporting the flow from behind with sufficient overdrive). Besides being the detonation with the minimum allowable shock velocity, it has many other interesting properties that are explored in the exercises.

In many cases the pressure p_b behind the shock is so high that the pressure p_a ahead of the shock can be neglected. When we set $p_a = 0$ in

(12), (13), (17), and (20) we obtain the *strong shock conditions*

$$u_{\text{cj}} = \frac{D_{\text{cj}}}{\gamma + 1} \tag{21}$$

$$v_{\text{cj}} = \frac{\gamma v_a}{\gamma + 1} \tag{22}$$

$$p_{\text{cj}} = \frac{D_{\text{cj}}^2}{(\gamma + 1)v_a} \tag{23}$$

$$p_{\text{cj}} = \frac{2(\gamma - 1)Q}{v_a} \tag{24}$$

The detonation products behind the shock have been assumed to satisfy the ideal gas equation of state. From our analysis in Chapter 5 we know that signals in an ideal gas are propagated with sound speed c where $c^2 = \gamma p v$. Using the strong shock conditions one can easily show that

$$c_{\text{cj}} = D_{\text{cj}} - u_{\text{cj}} \tag{25}$$

which implies that the flow issuing from behind the shock is sonic in a CJ detonation, relative to the speed of the front. That is, an observer riding on the shock front will see unreacted material entering the shock at speed D_{cj} and reacted material leaving the shock at the speed of sound. Stated still differently, the speed of the burned gas relative to the shock equals the sound speed in the burned gas. One can show that in a strong detonation the flow behind is supersonic, and in a weak detonation the flow is subsonic.

Example 1 (A Piston Problem). In the preceding discussion we have asked only about conditions that hold across a reactive shock without regard to the initial conditions or the complete nature of the flow in the burned regime behind the shock. In gas dynamics, however, it is often assumed that there is a mechanism, usually a piston (which is an idealization of a detonator or ignitor) that initiates the flow. This piston, which is controlled by external forces, induces a shock that propagates into the reactants and initiates the chemical reaction. As we remarked in Section 7.2, the *piston problem* is to determine the entire flow characteristics between the shock and the following piston for all $t > 0$. Generally, this can be a highly complex problem, but in certain cases, which we now discuss, there is a simple solution. Consider the special case when the detonation velocity D and the piston velocity u_p are *constant*, and $u_p > u_{\text{cj}}$. Then we can take $u = u_p =$ constant in the entire region between the shock and the piston; the remaining two flow parameters, p and v, are taken to be constant and equal to their values immediately

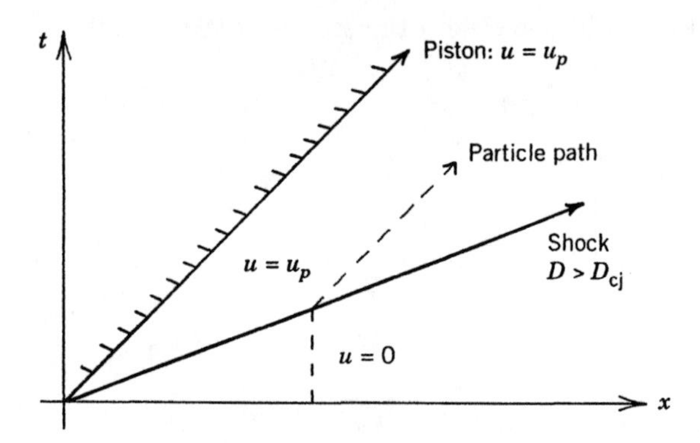

Figure 7.10. Overdriven detonation.

behind the shock, consistent with the Rankine–Hugoniot jump conditions. It is easy to see that $D > D_{cj}$, so this detonation is a strong detonation. Such a detonation is called an overdriven detonation. Figure 7.10 shows the schematic of the spacetime diagram, which includes a typical particle path. If $u_p = u_{cj}$, many of the same remarks apply. In the region between the piston and the shock front we have a constant flow given by $u = u_{cj}$, $p = p_{cj}$, $v = v_{cj}$, and the shock velocity is $D = D_{cj}$.

If $u_p < u_{cj}$ a simple flow is still possible. Figure 7.11 shows the spacetime diagram in this case. This shock front still moves with CJ velocity D_{cj}; it is followed by a rarefaction wave (where the flow parameters are constant on straight lines $x = kt$) that reduces the particle velocity u_{cj} at the front to the piston velocity u_p. Following the rarefaction is a constant state that connects to the piston.

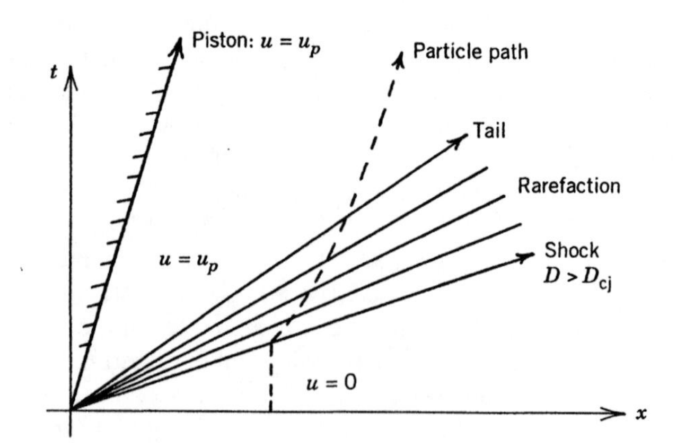

Figure 7.11. CJ detonation with $u_p < u_{cj}$.

ZND Model

During World War II there was obvious renewed interest in detonation processes. Independently, Zel'dovich (of the Soviet Union), von Neumann (of the United States), and Doering (of Germany) developed a model of detonation, now called the ZND theory, that resolves the chemistry over a finite region behind an inert shock, based on the reactive gas dynamic equations (1)–(4). So, an inert compression wave raises the temperature of the reactants and initiates a chemical reaction that takes place in a reaction zone where the state is governed by the equations of reactive flow.

In the discussion of the CJ model the shock was reactive; that is, in pv-space there was a jump from the initial state (v_a, p_a) ahead of the shock to a point on the reactive Hugoniot curve. For a ZND shock, which is nonreactive, there is a jump from the initial state to a point on the *nonreacted Hugoniot curve*

$$\left(\frac{p}{p_a} + m\right)\left(\frac{v}{v_a} - m\right) = 1 - m^2 \tag{26}$$

This is equation is (17) with $y_a = y_b = 1$ (because no reaction occurs in the shock front). The final state behind the shock must also lie on the Rayleigh line, so the state just behind the shock must be at a point of intersection of the Rayleigh line and the nonreacted (Hugoniot (see Figure 7.12). For the ZND model we assume that the shock velocity is $D = D_{\text{cj}}$, the constant CJ detonation velocity. Thus a steady ZND process is an inert shock represented by a jump up to the point N on the unreacted Hugoniot, followed by a smooth path, as yet undetermined, to a point on the reacted Hugoniot where the chemical reaction is complete ($y = 0$). One can regard (17) as an equation defining a family of partial reacted Hugoniot curves depending on the parameter $y = y_b$, with $0 \le y \le 1$. When $y = 1$ we obtain the unreacted Hugoniot passing through the initial state, and when $y = 0$ we obtain the Hugoniot for the burned or reacted materials. For y values in between we obtain a family of curves representing partially reacted material.

Thus the shock conditions for a ZND detonation are

$$\frac{v_N}{v_a} = 1 - \frac{u_N}{D} \tag{27}$$

$$p_N - p_a = \frac{Du_N}{v_a} \tag{28}$$

$$(p_N - mp_a)(v_N - mv_a) = (1 - m^2)v_a p_a \tag{29}$$

where $D = D_{\text{cj}}$, $m = (\gamma - 1)/(\gamma + 1)$, and the subscript N denotes the state just behind the shock (the point on the nonreacted Hugoniot in Figure 7.12).

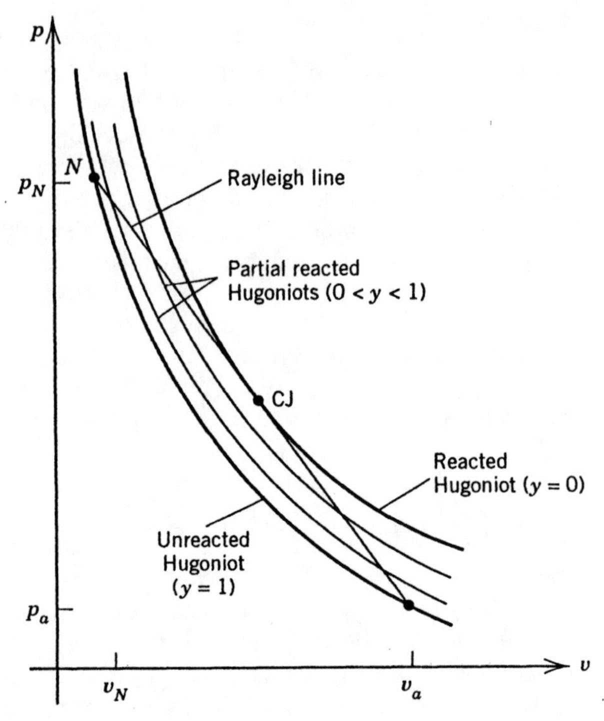

Figure 7.12. Rayleigh line corresponding to $D = D_{cj}$ traversing the one-parameter family of Hugoniot curves.

The point N is called the von Neumann point. Equations (27), (28), and (29) are just the jump conditions for mass, momentum, and energy across the inert shock. They uniquely determine the point N. If we assume strong shocks and set $p_a = 0$, we obtain the *inert strong shock conditions*

$$\frac{v_N}{v_a} = \frac{\gamma - 1}{\gamma + 1} \tag{30}$$

$$u_N = \frac{2D}{\gamma + 1} \tag{31}$$

$$p_N = \frac{2D^2}{(\gamma + 1)v_a} \tag{32}$$

where $D = D_{cj}$. We note that these conditions differ from the reactive strong shock conditions in the CJ model.

Now, to see how the chemically reacting flow behind the shock behaves, we seek traveling wave solutions of the detonation equations (1)–(4) of the form

$$\rho = \rho(z), \qquad z = x - Dt, \qquad D = D_{\mathrm{cj}}$$

and similarly for the state parameters u, v, p, T, and y. Substituting these forms into (1)–(4), we obtain a system of autonomous ordinary differential equations

$$\rho w' + w\rho' = 0 \tag{33}$$

$$ww' + \rho^{-1}p' = 0 \tag{34}$$

$$c_p \rho w T' - wp' = Qr \tag{35}$$

$$\rho w y' = -r \tag{36}$$

where $w = u - D$ is the particle velocity relative to D and a prime denotes d/dz. We now show how to solve these equations and determine wave profiles behind the shock. Because the system is autonomous, we may take the shock to be at $z = 0$ (i.e., at $t = 0$ the shock is at $x = 0$); the region behind the shock is then given by $z < 0$, and boundary conditions at $z = 0$ are given by the jump conditions as

$$\begin{aligned} \rho(0) &= \frac{\rho_a(\gamma+1)}{\gamma-1}, \quad u(0) = \frac{2D}{\gamma+1} \\ p(0) &= \frac{2\rho_a D^2}{\gamma+1}, \quad w(0) = -\frac{D(\gamma-1)}{\gamma+1}, \quad y(0) = 1 \end{aligned} \tag{37}$$

These conditions follow from (30)–(32) and the definition of w. The temperature T will be eliminated, so no condition is required. If we use the equation of state $p/\rho = RT$ to eliminate T from (35) and then use the relations

$$\frac{c_p}{R} = \frac{\gamma}{\gamma-1}, \qquad \frac{c_p}{R} - 1 = \frac{1}{\gamma-1}$$

while introducing the sound speed $c^2 = \gamma p/\rho$, we obtain

$$wp' - c^2 w\rho' = (\gamma - 1)Qr$$

Using (33), (34), and (36) allows us to write this equation as

$$(c^2 - w^2)w' = -(\gamma - 1)Qwy' \tag{38}$$

consequently, we may eliminate the independent variable z and write

$$(c^2 - w^2)\,dw = -(\gamma - 1)Qw\,dy \tag{39}$$

This equation can be integrated provided that we can write c^2 as a function of w, which we now do. First, equation (33) can be integrated directly to obtain

$$\rho w = \text{const} = -\rho_a D \tag{40}$$

where the constant was evaluated using the boundary conditions (37). Then (34) yields, upon integration,

$$p = \rho_a D w + D^2 \rho_a \tag{41}$$

where, again, (37) was used to evaluate the constant of integration. Finally, therefore,

$$c^2 = \frac{\gamma p}{\rho} = -\gamma D w - \gamma w^2 \tag{42}$$

so (39) becomes

$$[(\gamma + 1)w + \gamma D]\,dw = (\gamma - 1)Q\,dy \tag{43}$$

Equation (43) may now be integrated to give

$$\frac{\gamma + 1}{2} w^2 + \gamma D w = (\gamma - 1)Qy + K \tag{44}$$

where the constant K of integration is given by [using (37)]

$$K \equiv \frac{(1 - \gamma)^2 D^2}{2(\gamma + 1)} + \frac{\gamma D^2 (1 - \gamma)}{\gamma + 1} + (1 - \gamma)Q$$

Equation (44) is a quadratic equation and can be solved for w. After some tedious algebra we obtain

$$w = -\frac{D}{(\gamma + 1)}(\gamma - y^{1/2}) \tag{45}$$

where we used the equation

$$Q = \frac{D^2}{2(\gamma^2 - 1)}$$

The latter equation is obtained easily from (23) and (24). Using (45), we may then determine ρ, p, and u in terms of the reaction coordinate y from (40) and (41). We obtain

$$\rho = \frac{1}{v} = \frac{\rho_a(\gamma + 1)}{\gamma - y^{1/2}} \tag{46}$$

$$p = \frac{\rho_a D^2}{\gamma + 1}(1 + y^{1/2}) \tag{47}$$

$$u = w + D \tag{48}$$

Therefore, we have formulas for the density, pressure, and particle velocity in terms of the composition y. To determine these quantities as functions of the traveling wave variable z we must solve the species equation (36) which we catalog again as

$$\rho w y' = -r \tag{49}$$

This, of course requires that we now specify a reaction rate r.

We make one important observation before considering an example. If we substitute p and v from (46) and (47) into the equation for the Rayleigh line (with $p_a = 0$)

$$p = -\left(\frac{D}{v_a}\right)^2 (v - v_a)$$

we see that the equation is satisfied identically. Thus the pv-states of the traveling wave solution that we obtained must lie on the Rayleigh line. Geometrically, therefore, a steady ZND detonation, or ZND wave, can be described as follows. The reactant at state $v_a, p_a = 0$, is shocked up the von Neumann point (v_N, p_N) on the nonreacted Hugoniot; then the state moves continuously down the Rayleigh line, crossing the partially reacted Hugoniot curves, to the CJ point on the reacted Hugoniot. At the CJ point the reaction is complete since $y = 0$. We can infer from this description that, qualitatively, the pressure falls and the volume increases as we go back into the reaction zone.

Example 2. Suppose that the reaction rate is simply proportional to the concentration of the reactant, or

$$r = k\rho y \tag{50}$$

where k is a rate constant. Then (49) becomes

$$wy' = -ky$$

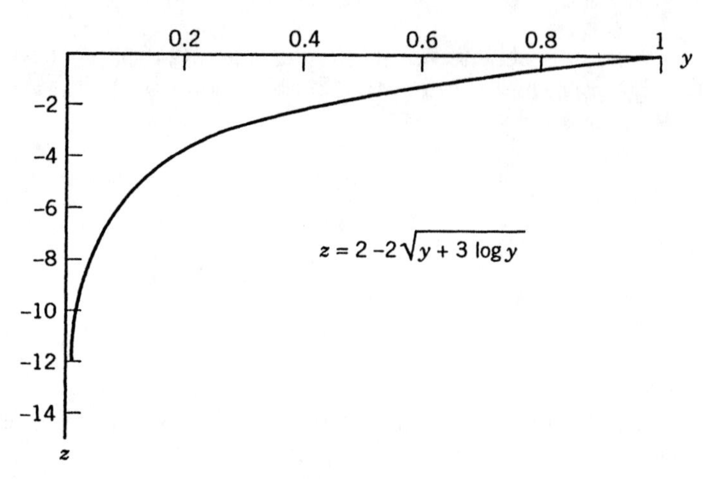

Figure 7.13

or, using (45),

$$\frac{\gamma - \gamma^{1/2}}{y}\,dy = (\gamma + 1)kD^{-1}\,dz$$

Integrating and noting that $y = 1$ at $z = 0$, we get

$$z = \frac{D}{k(\gamma + 1)}(2 - 2y^{1/2} + \gamma \ln y)$$

We may now graph z versus y and then invert the picture to observe how y changes with z (see Figure 7.13). In this case the reaction zone is infinitely long since $y \to 0$ as $z \to -\infty$.

Lagrangian Coordinates

The detonation equations (1)–(4) are written in a fixed laboratory coordinate x, also known as an Eulerian coordinate. It is common and sometimes advantageous in detonation processes, and in fluid mechanics in general, to work in a coordinate system that moves with the particles. Such a system is called a Lagrangian or material system. We now formulate the detonation equations in a Lagrangian system.

Let h be a particle label that is specified at time $t = 0$ by taking $h = x$ at time $t = 0$. So each particle or cross section always carries its label even though its position x in the laboratory changes with time. We assume that

$$x = x(h, t) \tag{51}$$

is a twice continuously differentiable function that gives the position x of the particle h at time t. By construction $x(h, 0) = h$. Further, we assume that for each x, equation (51) uniquely determines a value h, for each time t. In other words, the transformation (51) is invertible and we can obtain

$$h = h(x, t) \tag{52}$$

Because (51) and (52) are then inverse transformations

$$h(x(h, t), t) = h \quad \text{and} \quad x(h(x, t), t) = x \tag{53}$$

Equation (51) is said to define the *flow* because it gives the location of each particle at any time. Now we define the *Lagrangian velocity* of the flow by the equation

$$U(x, t) = x_t(h, t) \tag{54}$$

We make the plausible assumption that the fluid velocity $u(x, t)$ is related to the Lagrangian velocity $U(h, t)$ by the formulas

$$u(x, t) = U(h(x, t), t), \qquad U(h(x, t), t) = u(x, t)$$

In other words, the velocity measured in the laboratory by an observer located at x is the same as the velocity measured by an observer riding on the particle h when their positions coincide. In general, we assume that any laboratory (or Eulerian) measurement $f(x, t)$ made at the location x is related to a Lagrangian measurement $F(h, t)$ of the same quantity, made on a particle h, by

$$f(x, t) = F(h(x, t), t), \qquad F(h, t) = f(x(h, t), t) \tag{55}$$

Our goal is to rewrite the detonation equations in a Lagrangian system with independent variables t and h, and dependent parameters U, V, P, T, and Y, all functions of h and t, representing the particle velocity, specific volume, pressure, temperature, and composition, measured by observers riding with the particles.

To this end we require some simple formulas for various derivatives. We introduce the Jacobian of the transformation (51) defined by

$$J(h, t) = x_h(h, t) \tag{56}$$

It is easy to show that

$$J_t(h, t) = J(h, t) u_x(x(h, t), t) \tag{57}$$

which is known as the *Euler expansion formula*. Next, we define the *convective derivative* Df/Dt, where $f = f(x, t)$ is an Eulerian measurement, by the equation

$$\frac{Df}{Dt} = F_t(h, t)|_{h=h(x,t)} \tag{58}$$

Thus Df/Dt is the time rate of change of a quantity $F(h, t)$ measured by an observer riding with the particle, then frozen at the location x. Using the chain rule it is straightforward to show that

$$\frac{Df}{Dt} = f_t + uf_x \tag{59}$$

Again using the chain rule we can show that

$$f_x(x(h,t),t) = \frac{F_h(h,t)}{J(h,t)} \tag{60}$$

which completes the equations necessary to perform the transformation to a Lagrangian system.

In terms of the specific volume $v = 1/\rho$, the conservation of mass equation (1) can be written

$$v_t + uv_x - vu_x = 0$$

or

$$\frac{Dv}{Dt} - vu_x = 0 \tag{61}$$

To transform to a Lagrangian system we evaluate this equation at $x = x(h, t)$. Then (61) becomes

$$V_t - \frac{VJ_t}{J} = 0 \tag{62}$$

Then

$$\left(\frac{V}{J}\right)_t = 0 \quad \text{or} \quad \frac{V(h,t)}{J(h,t)} = \frac{V(h,0)}{J(h,0)} = V(h,0) \tag{63}$$

We note that (63) gives an interpretation of the Jacobian J as the ratio of the volume at time t to the volume at time zero; thus J is the volume expansion. Using (60) and (57) we can write

$$V_t - V(h,0)U_h = 0 \qquad \text{(mass balance—Lagrangian form)} \tag{64}$$

In the same way we may easily rewrite (2), (3), and (4) as

$$Y_t = -Vr \quad \text{(species equation—Lagrangian form)} \tag{65}$$

$$U_t + V(h,0)P_h = 0 \quad \text{(momentum balance—Lagrangian form)} \tag{66}$$

$$c_p T_t - VP_t = QVr \quad \text{(energy balance—Lagrangian form)} \tag{67}$$

Equations (64)–(67) are the detonation equations in Lagrangian form. We note that they are simpler than the Eulerian form (1)–(4), especially if at time $t = 0$ the density is constant. In that case $V(h,0) = v_a$, where v_a is a constant, and (64) and (66) become linear equations in which U can be eliminated to obtain the single equation

$$V_{tt} + v_a^2 P_{hh} = 0 \tag{68}$$

If the ideal gas law is used to eliminate T from the energy equation (67), we obtain

$$P_t + \frac{\gamma P}{V} V_t = (\gamma - 1)Qr \tag{69}$$

Equations (65), (68), and (69) are three equations governing the pressure P, the volume V, and the composition Y, the latter coupled to (69) through the reaction rate r.

EXERCISES

1. (a) For a CJ detonation prove that $D_{cj} = [2Q(\gamma^2 - 1)]^{1/2}$.
   (b) If $v_a = 0.5263\ \text{cm}^3/\text{g}$, $\gamma = 3$, and $Q = 5.776(10^{10})\ \text{cm}^2/\text{s}^2$ ($= 1.38$ kcal/g), find D_{cj}, p_{cj}, u_{cj}, and v_{cj} for a CJ detonation. How much energy is released in the shock front?
2. For any shock prove that $u_b^2 = (p_b - p_a)(v_a - v_b)$.
3. Prove Proposition 2.
4. Verify (25).
5. Determine the steady ZND detonation when the reaction rate is given by $r = -k\rho y w$ (note that w is negative). Give a qualitative description of the traveling waveforms for ρ, u, p, and y.
6. Rework Exercise 5 when $r = -k\rho w y^{1/2}$.
7. For the Arrhenius rate $r = k\rho y \exp(-E/R^0 T)$, show that for a steady ZND wave the composition y and the traveling wave variable z are related by the formula

$$z = -\frac{D}{(\gamma + 1)k} \int_1^y (\gamma - \eta^{1/2}) e^{f(\eta)} \eta^{-1}\, d\eta$$

   and determine $f(\eta)$.

8. For a steady ZND detonation prove that $w^2/2 + c^2/(\gamma - 1) + Qy =$ constant. This is an analog of Bernoulli's law for reactive flows.

9. For an ideal gas $e = c_v T +$ const and $pv = RT$, where c_v is the specific heat at constant volume. For any gas the first law of thermodynamics $de = T\,ds - p\,dv$ holds true, where s is the entropy.
   (a) For an ideal gas show that $p = k\rho\gamma e^{s/c_v}$, where k is a constant.
   (b) The sound speed c for an ideal gas is defined by $c^2 = \partial p/\partial\rho$, where $p = p(\rho, s)$. Show that $c^2 = \gamma p/\rho$.
   (c) In this exercise we denote directional derivatives along the Hugoniot curve, the Rayleigh line, and a constant-entropy curve (isentrope) by the subscripts H, R, and s, respectively. At the CJ point prove that

$$\left(\frac{dp}{dv}\right)_s = \left(\frac{dp}{dv}\right)_H \quad \text{and} \quad \left(\frac{ds}{dv}\right)_R = 0$$

   Thus the entropy is a local extremum at the CJ point. (See Courant and Friedrichs [1948] for additional properties of the CJ detonation.)

10. Prove that $u_b > 0$ for a detonation and $u_b < 0$ for a deflagration.

11. In the piston problem indicated in Figure 7.12 (where $u_p < u_{\text{cj}}$), calculate the speed of the tail of the rarefaction that separates the rarefaction fan from the constant state in front of the piston.

12. Take the reaction rate to be proportional to pressure—that is, $r = kP$—and assume that $V = v_a$ and $P = p_a$ at $t = 0$. Show that (68) and (69) can be reduced to a single partial differential equation for V.

7.4 COMBUSTION

We first summarize to get our bearing. In Section 7.1 we obtained a general set of nonlinear partial differential equations that model the flow of a two-species reacting fluid. We catalog those equations again as

$$\frac{D\rho}{Dt} + \rho u_x = 0 \tag{1}$$

$$\rho\frac{Du}{Dt} + p_x = \frac{4}{3}ku_{xx} \tag{2}$$

$$c_p\rho\frac{DT}{Dt} - \frac{Dp}{Dt} = Qr + \lambda T_{xx} + \frac{4}{3}ku_x^2 \tag{3}$$

$$\rho\frac{Dy}{Dt} = -r + Dy_{xx} \tag{4}$$

where $p = R\rho T$ and r is the chemical reaction rate, which we now assume has the form

$$r = K\rho y \exp\left(-\frac{E}{R^0 T}\right) \tag{5}$$

Here D/Dt is the convective derivative. Equations (1)–(4) represent conservation of mass, balance of momentum, balance of energy, and chemical species balance, respectively, with y being the mass fraction of the reactant A in a one-step chemical reaction A → B. These general equations include *viscous effects* (the term $\frac{4}{3}ku_{xx}$ represents the momentum transfer due to viscous forces, and the term $\frac{4}{3}ku_x^2$ represents the mechanical work done by the viscous forces), and *diffusion effects* (the term λT_{xx} is the heat diffusion and the term Dy_{xx} is mass diffusion). For detonation processes, discussed in Section 7.3, we neglected the viscous and diffusion terms, arguing that the time scale over which these processes occur is long compared to the much shorter time scale for convection of mass, momentum, and energy. Stated differently, in a detonation process all the interesting action is over before viscous and diffusion effects can take place. The absence of these terms in the governing equations leads to a hyperbolic system, equations (1)–(4) in Section 7.3, which describe detonations. For combustion processes, however, viscous and diffusion effects are important and they often occur on the same time scale as that of other processes in the problem. In fact, in combustion, diffusion often controls the process. For example, in a hot flame, it is diffusion that heats up the material in front of the flame, initiating reaction and subsequently causing the flame to propagate forward. Therefore, in combustion phenomena, the viscous and diffusion terms can play a decisive role and we must deal with the entire parabolic system (1)–(4), however complicated it may appear.

Scaling in Combustion

To make analytical progress in dealing with the general combustion equations (1)–(4) we adopt the strategy of studying the equations, which have a very rich mathematical structure, in special limits. For example, in many combustion problems the ratio of the particle velocity u to the velocity c with which sound travels in the medium is small. This ratio is called the *Mach number*, and thus we may examine flows in the limit of a small Mach number. In this limit certain terms in the equations become negligible compared to other terms, and a simplification occurs. To know precisely which terms may be ignored in a given limit, it is necessary to normalize the problem by introducing dimensionless variables and reducing the problem to dimensionless form. Only then will we be able to compare the order of magnitude of the various terms in the equations. This process is called *scaling* the problem, and we have alluded to it in previous discussions; it is one of the important concepts in modeling. The role of scaling in combustion problems, and for that matter in any problem, is crucial and cannot be overstated. We recall that the process of scaling involves selecting representative values for the dependent and independent variables and then redefining new dimensionless variables, of order 1, by dividing the original dimensioned variables by those representative values or scales (A detailed discussion of scaling can be found in Lin and Segel [1974] or Logan [1987].)

Therefore, to scale the equations (1)–(4) we let T_0, p_0, ρ_0, u_0, and y_0 be representative values of the temperature T, pressure p, density ρ, particle velocity u, and composition y. Later, when examining a specific problem, we shall state precisely what these values are; for example, T_0 might be the average value of an initial temperature profile in an initial value problem under consideration. Since we have the ideal gas law, we assume that

$$p_0 = \frac{c_0^2 \rho_0}{\gamma} \qquad c_0^2 = R\gamma T_0 \tag{6}$$

where c is the sound speed defined by $c^2 \equiv \gamma p/\rho$, and c_0 is the scale for c. Then we may define new, dimensionless dependent variables T^*, p^*, ρ^*, u^*, and y^* by $T^* = T/T_0$, $p^* = p/p_0$, and similarly for ρ^*, u^*, and y^*. We also require a time scale t_0 and a length scale x_0 for the independent variables. It is these two scales that are important in combustion problems, and their choice leads to certain aspects of the model in which we are interested. For example, a problem may be studied on a diffusion time scale or a chemical reaction time scale; x_0 could be the size of the container or the size of a spatial disturbance in the fluid. Different choices will specialize the problem to different limits. For combustion we can identify three different possible time scales:

$$t_a = \frac{x_0}{c_0}, \qquad t_c = \frac{\rho_0 c_p x_0^2}{\lambda}, \qquad t_r = K^{-1} \exp\left(\frac{E}{R^0 T_0}\right) \tag{7}$$

These are the *acoustic time scale* (roughly the time required for an acoustic signal to propagate), the *conduction time scale* (roughly the time required for diffusion), and the *reaction time scale* (roughly the time for the chemistry to occur). At the present time we will not specify the time scale and use the generic t_0 to denote one of the scales in (7). For x_0 we take $u_0 t_0$. Then, dimensionless independent variables can be defined by

$$x^* = \frac{x}{u_0 t_0}, \qquad t^* = \frac{t}{t_0}$$

Rewriting the governing equations (1)–(4) in terms of the starred, dimensionless quantities, and then for notational simplicity dropping the star, we obtain

$$\rho_t + u\rho_x + \rho u_x = 0 \tag{8}$$

$$\rho(u_t + uu_x) + \frac{1}{\gamma M^2} p_x = \frac{4}{3} \Pr\left(\frac{\delta}{M}\right) u_{xx} \tag{9}$$

$$\rho(T_t + uT_x) - \frac{\gamma - 1}{\gamma}(p_t + up_x)$$

$$= \beta \Delta \rho\, y e^{\theta - \theta/T} + \frac{\delta}{M} T_{xx} + \frac{4}{3}(\gamma - 1) \Pr M \delta u_x^2 \tag{10}$$

$$\rho(y_t + uu_x) = -\Delta \rho\, y e^{\theta - \theta/T} + \mathrm{Le}\left(\frac{\delta}{M}\right) y_{xx} \tag{11}$$

where

$$\Pr = \frac{kc_p}{\lambda}, \qquad \mathrm{Le} = \frac{Dc_p}{\lambda}, \qquad \beta = \frac{Qy_0}{c_p T_0}, \qquad \theta = \frac{E}{R^0 T_0}$$

are dimensionless constants, and

$$M = \frac{t_a}{t_0}, \qquad \delta = \frac{t_a}{t_c}, \qquad \Delta = \frac{t_0}{t_r}$$

are ratios of various time scales. The constants Pr and Le are called the *Prandtl number* and the *Lewis number*, and they measure the ratios of viscosity to thermal diffusion and mass diffusion to thermal diffusion, respectively; β is the heat release parameter, θ is a dimensionless activation energy, M is the Mach number, and δ is the ratio of the acoustic to diffusive time scales. We shall always assume that δ is a small parameter; that is, acoustic signals travel much faster than diffusion occurs. Also, the ideal gas equation of state normalizes to

$$p = \rho T$$

Equations (8)–(11) are normalized equations of reactive flow, in dimensionless form. It is assumed by the principle of correct scaling that the coefficient of each term represents the order of magnitude of that term; that is, all of the dependent variables and their derivatives are, by scaling, order 1 quantities. We can now address the question of which terms dominate in various special limits.

Small Mach Number Equations

From our definitions it is easy to observe that the Mach number M is given by

$$M = \frac{t_a}{t_0} = \frac{u_0}{c_0}$$

That is, M is the ratio of a representative particle speed to the acoustic sound speed. In many combustion problems this ratio is small. In fact, if we assume that $M = O(\delta)$ (i.e., M and δ have the same order of smallness), only the pressure term in the momentum equation (9) survives, and to leading order we have

$$p_x = 0$$

which means that the pressure p depends only on t. Then the mechanical work term (involving u_x^2) in the energy equation (10) may be neglected, and the energy equation becomes

$$\rho(T_t + uT_x) - \frac{\gamma - 1}{\gamma'} p'(t) = \Delta\beta\, \rho y e^{\theta - \theta/T} + \frac{M}{\delta} T_{xx}$$

The species equation (11) becomes, to leading order,

$$\rho(y_t + uy_x) = -\Delta\rho\, y e^{\theta - \theta/T} + \frac{M}{\delta} \text{Le}\, y_{xx}$$

In fact, if we now specify the time scale to be the conduction time scale (i.e., $t_0 = t_c$), then $M = \delta$ and we have the so-called *small Mach number equations*

$$\rho_t + u\rho_x + \rho u_x = 0 \tag{12}$$

$$p = p(t) \tag{13}$$

$$\rho(T_t + uT_x) - \frac{\gamma - 1}{\gamma} p'(t) = \Delta\beta\, \rho y e^{\theta - \theta/T} + T_{xx} \tag{14}$$

$$\rho(y_t + uy_x) = -\Delta\rho\, y e^{\theta - \theta/T} + \text{Le}\, y_{xx} \tag{15}$$

The small Mach number equations are valid in the limit $t_a \ll t_c = t_r$. The order 1 ratio Δ, which is now the conduction time divided by the reactions time, is called the *Damköhler number*. In this simplified model acoustic signals travel through the medium and equilibrate the pressure, precluding pressure gradients; mechanical work is small, as are changes in momentum.

Additional assumptions can simplify the model still further. A constant pressure assumption ($p = 1$) and the assumption that the Lewis number is unity cause (14) and (15) to become

$$\rho(T_t + uT_x) - T_{xx} = \Delta\beta\,\rho y e^{\theta-\theta/T} \tag{16}$$

$$\rho(y_t + uy_x) - y_{xx} = -\Delta\rho\, y e^{\theta-\theta/T} \tag{17}$$

in which the same differential operator appears on the left side of both (16) and (17). Multiplying (17) by β and adding the result to (16) gives

$$\rho(H_t + uH_x) - H_{xx} = 0 \tag{18}$$

where

$$H \equiv T + \beta y \tag{19}$$

H is called the *Shvab–Zel'dovich* variable, and (18) states that H is convected and diffused as in a reactionless medium.

Example 1. Consider a flame holder (a burner) at $x = 0$ that spews out reactant ($y = 1$) at temperature $T = T_0$ with velocity $u = u_0$ and density $\rho = \rho_0$ in the positive x direction. As the material moves to the right it reacts, giving off heat (see Figure 7.14). We use these constant values at the burner as scales to nondimensionalize the problem, arriving at the scaled equations (8)–(11). At $x = \infty$ we assume the boundary conditions $y = 0$ (no reactant remains) and $T = T_b > 1$, for some final dimensionless temperature T_b yet to be determined. We can ask if there is a low Mach number, constant pressure, steady flame satisfying these boundary conditions. By steady we mean time independent, or a solution that depends only on x. From (12) we have $(\rho u)_x = 0$ or ρu = constant = 1, and thus from (18) we obtain

$$H'(x) - H''(x) = 0, \qquad x > 0 \tag{20}$$

The boundary conditions on H are

$$H(0) = 1 + \beta, \qquad H(\infty) = T_b \tag{21}$$

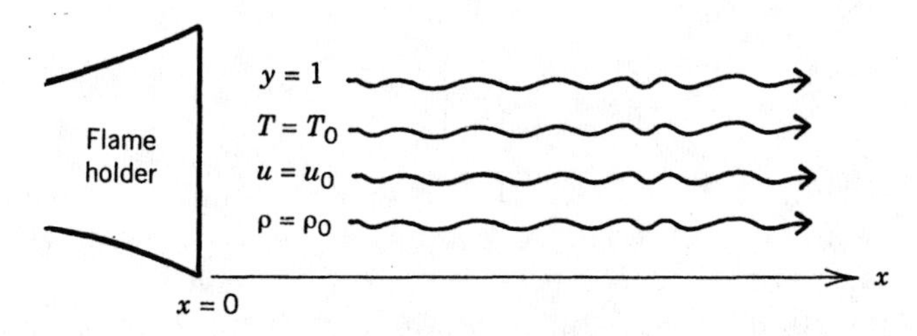

Figure 7.14. Flame holder.

Equation (20) is a second-order constant-coefficient equation with general solution

$$H(x) = A + Be^{x}$$

where A and B are constants. The boundary conditions (21) force $A = 1 + \beta$ and $B = 0$, and thus the Shvab–Zel'dovich variable is constant, that is,

$$H(x) = 1 + \beta, \qquad x > 0$$

and T_b, the temperature of the burned reactant at infinity, must be given by $T_b = 1 + \beta$. It now follows from (19) that $y = \beta^{-1}(T_b - T)$, which allows us to eliminate the the composition y from the temperature equation (16), giving

$$T'(x) - T''(x) = \Delta T^{-1}(T_b - T)e^{\theta - \theta/T}, \qquad x > 0 \tag{22}$$

$$T(0) = 1, \qquad T(\infty) = T_b = 1 + \beta \tag{23}$$

Here we have also used the equation of state, which is $\rho = T^{-1}$, to eliminate ρ. We can easily observe that the boundary value problem (22)–(23) has a solution if we introduce the phase variable $W = T'$ and write (22)–(23) as a system

$$T' = W \tag{24}$$

$$W' = W - \Delta T^{-1}(T_b - T)e^{\theta - \theta/T} \tag{25}$$

This system can be analyzed by the methods presented in the appendix to Chapter 4. Figure 7.15 shows the phase portrait of (24)–(25). The nullcline $W' = 0$, where the vector field is horizontal, is shown as the dashed line; the vector is vertically down along $W = 0$. The only critical point is $T = T_b$, $W = 0$, which is easily seen by linearization to be a saddle; the Jacobi matrix at $(T_b, 0)$ is

$$J = \begin{pmatrix} 0 & 1 \\ \Delta T_b^{-1} e_b^{\theta - \theta/T_b} & 1 \end{pmatrix}$$

which has eigenvalues

$$\lambda_{\pm} = \tfrac{1}{2}\left[1 \pm (1 + 4a)^{1/2}\right], \qquad a \equiv 4\Delta T_b^{-1} e^{\theta - \theta/T_b}$$

and eigenvectors $(1, \lambda)$ and $(1, \lambda_+)$. The stable manifold associated with the negative eigenvalue is a separatrix entering $(T_b, 0)$ as x increases to infinity; followed backward in x this separatrix crosses the nullcline $W' = 0$ and then crosses the vertical line $T = 1$ at some value x^* that can be taken to be zero because of the autonomous nature of the system (24)–(25). Therefore, there

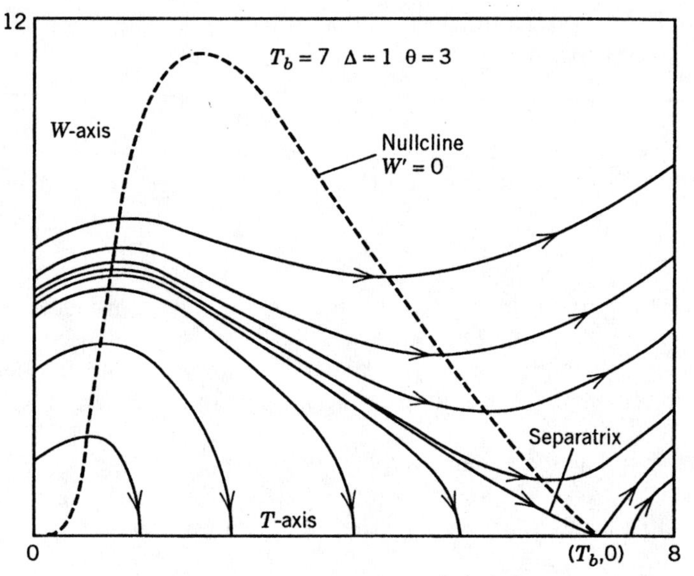

Figure 7.15. Phase portrait of (24)–(25).

is a trajectory satisfying the boundary value problem (22)–(23), and a steady solution has been found for any value of the Damköhler number Δ.

High Activation Energy

Another interesting limit of the governing equations (8)–(11) is the limit of high activation energy. For many chemical reactions the activation energy E is large, making $\varepsilon = 1/\theta$ a small parameter in the problem. If we assume expansions of the form

$$\rho = 1 + \varepsilon m, \qquad p = 1 + \varepsilon P, \qquad T = 1 + \varepsilon \phi, \qquad u = \varepsilon U, \qquad y = 1 - \varepsilon Y \tag{26}$$

and substitute into (8)–(11), we obtain

$$(m_t + U_x)\varepsilon + O(\varepsilon^2) = 0 \tag{27}$$

$$\left[U_t + \frac{1}{\gamma M^2} P_x - \frac{4}{3} \Pr\left(\frac{\delta}{M}\right) U_{xx}\right]\varepsilon + O(\varepsilon^2) = 0 \tag{28}$$

$$\left(\phi_t - \frac{\gamma - 1}{\gamma} P_t - \frac{\delta}{M}\phi_{xx}\right)\varepsilon = \Delta\beta e^{\phi} + O(\varepsilon^2) + O(\varepsilon\Delta) + O(M\delta\varepsilon) \tag{29}$$

$$\left(Y_t - \frac{\delta}{M}\mathrm{Le}\, Y_{xx}\right)\varepsilon = \Delta e^{\phi} + O(\varepsilon^2) + O(\varepsilon\Delta) \tag{30}$$

where we have yet to fix the orders of time-scale ratios M, δ, and Δ. Here we have used the expansion

$$e^{\theta-\theta/T} = e^{[1-1/(1+\varepsilon\phi)]/\varepsilon} = e^{\phi}[1 + O(\varepsilon)]$$

Substituting (26) into the equation of state $p = \rho T$ gives, to leading order,

$$P = m + \phi \tag{31}$$

The Ansatz (26) really means that we are looking for disturbances that differ only slightly from a constant state; physically, this could occur during an *induction period* when these perturbations of the dependent variables remain small. Equations (27)–(31) form what is sometimes called the *induction model.*

If chemical reaction is going to play a role in determining how these small perturbations evolve, the reaction terms on the right sides of equations (29) and (30) must come in at order ε. Therefore, we assume that

$$\Delta = \frac{t_0}{t_r} = \varepsilon \tag{32}$$

which implies that the time scale is given by

$$t_0 = \varepsilon K^{-1} e^{\theta} = K^{-1}\theta e^{\theta} \tag{33}$$

Now, depending on the relative sizes of the time scale ratios M and δ, various reduced ignition models can be obtained. These models give rise to interesting sets of reaction–diffusion equations, as well as reactive–nondiffusive equations, that have been the subject of much analysis in applied mathematics. We particularly refer the reader to Bebernes and Eberly [1989] for an excellent introduction to this analysis as well as a complete bibliography.

Example 2. If we assume that the time scale for heat conduction is defined by

$$\frac{t_0}{t_c} = a = O(1) \tag{34}$$

then it follows that

$$M = \frac{t_a}{(at_c)} \qquad \frac{\delta}{M} = a, \qquad \delta M = a^{-1}\left(\frac{t_a}{t_c}\right)^2$$

Now, from our basic assumption that $\delta = t_a/t_c \ll 1$, equations (27)–(30)

become, to leading order,

$$m_t + U_x = 0 \tag{35}$$

$$P_x = 0 \tag{36}$$

$$\phi_t - \frac{\gamma - 1}{\gamma} P'(t) - a\phi_{xx} = \beta e^{\phi} \tag{37}$$

$$Y_t - a \,\mathrm{Le}\, Y_{xx} = e^{\phi} \tag{38}$$

If we now assume that the reaction occurs in a bounded domain $0 < x < l$, the total mass in the domain is constant and given by

$$\int_0^l \rho(x,t)\, dx = \text{const}$$

Then

$$\int_0^l m(x,t)\, dx = 0$$

and equation (31) forces

$$\int_0^l P(t)\, dx = \int_0^l \phi(x,t)\, dx \quad \text{or} P(t) = l^{-1} \int_0^l \phi(x,t)\, dx$$

Then the pressure perturbation P may be eliminated from the energy equation (37) and we can obtain an equation for the temperature perturbation ϕ, namely,

$$\phi_t - \phi_{xx} = \beta e^{\phi} + \frac{\gamma - 1}{\lambda l} \int_0^l \phi_t\, dx \tag{39}$$

Equation (39) can be accompanied by an initial condition

$$\phi(x,0) = \phi_0(x), \qquad 0 < x < l \tag{40}$$

and boundary conditions of the form

$$\phi(0,t) = \phi(l,t) = 0, \qquad t > 0 \tag{41}$$

The problem (39)–(41) is called the *gaseous reaction–diffusion ignition model*. In the limit as $\gamma \to 1$ equation (39) becomes

$$\phi_t - \phi_{xx} = \beta e^{\phi} \tag{42}$$

which is the *solid-fuel ignition model.* In Chapter 8 we discuss some of the mathematical aspects of this model.

EXERCISES

1. Consider the steady-state solid-fuel ignition model (Gelfand problem)

$$-u'' = \beta e^{u}, \quad -1 < x < 1; \qquad u(-1) = u(1) = 0 \tag{43}$$

Letting $\alpha = u(0)$, show that

$$\sqrt{2\beta}\,(1-x) = \int_0^u (e^{\alpha} - e^{w})^{-1/2}\, dw$$

and hence

$$\beta = 2e^{-\alpha} \operatorname{arctanh}^2 \sqrt{1 - e^{-\alpha}}$$

Conclude that there is no solution to (43) if $\beta > \beta_{cr}$, for some positive number β_{cr}, and there are two solutions of (43) if $0 < \beta < \beta_{cr}$. *Hint*: Notice that the problem is symmetric on $[-1, 1]$, and thus the problem may be restricted to the interval $[0, 1]$.

2. Consider the high activation equations (27)–(30) and assume the order relation (32). Derive a set of model equations that describe a combustion process in the limit where $O(t_r) = t_a \ll t_c$, that is, where the reaction time is on the same order as the acoustic time, and both are small compared to the conduction time. (This model is known as the induction period reaction period reactive Euler model.)

REFERENCES

The standard reference on detonations is Fickett and Davis [1979], a book that incorporates both theory and experimental results. The standard references on combustion are Williams [1985] and Zel'dovich et al. [1985]; Buckmaster and Ludford [1982] can be recommended for sophisticated readers. Because these areas have attracted the attention of both mathematicians and engineers, the standards of rigor, the writing styles, and the motivating factors in the various expositions are often quite diverse. For the mathematics-oriented student the books by Bebernes and Eberly [1989] and Kapila [1983] are highly recommended for the analysis of deeper questions in the study of reaction–diffusion equations and asymptotics, respectively, related to combustion theory. The infrequently referenced Engineering Design Handbook [1972], available from the U.S. Government Printing

Office, contains an excellent introduction to the theory of detonation up to 1970. References to other short monographs on various aspects of combustion and detonation can be found below.

Bebernes, J., and D. Eberly, 1989. *Mathematical Problems from Combustion Theory, Springer-Verlag, New York.*

Buckmaster, J. D., Editor, 1985. *The Mathematics of Combustion*, Society for Industrial and Applied Mathematics, Philadelphia.

Buckmaster, J. D., and G. S. S. Ludford, 1982. *Theory of Laminar Flames*, Cambridge University Press, Cambridge.

Buckmaster, J. D., and G. S. S. Ludford, 1983. *Lectures on Mathematical Combustion*, Society for Industrial and Applied Mathematics, Philadelphia.

Courant, R., and K. O. Friedrichs, 1948. *Supersonic Flow and Shock Waves*, Wiley-Interscience, New York. Reprinted by Springer-Verlag, New York, 1976.

Engineering Design Handbook, 1972. *Principles of Explosive Behavior*, AMC Pamphlet 706–180, U.S. Army Material Command, Washington, D.C.

Fickett, W., 1985. *Introduction to Detonation Theory*, University of California Press, Berkeley.

Fickett, W., and W. C. Davis, 1979. *Detonation*, University of California Press, Berkeley.

Kapila, A. K., 1983. *Asymptotic Treatment of Chemically Reacting Systems*, Pitman, Boston.

Lin, C. C., and L. A. Segel, 1974. *Mathematics Applied to Deterministic Problems in the Natural Sciences*, Macmillan, New York. Reprinted by the Society for Industrial and Applied Mathematics, Philadelphia.

Logan, J. D., 1987. *Applied Mathematics: A Contemporary Approach*, Wiley-Interscience, New York.

Ludford, G. S. S., Editor, 1986. *Reacting Flows: Combustion and Chemical Reactors, Parts* 1 *and* 2, American Mathematical Society, Providence, R.I.

Nauman, E. B., 1987. *Chemical Reactor Design*, Wiley, New York.

Truesdell, C., 1965. On the foundations of mechanics and energetics, in *Rational Mechanics of Materials: Continuum Mechanics II*, edited by C. Truesdell, Gordon and Breach, New York, pp. 292–305.

Williams, F. A., 1985. *Combustion Theory*, 2 ed., Benjamin/Cummings, Menlo Park, Calif.

Zel'dovich, Ya. B., G. I. Barenblatt, V. B. Librovich, and G. M. Makhviladze, 1985. *The Mathematical Theory of Combustion and Explosion*, Consultants Bureau, Division of Plenum Press, New York.

8

ELLIPTIC EQUATIONS

In this final chapter we address some basic questions in the theory of nonlinear elliptic partial differential equations. Consistent with our pedagogical approach, we consider only some representative examples, most of which are elementary in nature; we do not lay out a general theory. The idea is to equip the reader with some concrete examples that will hopefully provide the motivation for further intensive study.

We may think of elliptic equations as equilibrium problems being born out of the time-asymptotic limit of reaction–diffusion equations. For example, steady-state solutions, if they exist, of the parabolic, reaction–diffusion equation

$$u_t - \Delta u = f(u) \tag{1}$$

must satisfy the equilibrium equation

$$-\Delta u = f(u) \tag{2}$$

The latter semilinear equation is a steady-state equation in that time does not appear, and $u = u(x)$, a function of the spatial variable only. So one may think of the solutions $u(x, t)$ of (1) as evolving into the solutions $u = u(x)$ of (2) as t gets large. One of the major problems in the study of reaction–diffusion equations is to confirm this observation in special cases, and it leads to questions of existence of solutions of both (1) and (2) and also to questions of stability and bifurcation.

Equilibrium, or steady-state equations such as (2) are elliptic, and they are fundamentally different from parabolic and hyperbolic problems. As we should expect, initial conditions for elliptic equations are inappropriate; rather, only boundary conditions lead to well-posed elliptic problems. Thus concepts such as characteristics, which are fundamental for hyperbolic problems, play no role in the steady state. For elliptic equations we look for solutions on a spatial domain that match given conditions on the boundary of that domain; we do not regard, as in the case of hyperbolic or parabolic equations, initial or boundary data as being carried into the domain along, for example, characteristic curves.

In Section 8.1 the question of classification is revisited and some basic elliptic models are discussed. In Section 8.2 we present two of the fundamental tools in the study of nonlinear elliptic equations; these are the maximum principle and the basic existence theorem involving upper and lower solutions. Section 8.3 focuses on a nonlinear eigenvalue problem, the Gelfand problem, which was introduced in Chapter 7 as a model for solid-phase combustion. In Section 8.4 we address the question of the stability of equilibrium solutions to (1) and the related question of the bifurcation of solutions of (2) when a parameter is present. We also discuss nonlinear stability analysis based on normal modes.

The field of nonlinear elliptic boundary value problems is a vast area of intensive study, and our treatment is parsimonious compared to our development of evolution equations. The reader is invited to explore this field further by examining one or more of the references listed at the end of the chapter.

8.1 ELLIPTIC EQUATIONS AND MODELS

Classification Revisited

The focus of all the analysis in the first seven chapters has been on parabolic and hyperbolic equations. These equations describe evolutionary processes that are changing in time, and problems associated with these types of equations have the property that signals or data are carried into regions of spacetime from the boundaries; for example, in hyperbolic problems, characteristics carry initial and boundary data forward into regions of space and time. Now we wish to discuss a different type of problem that is nonevolutionary; these are problems given by elliptic equations. Physically, in many cases one may regard these steady-state problems as a limiting problem of an evolutionary system, when all the transients have disappeared. That is, over a long period of time the transients caused by initial or boundary data decay away, and an equilibrium situation results. This asymptotic equilibrium state is time independent and is generally governed by an elliptic system where all the independent variables are spatial variables and the proper auxiliary conditions are boundary conditions that are independent of time. Consequently, physical intuition suggests that the initial value problem for elliptic equations should not be well posed.

Let us return to the question of classification of PDEs. In Chapter 1 we discussed briefly classification of PDEs, with the idea of determining transformations under which PDEs could be simplified to certain canonical forms. There the notion of ellipticity was introduced. With more care, in Chapter 5 we classified first order system of PDEs; hyperbolic systems were characterized by a matrix having real eigenvalues and a full complement of independent eigenvectors. The eigenvalues defined special characteristic directions, the integral curves of which were the characteristics. An elliptic equation was one in which there were no (real) characteristics; that is, the eigenvalues of the associated matrix for the system were all complex. Now we want to examine this situation again in a different context. To this end, we consider the second order partial differential equation

$$au_{xx} + 2bu_{xy} + cu_{yy} = d \tag{1}$$

where $u = u(x, y)$ and the coefficients a, b, c, and d are continuous and depend on the independent variables x and y, and possibly even on u, u_x, and u_y. We assume that the problem is defined in some open, connected domain D of the xy-plane. To approach the classification question in a different way we consider the *Cauchy problem*, where we require u to satisfy the PDE (1) along with *Cauchy data*

$$u = f(s), \qquad u_x = g(s), \qquad u_y = h(s) \tag{2}$$

given along some given, smooth curve C lying in D with parametric equations

$$C: x = x(s), \quad y = y(s) \tag{3}$$

where s ranges over some parameter interval I. It is easy to observe that f, g, and h in (2) are not all independent, since

$$\begin{aligned} f'(s) &= \frac{d}{ds}u(x(s), y(s)) \\ &= u_x(x(s), y(s))x'(s) + u_y(x(s), y(s))y'(s) \\ &= g(s)x'(s) + h(s)y'(s) \end{aligned} \tag{4}$$

Thus we may not specify u, u_x, and u_y independently along C, and therefore we assume the compatibility condition (4). Alternatively, one may specify u and its normal derivative along C.

We know the values of u and its first derivatives on the curve C, and we might inquire what it would take to determine the values of u off the curve C. Let us ask this question locally by fixing a point P on the curve C and then asking if the solution u to (1)–(2) could be determined in a small neighborhood of P. Evidently, if u is sufficiently smooth, then u at some point Q in this neighborhood could be found from its Taylor series expansion around the point P on C. This would in turn require knowledge of all the higher derivatives of u at P [i.e., $u_{xx}(P)$, $u_{xy}(P)$, $u_{yy}(P)$, etc.]. Consequently, let us pose the problem of how to determine these second derivatives at P. The answer comes from differentiating u_x and u_y along the curve C. We have, by the chain rule,

$$\frac{du_x}{ds} = u_{xx}x'(s) + u_{xy}y'(s) = g'(s) \tag{5}$$

where we have conserved notation by not writing down the arguments $(x(s), y(s))$ of the partial derivatives. Similarly,

$$\frac{du_y}{ds} = u_{yx}x'(s) + u_{yy}y'(s) = h'(s) \tag{6}$$

Equations (5) and (6), along with the PDE (1) itself, when evaluated at P, give three equations in the three unknowns u_{xx}, u_{yy}, and u_{xy} at P. In matrix form,

$$\begin{pmatrix} a & 2b & c \\ x' & y' & 0 \\ 0 & x' & y' \end{pmatrix} \begin{pmatrix} u_{xx} \\ u_{xy} \\ u_{yy} \end{pmatrix} = \begin{pmatrix} d \\ g' \\ h' \end{pmatrix}$$

This nonhomogeneous system has a unique solution if the coefficient determinant is nonzero. Notice that the determinant is given by

$$\delta \equiv cx'^2 - 2bx'y' + ay'^2 \tag{7}$$

Therefore, if δ is nonzero at P, the second derivatives of u at P can be uniquely determined. In a similar fashion, by differentiating the PDE and the second derivatives u_{xx} and u_{xy} at P along the curve C, we can obtain a system of three equations for the third derivatives of u, and so on, and each will have the same coefficient matrix as the system above. Therefore, if δ is nonzero at P, a Taylor series expansion of u about P can be determined (so far as u is differentiable), thereby determining the solution u in a neighborhood of P.

The preceding discussion leads us to make the following definition. A curve C is said to be *characteristic* if $\delta = 0$ at each point of C. Thus, if C is characteristic, one would *not* expect to be able to solve the Cauchy problem (1)–(2), where the Cauchy data are given along C. If Cauchy data are given along a curve that is not characteristic, one would expect to determine u in a neighborhood of the curve.

If a curve C is characteristic, its differentials dx and dy satisfy the relation

$$c\,dx^2 - 2b\,dx\,dy + a\,dy^2 = 0 \tag{8}$$

or

$$a\left(\frac{dy}{dx}\right)^2 - 2b\frac{dy}{dx} + c = 0$$

Thus

$$\frac{dy}{dx} = \frac{b \pm \sqrt{b^2 - ac}}{a} \tag{9}$$

Let us assume for a moment that a, b, and c depend only on x and y. Then (9) is a differential equation whose solutions determine the characteristic curves. Now we can see, in a different context, the origin of classification. Clearly, the sign of $b^2 - ac$ determines if there are real characteristics, and we make the following definition: In a region D the PDE (1) is *hyperbolic* if $b^2 - ac > 0$ in D, *elliptic* if $b^2 - ac < 0$ in D, and *parabolic* if $b^2 - ac = 0$. Consequently, in the case of hyperbolic equations there are two families of characteristic curves, corresponding to the choice of + or − in (9), and for parabolic equations there is only one family of characteristic curves. *Elliptic equations have no real characteristics.*

We remark that in the case when a, b, and c depend on u, u_x, or u_y the characteristic curves cannot be determined a priori since the solution u is unknown.

It would appear from the preceding heuristic discussion that the Cauchy problem is ideally suited for elliptic equations because all curves are noncharacteristic, and therefore u and all of its derivatives can be determined along any such curve, giving the solution (via Taylor's expansion) in a neighborhood of a curve where Cauchy data are prescribed. However, the situation is more complicated than it would appear. There is another issue in this case, namely continuous dependence of the solution on Cauchy data. As the following example shows, the Cauchy problem is not well posed for elliptic equations, in that small changes in the Cauchy data lead to arbitrarily large changes in the solution.

Example 1 (*Hadamard's Example*). Consider the Cauchy problem

$$u_{xx} + u_{yy} = 0,$$

$$u(x,0) = 0, \qquad u_y(x,0) = 0$$

where C is the x-axis. Clearly, $u = 0$ is a solution to this problem. Now change the Cauchy data to

$$u(x,0) = 0, \qquad u_y(x,0) = n^{-1} \sin nx$$

which for large n is only a small change in the data. It is straightforward to verify that the solution is given by

$$u(x,y) = n^{-2} \sin nx \sinh ny$$

For large values of n the exponentially growing sinh function dominates the n^{-2} factor and hence the solution deviates from the original solution $u = 0$ by an arbitrarily large amount. Thus a small change in the Cauchy data produces an arbitrarily large change in the solution.

In summary, ellipticity for second order equations of the form (1) means that there are no characteristic curves as there are for hyperbolic equations, to carry data into a region from the boundary. Therefore, we expect to impose boundary conditions on an elliptic equation and consider a pure boundary value problem where we specify u *or* its normal derivative on the boundary of a region where we wish to solve the problem. For infinite domains this may mean imposing conditions on u at infinity (e.g., a boundedness condition or a limit condition). Generally, elliptic problems can be more difficult than their hyperbolic and parabolic counterparts because a solution must exist in the large, over an entire domain, whereas for the latter it still may be of interest to obtain a local solution in some small interval of time.

Model Problems

Example 2 *(Laplace's Equation)*. Let D be an open, bounded, connected region in R^2 with boundary ∂D, and let $u = u(x, t)$ denote the temperature in D at any position $x = (x_1, x_2)$ at time t. Suppose that at time $t = 0$ we specify the initial temperature to be $u(x, 0) = u_0(x)$, and suppose that for all $t \geq 0$ we impose the boundary condition

$$u(x,t) = h(x), \qquad x \in \partial D, \quad t > 0 \tag{10}$$

where h is a given function. Then from Chapter 1 we know that conservation of energy and Fourier's heat law imply that the temperature evolves according to the diffusion equation

$$u_t - k\,\Delta u = 0, \qquad x \in D, \quad t > 0 \tag{11}$$

where we have assumed that the region is homogeneous with constant specific heat, constant density, and constant thermal conductivity. Δ is the two-dimensional Laplacian, and k is the diffusivity. If we imagine that the system evolves so long that the initial condition at $t = 0$ no longer affects changes in the system, we reach an asymptotic temperature state that is independent of time [i.e., $u = u(x)$]. Thus the equilibrium temperature $u(x)$ must satisfy

$$\Delta u = 0, \qquad x \in D \tag{12}$$

along with the boundary condition (10). The elliptic equation (12) is called

Laplace's equation, and the boundary value problem consisting of Laplace's equation (12) and the boundary condition (10) is called the *Dirichlet problem* on the domain D. If the boundary condition (10) is replaced by a flux condition of the form

$$\frac{du}{dn} = g(x), \qquad x \in \partial D, \quad t > 0 \tag{13}$$

where $du/dn = \operatorname{grad} u \cdot \mathbf{n}$, where $\mathbf{n}$ is the outward unit normal on ∂D, then the boundary value problem consisting of Laplace's equation (12) and the flux condition (13) is called the *Neumann problem*. More generally, one could consider a combined boundary condition of the form

$$a(x)u + b(x)\frac{du}{dn} = c(x), \qquad x \in \partial D, \quad t > 0 \tag{14}$$

which is called a mixed condition (or a Robin condition).

The Dirichlet problem and the Neumann problem are two of the most important problems in analysis. It is clear that we can generalize these two problems to three spatial dimensions, or even to n spatial dimensions.

Example 3 (*Poisson's Equation*). Returning to the discussion in Example 2, if a time-independent heat source $F(x)$ is present in the domain D, then the transient temperatures $u = u(x, t)$ are governed by the parabolic nonhomogeneous diffusion equation

$$u_t - k\,\Delta u = F(x), \qquad x \in D, \quad t > 0$$

Equilibrium temperatures $u = u(x)$ are therefore solutions to the elliptic equation

$$-\Delta u = f(x), \qquad x \in D \tag{15}$$

which is known as *Poisson's equation*. Here $f(x) \equiv F(x)/k$. For boundary conditions on ∂D we may append either a Dirichlet, a Neumann, or a mixed condition. If the heat source depends on the temperature u, we obtain a semilinear elliptic equation of the form

$$-\Delta u = f(x, u), \qquad x \in D \tag{16}$$

Equation (16) is one of the basic nonlinear elliptic equations that is frequently discussed in the literature; the nonlinearity occurs in the source term and not in the differential operator. Such equations are called *semilinear*.

Example 4 (*Reaction–Diffusion Equations*). As we noted in Chapter 6, there has been considerable interest in the last several years in nonlinear reactive–diffusive systems. These systems are governed, for example, by equations of the form

$$u_t - d_1 \Delta u = f(u,v), \qquad v_t - d_2 \Delta v = g(u,v) \tag{17}$$

where u and v represent the unknown density functions (populations, chemical concentrations, etc.), and d_1 and d_2 are diffusion constants. These equations provide qualitative models for a number of biological, chemical, and physical phenomena: predator–prey interactions, chemical reactions, combustion phenomena, morphogenesis, and nerve-pulse conduction, to mention a few. Just as in heat conduction, discussed in the preceding examples, we are often interested in the equilibrium situation where u and v are time independent. In this case, $u = u(x)$ and $v = v(x)$ satisfy the elliptic system

$$-d_1 \Delta u = f(u,v), \qquad -d_2 \Delta v = g(u,v) \tag{18}$$

subject, of course, to the boundary conditions imposed on the problem. We may think of the solutions of (18) as the asymptotic limit of the solutions of (17). Here we are assuming that solutions exist. This interpretation leads to several interesting questions. For example, what kinds of initial data imposed on (17) cause its solutions to evolve into the stationary solutions of the equilibrium problem (18)? This question becomes even more exciting when one discovers that (18) may admit multiple equilibrium solutions. So, we can liken equilibrium solutions of (18) to critical points for plane autonomous systems of ODEs and then ask similar stability questions for the stationary solutions of the PDEs (18). For example, is a given stationary solution of (18) asymptotically stable, that is, does it attract all solutions of (17) that are sufficiently close to it (in some norm) at time $t = 0$? These are important questions in the nonlinear analysis associated with reaction–diffusion equations and the resulting elliptic systems that arise in a time-asymptotic limit.

Another important exercise is to investigate bifurcation, or branching, phenomena in systems such as (17) and (18). For example, the system (17) may contain a parameter such that below a critical value there is a single, asymptotically stable equilibrium solution of (18); then, as the parameter increases beyond the critical value, additional equilibria appear and the previous asymptotically stable solution may become unstable. The study of multiple equilibria and their stability properties, as functions of parameters in the system, lies at the foundation of what is called bifurcation theory.

Yet another class of problems associated with elliptic equations are eigenvalue problems. These are discussed in Section 8.4.

Example 5 (*Schrödinger Equation*). In the doctrine of classical mechanics the position of a particle of mass m moving under the influence of a potential

$V(x)$, $x \in R^3$, is determined precisely by solving the differential equations of motion given by Newton's second law, subject to initial conditions. After the turn of the twentieth century it was recognized that the classical theory did not apply in all cases, and quantum mechanics was developed. In quantum theory a statistical interpretation is advanced where all that can be known about a particle's location is the probability of it being in some region of space. The information about a quantum mechanical system is contained in a state function $\psi(x, t)$, called the wave function, where $|\psi|^2$ represents the probability density. That is,

$$\int_D |\psi|^2 \, dx$$

represents the probability of finding the particle in the volume D. It is a fundamental postulate of quantum mechanic that the wave function satisfies the *time-dependent Schrödinger equation* given by

$$i\hbar\psi_t = \left[-\hbar^2 \frac{\Delta}{2m} + V(x) \right] \psi \tag{19}$$

where Δ is the three-dimensional Laplacian and $h = 2\pi\hbar$ is Planck's constant. If we assume that $\psi(x, t) = T(t)\Psi(x)$, then (19) separates into the two equations

$$T' = -\frac{iE}{\hbar} T \quad \text{and} \quad \Delta\Psi + \frac{2m}{\hbar^2} [E - V(x)] \Psi = 0$$

where E is a separation constant. This equation for Ψ is called the *time-independent Schrödinger equation*, and it is elliptic. Typically, when boundary conditions are imposed, the time-independent problem is regarded as an eigenvalue problem where the eigenvalues E are the possible energy levels of the particle.

EXERCISES

1. The *divergence theorem* in R^n is contained in the statement

$$\int_D D_k u(x) \, dx = \int_{\partial D} u(x) n_k(x) \, dS, \qquad D_k = \frac{\partial}{\partial x_k}$$

where D is an open, bounded region in R^n with smooth boundary ∂D, and n_k is the kth component of the outward unit normal on ∂D, and u is a scalar function with $u \in C(\bar{D}) \cap C^1(D)$.

(a) Use the divergence theorem to derive the *integration-by-parts formula*

$$\int_D vD_k u\,dx = \int_{\partial D} uvn_k\,dS - \int_D uD_k v\,dx$$

(b) Assume that $u, v \in C(\bar{D}) \cap C^2(D)$. Use the integration-by-parts formula in part (a) to derive the two Green's identities:

$$\int_D v\,\Delta u\,dx = \int_{\partial D} v\frac{du}{dn}\,dS - \int_D \operatorname{grad} u \operatorname{grad} v\,dx$$

$$\int_D v\,\Delta u\,dx = \int_D u\,\Delta v\,dx + \int_{\partial D}\left(v\frac{du}{dn} - u\frac{dv}{dn}\right)dS$$

where D is the n-dimensional Laplacian, grad $= (D_1, \ldots, D_n)$ is the gradient operator, and du/dn is the normal derivative given by

$$du/dn = \operatorname{grad} u \cdot \mathbf{n}$$

where $\mathbf{n}$ is the normal vector.

(c) Use part (b) to prove that classical solutions [in $C(\bar{D}) \cap C^2(D)$] to the Dirichlet problem

$$-\Delta u = g(x), \qquad x \in D$$

$$u = f(x), \qquad x \in \partial D$$

are unique, provided that they exist. [Assume that $f, g \in C(\bar{D})$.]

2. Use an *energy method* to show that the parabolic problem

$$u_t - \Delta u = 0, \qquad x \in D, \quad t > 0$$

$$u(x,t) = 0, \qquad x \in \partial D, \quad t > 0$$

$$u(x,0) = 0, \qquad x \in D$$

has only the trivial solution $u = 0$ on $D \times R^+$ [Consider $E(t) = \int_D u(x,t)^2\,dx$ and use Exercise 1b.]

3. Determine all solutions of Laplace's equation $\Delta u = 0$ of the form $u = \psi(r)$, $r = |x - \xi|$, where $x = (x_1, \ldots, x_n)$, where $\xi = (\xi_1, \ldots, \xi_n)$ is a fixed point in R^n, and where r is the Euclidean distance from x to ξ. The answer is

$$\psi(r) = A + C\ln r \quad \text{if } n = 2; \qquad \psi(r) = A + \frac{Cr^{2-n}}{2-n} \quad \text{if } n > 2$$

where A and C are constants.

4. Consider the PDE

$$u_{xx} + x^2 u_{yy} = y u_y$$

Show that the equation is elliptic. Determine characteristic coordinates and reduce the equation to canonical form (see Section 1.1).

5. Consider the initial value problem for the Schrödinger equation for a free particle,

$$u_t = i u_{xx}, \qquad x \in R, \quad t > 0$$

$$u(x,0) = u_0(x), \qquad x \in R$$

where u_0 is continuous and square integrable on R. Prove that if a solution $u(x,t)$ exists and vanishes, along with all of its derivatives, at infinity, then

$$\int_R |u(x,t)|^2 \, dx = \int_R |u_0(x)|^2 \, dx \qquad \text{for all } t > 0$$

(This form of the Schrödinger equation, without the accompanying constants, is the one that is often discussed in mathematics literature.)

6. Use the result for the solution to the initial value problem for the heat equation (see Section 4.1) to determine a formal solution to the initial value problem in Exercise 5. Compare and contrast the two kernels in the integral representations of the solutions to these two initial value problems.

7. Consider the Helmholtz equation in R^3,

$$\Delta u + cu = 0$$

where c is a positive constant and Δ is the three-dimensional Laplacian. Find all solutions of the form $u(x) = \Psi(|x - \xi|)$, where ξ is some fixed point in R^3.

8.2 BASIC RESULTS FOR ELLIPTIC OPERATORS

We have observed that elliptic equations are associated with boundary value problems which are independent of time. A common feature of elliptic problems is that their solutions are determined by their values on the boundary of the domain. Another feature of elliptic problems is that a solution must satisfy some sort of maximum or minimum principle requiring that it take on its maximum or minimum value on the boundary of the domain over which the problem is defined. In this section we present two of the fundamental tools used in the analysis of elliptic problems, the maximum principle and the basic existence theorem.

Maximum Principle

In the sequel we assume that D is an open, bounded, connected subset of R^n, and we denote the points of D by $x = (x_1, \ldots, x_n)$. The symbol $\bar{D}$ will denote the *closure* of D, that is, $\bar{D} = D \cup \partial D$, where ∂D is the boundary of D. We will not deal with questions of regularity of the boundary; generally, we assume that the boundary of D is sufficiently smooth to give validity to our results. For elliptic equations it is not much more difficult to work in n-dimensional space, and then our results will be valid when specialized to either R^2 or R^3. The Laplacian in R^n is the second order differential operator Δ defined by

$$\Delta u = u_{x_1x_1} + \cdots + u_{x_nx_n}$$

We say that a function $u \in C^2(D)$ is *subharmonic* (*superharmonic*) in D if $-\Delta u \leq 0$ in D (correspondingly, $-\Delta u \geq 0$ in D). A function $u \in C^2(D)$ that satisfies Laplace's equation $\Delta u = 0$ in D is said to be *harmonic* in D.

The maximum principle, in its simplest form, requires that a subharmonic function on D that is continuous in $\bar{D}$ assume its maximum value on the boundary ∂D. Correspondingly, the minimum principle insists that a superharmonic function assume its minimum value on ∂D. It is clear that these results should be expected if one restricts the analysis to $n = 1$, where D is an interval in R^1 and u is a function on D. In this case the Laplacian is just the second derivative operator and $u'' \geq 0$ means that u is concave up, and $u'' \leq 0$ means that u is concave down. Respectively, such functions have their maximum and their minimum at the endpoints (assuming that they are not constant functions). In fact, the one-dimensional case is a convenient mnemonic device. Therefore, we have the following weak form of the maximum principle.

Theorem 1 (*Weak Maximum Principle*). If $u \in C(\bar{D}) \cap C^2(D)$ and $\Delta u \geq 0$ on D, then $\max_{\bar{D}} u = \max_{\partial D} u$.

We remark that because u is a continuous function on a compact (closed and bounded) set, the function u must assume its maximum somewhere on $\bar{D}$. The theorem states that the maximum must occur at least somewhere on ∂D, but it does not preclude the maximum also occurring at an interior point. A stronger version of the maximum principle, stated later, will preclude this, unless the function is constant. Now the proof of the weak maximum principle (this proof is given in most books on PDEs). First, we observe that if $\Delta u > 0$ on D (strict inequality), then a maximum cannot occur in D. For, at such a point we would have $u_{x_ix_i} \leq 0$ for all i, which would violate $\Delta u > 0$ at that point. Next consider the case $\Delta u \geq 0$ on D. For $\varepsilon > 0$ we define the auxiliary function $v = u + \varepsilon |x|^2$ for which $\Delta v > 0$. Thus, from the strict inequality case, we know that v cannot assume its maximum in D. Then

$\max_{\overline{D}} v = \max_{\partial D} v$. Consequently,

$$\max_{\overline{D}} u \le \max_{\overline{D}} \left(u + \varepsilon |x|^2\right) = \max_{\partial D} \left(u + \varepsilon |x|^2\right) \le \max_{\partial D} u + \varepsilon \max_{\partial D} |x|^2$$

Because ε is arbitrary, $\max_{\overline{D}} u \le \max_{\partial D} u$. The opposite inequality is automatically true, and therefore the theorem is proved.

Upon replacing u by $-u$ in the statement of the theorem, we obtain a weak form of the minimum principle.

Theorem 2 (*Weak Minimum Principle*). If $u \in C(\overline{D}) \cap C^2(D)$ and $\Delta u \le 0$ on D, then $\min_{\overline{D}} u = \min_{\partial D} u$.

It is obvious that if u is harmonic in D and continuous on $\overline{D}$, then u must assume both its maximum and minimum on ∂D. In particular, if $u = 0$ on ∂D and $\Delta u = 0$ in D, then u must vanish identically in D. Therefore, a harmonic function on D that is continuous in $\overline{D}$ is uniquely determined by its values on the boundary.

We now state, without going through the proof, the strong version of the maximum principle. A proof can be found in Protter and Weinberger [1967], which is an excellent general reference for maximum principles in partial differential equations. We formulate the theorem for more general elliptic operators than the Laplacian. To this end, let us consider an operator L defined by

$$Lu = \sum_i \sum_j a_{ij}(x) u_{x_i x_j} + \sum_j b_j(x) u_{x_j} \tag{1}$$

where the coefficients a_{ij} and b_j are continuously differentiable functions on D which are continuous on $\overline{D}$, and where

$$\sum_i \sum_j a_{ij}(x) \xi_i \xi_j \ge \mu \sum_j \xi_j^2 \tag{2}$$

for all x in D, for some positive real number μ. An operator L satisfying these conditions is called *uniformly elliptic* in D. Of course, the Laplacian Δ is uniformly elliptic. Then one version of the strong form of the maximum principle can be stated as follows.

Theorem 3 (*Strong Maximum Principle*). Let $u \in C(\overline{D}) \cap C^2(D)$ and assume that $Lu + c(x)u \ge 0$ for $x \in D$, where L is uniformly elliptic in D, and where the function c is continuous, bounded, and $c \le 0$ on D. If u attains a nonnegative maximum M in D, then $u = M$ for all $x \in D$.

It is possible to slightly loosen the hypotheses in this theorem, and we refer the reader to Protter and Weinberger [1967]. One can also draw some conclusions about the normal derivative of u at the boundary. Recall that the normal derivative is $du/dn = \mathbf{n} \cdot \operatorname{grad} u$, where $\mathbf{n}$ is the outward unit normal.

Theorem 4. Under the same hypotheses as in Theorem 3, assume that $u \le M$ in D, $u = M$ at a boundary point x_0, and $M \ge 0$. If x_0 lies on a boundary of some sphere inside D, then $du/dn > 0$ at x_0, provided that u is not constant in D.

We note that if the boundary of D is smooth, then an interior ball, or sphere, can always be placed at x_0. We also remark that it can be shown that one need not restrict the conclusion of the theorem to the normal derivative; more generally, the directional derivative of u in any outward direction is positive.

The corresponding minimum principle may be stated as follows. The proof comes from replacing u by $-u$ in the weak maximum principle, and we leave it as a simple exercise.

Theorem 5 (*Strong Minimum Principle*). Let $u \in C(\overline{D}) \cap C^2(D)$ and assume that $Lu + c(x)u \le 0$ for $x \in D$, where L is uniformly elliptic in D, and where c is continuous, bounded, and $c \le 0$ on D. If u attains a nonpositive minimum m in D, then $u = m$ for all $x \in D$.

Once we have maximum and minimum principles, it is straightforward to obtain comparison theorems. For example, we have the following:

Theorem 6. Let $u, v \in C(\overline{D}) \cap C^2(D)$ and assume that $Lu \ge 0$ and $Lv = 0$ in D, where L is a uniformly elliptic operator given by (1). If $u \le v$ on ∂D, then $u \le v$ for all $x \in D$.

Proof. Let $w = v - u$. Then $w \ge 0$ on ∂D and $Lw \le 0$ on D. Now, by way of contradiction, assume that $w(x_0) < 0$ for some x_0 in D. Then w has a negative minimum in D, and therefore by the strong minimum principle the function w must be constant and negative in D. But this contradicts the fact that w is nonnegative on ∂D. Hence $w \ge 0$ on D, giving the result. Our assumption that D is a bounded set is important in this theorem; it is not true, in general, for unbounded domains.

The exercises contain several applications of the maximum and minimum principles.

Existence Theorem

Let us consider the semilinear boundary value problem

$$-Lu = f(x, u), \qquad x \in D \tag{3}$$

$$u = h(x), \qquad x \in \partial D \tag{4}$$

where L is the uniformly elliptic operator defined by (1). The function h in the Dirichlet boundary condition, and the function f, defined in $D \times R$, are both assumed to be smooth (class C^1) on their respective domains. A minus sign appears in front of the operator L on the left side of (3) so that subsequent inequalities will have a symmetric form (in fact, the maximum and minimum principles stated above are often presented with negative signs on the operator and the inequalities reversed).

One of the fundamental tools in the analysis of elliptic equations and eigenvalue problems for elliptic operators is the notion of upper and lower solutions to (3)–(4), and the fact that if such solutions exist, a classical solution can be sandwiched in between. In the next section, where we discuss eigenvalue problems, we will experience the power of this method to prove existence results. A function $\bar{u}$ in $C(\bar{D}) \cap C^2(D)$ is called an *upper solution* to (3)–(4) if

$$-L\bar{u} \geq f(x, \bar{u}), \qquad x \in D \tag{5}$$

$$\bar{u} \geq h(x), \qquad x \in \partial D \tag{6}$$

A *lower solution* $\underline{u}(x)$ is defined similarly, with the two inequalities reversed. The fundamental theorem can now be stated.

Theorem 7 (*Existence*). If the boundary value problem (3)–(4) has a lower solution $\underline{u}(x)$ and an upper solution $\bar{u}(x)$ with $\underline{u}(x) \leq \bar{u}(x)$ in D, then the boundary value problem (3)–(4) has a solution $u(x)$ with the property that $\underline{u}(x) \leq u(x) \leq \bar{u}(x)$ in D.

We shall not prove this theorem. It is not difficult to prove, and accessible proofs can be found in Sattinger [1973] and Smoller [1983]. One method of proof uses the idea of monotone iteration schemes; that is, monotone sequences of iterates bounded by the upper and lower solutions are constructed and can be shown to converge a solution from above and from below. There are other proofs using topological, or degree, methods. Upper and lower solutions were introduced in the late 1960s. The proof of the general existence theorem using monotone iteration methods and upper and lower solutions was given by H. Amann in 1971 and D. Sattinger in 1972. We point out that the result is valid for boundary conditions more general than (4). Generalizations have been offered in other directions as well. The reader can consult the complete bibliography in Pao [1992] for specific references to these cited works as well as other treatments of the subject.

Example 1. Let D be the square $0 < x, y < a$ in R^2 and consider the boundary value problem

$$-\Delta u = u(1-u) \qquad \text{in } D \tag{7}$$

$$u = 0 \qquad \text{on } \partial D \tag{8}$$

where Δ is the two-dimensional Laplacian. We use the existence theorem to prove that a solution exists to this BVP, provided that the square D is large enough. The reader should be alert to the notation change, where we are using x and y in lieu of x_1 and x_2. First, define $\bar{u}(x)$ as the solution to the ordinary differential equation $-\bar{u}'' = \bar{u}(1-\bar{u})$ on $0 < x < a$, with $\bar{u}(0) = \bar{u}(a) = 0$. Then $\bar{u}(x) \geq 0$ on ∂D and $-\Delta \bar{u} = -\bar{u}'' = \bar{u}(1-\bar{u}) \geq 0$. Therefore, $\bar{u}(x)$ is an upper solution of the BVP (7)–(8). To find a lower solution we proceed as follows. First, consider the eigenvalue problem

$$-\Delta u = \lambda u \qquad \text{in } D$$
$$u = 0 \qquad \text{on } \partial D$$

The eigenvalues are given by $\lambda = \pi^2(n^2+m^2)/a^2$ for $m, n = 1, 2, 3, \ldots$ (see, e.g., Courant and Hilbert [1953]). Thus the smallest eigenvalue is $\lambda_1 = 2\pi^2/a^2$, and an associated eigenfunction is given by $\phi(x, y) = \sin(\pi x/a)\sin(\pi y/a)$, which is positive on D. Then $-\Delta\phi = \lambda_1\phi$ and $\phi = 0$ on ∂D. Now choose a such that $\lambda_1 < 1$, and take $\underline{u} = \varepsilon\phi$, where ε is a positive constant to be selected later. Then

$$\Delta\underline{u} + \underline{u}(1-\underline{u}) = \varepsilon\phi(-\lambda_1 + 1 - \varepsilon\phi) \geq 0$$

provided that ε is chosen small enough. Therefore, $\underline{u}$ is a lower solution. By the existence theorem there is a solution u to (7)–(8) with the property that $\varepsilon\phi(x, u) \leq u(x, y) \leq \bar{u}(x)$ for all (x, y) in D, provided that a is large enough.

We remark that a similar existence theorem using upper and lower solutions holds for parabolic equations as well. For example, consider the initial boundary value problem

$$u_t - \Delta u = f(x, u), \qquad x \in D, \quad 0 < t < T \tag{9}$$

$$u(x, t) = g(x, t), \qquad x \in \partial D, \quad 0 < t < T \tag{10}$$

$$u(x, 0) = u_0(x), \qquad x \in D \tag{11}$$

We assume that f is a smooth function of its arguments and that the initial and boundary conditions are continuous. We say that $\underline{u}(x, t)$ is a lower solution of (9)–(11) if (9)–(11) hold with the inequality $\leq$ rather than with equality. In the same way, $\bar{u}(x, t)$ is an upper solution if (9)–(11) hold with the inequality $\geq$. Here, both $\underline{u}$ and $\bar{u}$ are assumed to be twice continuously differentiable on $\Omega_T = D \times (0, T)$ and continuous on $\bar{\Omega}_T$. Then we state, without proof, the following existence theorem.

Theorem 8. Let $\underline{u}$ and $\bar{u}$ be lower and upper solutions of (9)–(11) with $\underline{u}(x, t) \leq \bar{u}(x, t)$ on Ω_T. Then the initial boundary value problem (9)–(11)

has a solution $u(x, t)$ with $\underline{u}(x, t) \leq u(x, t) \leq \bar{u}(x, t)$ on Ω_T that is continuous on $\bar{\Omega}_T$ and class $C^2(\Omega_T)$.

EXERCISES

1. Let $u \in C(\bar{D}) \cap C^2(D)$ be a solution of $\Delta u + \Sigma_{k=1}^n b_k(x)u_x + c(x)u = 0$ on D where $c(x) < 0$ in D. Prove that $u = 0$ on ∂D implies that $u = 0$ in D. *Suggestion*: Show that min $u \geq 0$ and max $u \leq 0$.
2. Use the maximum to prove that a harmonic function on D which is continuous in $\bar{D}$ is uniquely determined by its values on ∂D.
3. What can be deduced about solutions to the nonlinear Dirichlet problem

$$\Delta u = u^2, \qquad x \in D$$

$$u(x) = 0, \qquad x \in \partial D?$$

4. Let $D = \{(x, y) | 0 < x, y < \pi\}$ be a domain in R^2, and consider the Dirichlet problem

$$u_{xx} + u_{yy} + 2u = 0 \qquad \text{in } D$$

$$u = 0 \qquad \text{on } \partial D$$

Where does the maximum of u occur? Discuss.

5. Let Ω in R^n denote the unbounded domain $|x| > 1$, and let $u \in C^2(\bar{\Omega})$, $\lim_{x \to \infty} u(x) = 0$, and $\Delta u = 0$ in Ω. Prove that $\max_{\bar{\Omega}} |u| = \max_{\partial\Omega} |u|$.

8.3 EIGENVALUE PROBLEMS

Linear Eigenvalue Problems

The reader should be familiar with eigenvalue problems occurring in matrix theory (the algebraic eigenvalue problem) and in ordinary differential equations (Sturm–Liouville problems). In Section 8.1 we mentioned how an eigenvalue problem arises naturally for the Schrödinger equation in quantum mechanics. In this section we examine some eigenvalue problems for partial differential equations with the goal of obtaining some interesting results on the nature of the eigenvalues for nonlinear equations.

One of the most important problems for linear partial differential operators is the eigenvalue problem for the negative Laplacian on a given open, bounded, connected domain D in R^n. Such a problem was encountered in Example 1 in Section 8.2. The problem is to determine values of λ for which the BVP

$$-\Delta u = \lambda u, \qquad x \in D \tag{1}$$

$$u = 0, \qquad x \in \partial D \tag{2}$$

has a nontrivial solution. The values of λ for which nontrivial solutions exist are called *eigenvalues*, and the corresponding solutions are called *eigenfunctions*. The set of eigenvalues is called the *spectrum* of $-\Delta$. It is also of interest to replace the Dirichlet boundary condition (2) with a Neumann-type condition on the normal derivative, $du/dn = 0$ on ∂D. Generally, eigenvalues and the corresponding eigenfunctions depend on the boundary condition, the operator, and the domain D.

Example 1. The physical origin of eigenvalue problems for PDEs comes out of the study of vibration problems. For example, let D represent an elastic membrane in two dimensions. Under special assumptions, the vertical displacement $u(x, t)$ of the membrane from equilibrium at position $x = (x_1, x_2)$ at time t is governed by the two-dimensional wave equation

$$u_{tt} - a\,\Delta u = 0, \qquad x \in D, \quad t > 0 \tag{3}$$

where $a > 0$ is a physical constant. If the boundary of the membrane is held fixed (pinned), we impose the boundary condition

$$u = 0, \qquad x \in \partial D, \quad t > 0 \tag{4}$$

Of interest in many engineering applications are solutions of the form

$$u(x,t) = U(x)e^{i\omega t} \tag{5}$$

We observe that these solutions are periodic, or oscillatory, in time with frequency ω. The shape of the solution is $U(x)$. If (3)–(4) admits solution of the form (5), ω is called a *fundamental frequency* and $U(x)$ is called a *normal mode of oscillation*. To find such solutions we substitute (5) into (3) and (4) to obtain

$$-\Delta U = \lambda U, \qquad x \in D$$
$$U = 0, \qquad x \in \partial D$$

where $\lambda \equiv \omega^2/a$. Therefore, an eigenvalue problem must be solved to determine the fundamental frequencies and normal modes of oscillation of the membranes.

A lot is known about the eigenvalues and eigenfunctions for $-\Delta$, the negative Laplacian. For special regions such as rectangles and balls, explicit formulas are known. The classic book by Courant and Hilbert [1953] is one of the best references for properties of eigenvalues of linear partial differential operators. For the sequel we require only a few facts that are collected together in the following theorem on the *principal* (smallest) eigenvalue and its monotonicity. The domain D over which the problem is defined is assumed to be an open, bounded region with a sufficiently smooth boundary.

Theorem 1. For the eigenvalue problem (1)–(2) on the domain D, there are infinitely many eigenvalues λ_n, $n = 1, 2, 3, \ldots$, that can be arranged in a sequence $0 < \lambda_1 \leq \lambda_2 \leq \lambda_3 \leq \cdots$ with $\lim \lambda_n = +\infty$. The eigenfunction $\phi(x)$ associated with the *principal eigenvalue* λ_1 can be chosen to be positive on D. Moreover, if D^* is a subdomain of D, the principal eigenvalue λ_1^* for the eigenvalue problem (1)–(2) on D^* has the property that $\lambda_1^* \geq \lambda_1$.

Remark. This theorem can be extended to more general operators and boundary conditions. For example, if L is a uniformly elliptic operator on D, we can consider the eigenvalue problem

$$-Lu + c(x)u = \lambda r(x)u, \qquad x \in D$$

$$a\frac{du}{dn} + bu = 0, \qquad x \in \partial D$$

where $a = 1$ and $b \geq 0$ or $a = 0$ and $b = 1$, and where $c, r \in C(D)$, $c > 0$ and $r > 0$. The principal eigenvalue λ_1 is real, nonnegative, and has multiplicity 1; the corresponding principal eigenfunction can be taken to be positive on D.

Nonlinear Eigenvalue Problems

The first question to be decided is what kinds of nonlinear problems should be investigated. Here we shall consider one type of problem that has enjoyed considerable attention in the mathematical literature; it has its origins in combustion phenomena. Our analysis of this problem follows that of Bebernes and Eberly [1989]; see also Pao [1992].

***Example* 2.** We refer to the *solid-fuel model* developed in Section 7.4. In several dimensions the model equation is

$$u_t - \Delta u = \delta e^u, \qquad x \in D, \quad t > 0$$

with initial and boundary conditions

$$u(x,0) = u_0(x), \qquad x \in D$$

$$u(x,t) = 0, \qquad x \in \partial D, \quad t > 0$$

Here u is a dimensionless perturbed temperature variable. In the steady case we have the nonlinear boundary value problem

$$-\Delta u = \delta e^u, \qquad x \in D \tag{6}$$

$$u = 0, \qquad x \in \partial D \tag{7}$$

Problem (6)–(7) is called the *Gelfand problem*. An important problem in combustion theory is to determine values of the parameter δ for which the problem has a positive solution.

With the Gelfand problem as the motivation, we are led to consider a wide class of nonlinear eigenvalue problems of the form

$$-\Delta u = \lambda f(x,u), \qquad x \in D \tag{8}$$

$$u = 0, \qquad x \in \partial D \tag{9}$$

We assume that f is continuous on $D \times R$ and $f(x,u) \geq 0$. Notice that the linear eigenvalue problem for $-\Delta$ is not a special case of (8)–(9) since $f(x,u) = u$ does not satisfy the positivity requirement. Because we are interested in nonnegative solutions, we define the *spectrum* σ as the set of all real numbers λ for which a nonnegative solution of (8)–(9) exists. It is easy to show that if there is a positive number λ_1 in σ, then σ must also contain the interval $(0, \lambda_1)$. So for any λ in the interval $0 \leq \lambda \leq \lambda_1$, there is a nonnegative solution to (8)–(9). (Clearly, $0 \in \sigma$.) Consequently, the nonlinear problem (8)–(9) will not have a discrete spectrum as we have come to expect from linear boundary value problems. We record this as our first result.

Proposition 1. If $\lambda_1 > 0$ and $\lambda_1 \in \sigma$, then $[0, \lambda_1] \subseteq \sigma$.

The proof uses the fundamental existence theorem on upper and lower solutions for elliptic equations stated in Section 8.2. Let $0 < \lambda < \lambda_1$ be fixed. Then $\underline{u}(x) = 0$ is a lower solution to (8)–(9) since f is nonnegative. Now let $\bar{u}(x)$ be the nonnegative solution of (8)–(9) corresponding to λ_1. It is clear that

$$-\Delta \bar{u} = \lambda_1 f(x, \bar{u}) \geq \lambda f(x, \bar{u}), \qquad x \in D$$

and therefore $\bar{u}$ is an upper solution to (8)–(9). By the existence theorem there is a solution $u(x)$ to (8)–(9) satisfying $0 \leq u(x) \leq \bar{u}(x)$, $x \in D$, thereby completing the proof.

Now we prove an important result on the nonexistence of eigenvalues beyond a certain value, provided that f is bounded below by a linear function of u.

Proposition 2. Suppose that

$$f(x,u) \geq h(x) + r(x)u, \qquad (x,u) \in D \times [0,\infty)$$

where h and r are continuous on D, and $h, r > 0$ on D. Then the eigenvalue problem (8)–(9) has no nonnegative solutions for any $\lambda \geq \lambda_1(r)$, where $\lambda_1(r)$ is the principal eigenvalue of the problem

$$-\Delta u = \lambda r(x)u, \quad x \in D; \qquad u = 0, \quad x \in \partial D \tag{10}$$

For the proof we again use the idea of upper and lower solutions. By way of contradiction assume that $\bar{u}(x)$ is a nonnegative solution of (8)–(9) with $\lambda \geq \lambda_1(r)$. Then it is easy to see that $\bar{u}$ is an upper solution of the BVP

$$-\Delta u = \lambda(h(x) + r(x)u), \quad x \in D; \qquad u = 0, \quad x \in \partial D \tag{11}$$

Moreover, it is obvious that $\underline{u} = 0$ is a lower solution of (11). By the existence theorem on upper and lower solutions there is a solution $u(x)$ of (11) with the property that $0 \leq u(x) \leq \bar{u}(x)$ for $x \in D$. Moreover, by the maximum principle we must have $u(x) > 0$ on D [$-\Delta u > \lambda r(x)u \geq 0$ on D, and $u = 0$ on ∂D].

Now let ϕ be the positive, principal eigenfunction of (10) corresponding to the principal eigenvalue $\lambda_1(r)$. By Green's identity (see Exercise 1, Section 8.1) we have

$$0 = \int_D (u\,\Delta\phi - \phi\,\Delta u)\,dx = \int_D [\phi(\lambda h + \lambda ru) - \lambda_1(r) r\phi u]\,dx$$

Therefore,

$$[\lambda_1(r) - \lambda]\int_D ru\phi\,dx = \lambda\int_D \phi h > 0$$

which implies $\lambda_1(r) > \lambda$, a contradiction. Consequently, no nonnegative solution of (8)–(9) can exist for $\lambda \geq \lambda_1(r)$, completing the proof.

Example 3. Consider the Gelfand problem

$$\begin{aligned} -\Delta u &= \lambda e^u, \qquad & x \in D \\ u &= 0, & x \in \partial D \end{aligned}$$

Here we have $f(x,u) = e^u \geq u + 1$ for $u \geq 0$, so the last proposition applies with $h = r = 1$. We conclude that the Gelfand problem has no nonnegative solutions for any $\lambda \geq \lambda_1$, where λ_1 is the principal eigenvalue of the problem $-\Delta u = \lambda u$, $x \in D$, with $u = 0$ on ∂D. To compute λ_1 see Exercise 3.

Of course, the main question is the existence of a positive eigenvalue of (8)–(9).

Proposition 3. If $f(x,u) > 0$ for $x \in D$ and $u \geq 0$, there exists a real number $\lambda^* > 0$ such that the eigenvalue problem (8)–(9) has a positive solution on D for $\lambda = \lambda^*$.

The strong maximum principle shows that a necessary condition for a positive solution is $\lambda \geq 0$. For let $\lambda < 0$. Then $-\Delta u = \lambda f(x,u) < 0$ on D and $u \leq 0$ on D. By the maximum principle, $u \leq 0$ in D, a contradiction. To complete the proof we need to show that there is a positive solution for some $\lambda^* > 0$. Again the idea of upper and lower solutions is essential. We have already shown that $\underline{u} = 0$ is a lower solution for any $\lambda^* > 0$. Now let $\bar{u}$ be the solution to

$$-\Delta u = 1, \qquad x \in D$$
$$u = 0, \qquad x \in \partial D$$

By the maximum principle we have $\bar{u} > 0$ on D. For λ^* sufficiently small

$$1 = -\Delta \bar{u} \geq \lambda^* f(x, \bar{u}), \qquad x \in D$$

Therefore, $\bar{u}$ is an upper solution to (8)–(9) for $\lambda = \lambda^*$ (note that f is bounded). By the existence theorem there is a solution $u(x)$ of (8)–(9), with $\lambda = \lambda^*$, satisfying the condition $0 \leq u(x) \leq \bar{u}(x)$, $x \in D$. Again by the maximum principle we must have $u(x) > 0$. [Note that u cannot be an identically zero solution since $f(x,u) > 0$.]

EXERCISES

1. Consider the eigenvalue problem (8)–(9), where $f(x,u) \in C^2$, $f(x,0) > 0$, $f_u(x,0) > 0$, and $f_{uu}(x,u) \geq 0$, for all $x \in D$ and $u \geq 0$. If $\lambda > \lambda_1(f_u(x,0))$, prove that there are no nonnegative solutions.
2. Consider the nonlinear boundary value problem

$$\Delta u + \mu u = u^3, \qquad x \in D$$
$$u = 0, \qquad x \in \partial D$$

Prove that if μ is sufficiently small, then $u = 0$ is the only solution. *Suggestion*: Take $\mu < \lambda_1$, where λ_1 is the principal eigenvalue of $-\Delta$ on D.

3. Let D be the open ball $x^2 + y^2 < 1$ in the plane. Find a value of λ_1 for which the eigenvalue problem

$$-(u_{xx} + u_{yy}) = \lambda e^u, \quad x \in D; \qquad u = 0, \quad x \in \partial D$$

has no nonnegative solution for $\lambda \geq \lambda_1$.

8.4 STABILITY AND BIFURCATION

Physical systems governed by nonlinear partial differential equations suggest many interesting mathematical questions. One of these, which we now take up, is the existence of equilibrium solutions and how those solutions depend on parameters that occur in the problem. For example, equilibrium solutions may exist for some values of a parameter, yet not for other values. More interestingly, as a parameter is varied, there may be a critical value where an equilibrium solution may lose its stability properties and no longer exist as a physical possibility, even though it may continue to exist in a mathematical sense. At the same critical value, dividing may occur and additional equilibria may suddenly appear. The study of problems of this type lies in the domain of stability and bifurcation theory. By *bifurcation* we mean *branching* or dividing. First, we introduce such problems in the realm of ordinary differential equations, after which we study related questions for PDEs.

Ordinary Differential Equations

Example 1. Let us consider an animal population whose growth is determined by a logistics type of law, while at the same time the population is *harvested* at a constant rate. If $U = U(\tau)$ is the population at time τ, then the governing evolution equation is

$$U' = rU\left(1 - \frac{U}{K}\right) - h, \qquad \tau > 0 \tag{1}$$

where r is the growth rate, K the carrying capacity, and h the constant harvesting rate, with $h > 0$. Equation (1) can be nondimensionalized by introducing the scaled variables

$$u = \frac{U}{K}, \qquad t = r\tau$$

In other words, we scale the population by its carrying capacity, and we scale time by the reciprocal of the growth rate. Then, in dimensionless variables, equation (1) becomes

$$\frac{du}{dt} = u(1 - u) - \mu \qquad (2)$$

where μ is dimensionless harvesting parameter defined by $\mu = h/rK$, which is positive. Now, treating μ as control parameter, we may inquire if equilibrium populations exist. That is, are there any constant solutions of (2)? Easily, these are found by setting $du/dt = 0$. In the present cases, the equilibrium populations are values of u that satisfy

$$u(1 - u) - \mu = 0$$

or

$$u = \frac{1 \pm \sqrt{1 - 4\mu}}{2} \qquad (3)$$

Therefore, if the harvesting parameter μ exceeds $\frac{1}{4}$, there are no equilibrium populations; if $0 < \mu < \frac{1}{4}$, there are two possible equilibrium states in which the ecosystem could exist. It is common to sketch the locus (3) on a μu-coordinate system to graphically indicate the dependence of the equilibrium solutions on the parameter μ. Such a graph is shown in Figure 8.1 and is called a *bifurcation diagram*; the harvesting parameter μ is called a *bifurcation parameter*. At the critical value $\mu = \frac{1}{4}$, where equilibrium solutions appear, the point on the bifurcation diagram is called a bifurcation point. We can physically interpret the result in the following manner. If μ is too large, that is, if there is too much harvesting of the population, then (2) shows that du/dt is negative and therefore the population dies out. As the harvesting is decreased, there is a critical value of μ below which two sizes of equilibrium populations can exist; in these two populations there is a balance between growth and harvesting. We can now ask if nature prefers one of those

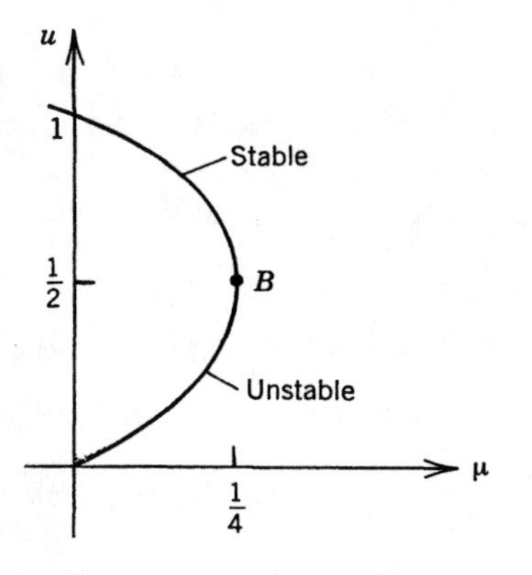

Figure 8.1. Bifurcation diagram showing the locus (3), a graph of the constant, equilibrium solutions versus the bifurcation parameter μ. The point B is a bifurcation point where the stability changes; the states on the upper branch are stable, and the states on the lower branch are unstable.

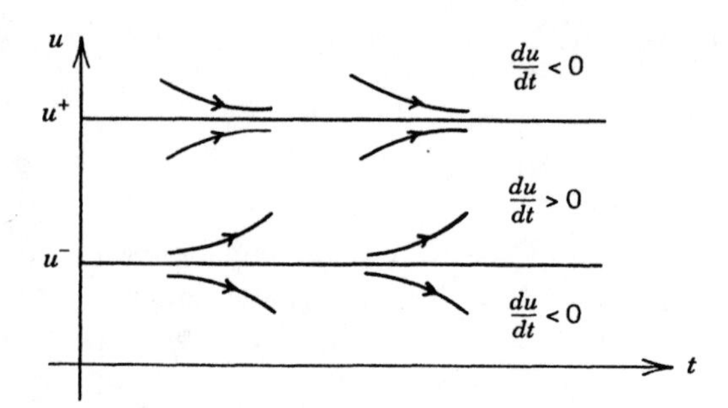

Figure 8.2. Diagram showing the direction of the integral curves of (2).

equilibrium populations over the other. This question is at the heart of the concept of stability, which has been discussed in earlier chapters. If the ecosystem is in one of these equilibrium states, and if a small perturbation is imposed, say by introducing a few more animals, does the system go out of balance or does it return to the original equilibrium? This question can be resolved easily for the present problem by examining the direction field of the differential equation (2) in the case $0 < \mu < \frac{1}{4}$. Letting u^+ and u^- denote the two equilibrium populations defined in (3), with the plus and minus sign, respectively, we observe that

$$\frac{du}{dt} < 0 \quad \text{if } u > u^+; \qquad \frac{du}{dt} > 0 \quad \text{if } u^- < u < u^+;$$

$$\frac{du}{dt} < 0 \quad \text{if } 0 < u < u$$

Therefore, the integral curves or solutions of (2) must behave as shown in Figure 8.2. Thus u^+ must be an attractor, or a stable equilibrium, and u^- must be a repeller, or unstable equilibrium. Therefore, nature would prefer the larger equilibrium population, so the upper branch of the equilibrium curve shown on the bifurcation diagram in Figure 8.1 is stable, while the lower branch is unstable.

The types of questions addressed in Example 1 form the basis of what is termed *bifurcation and stability theory*. Many problems arising in the physical and natural sciences have these ingredients, and it is of interest to determine how equilibrium solutions depend on parameters occurring in the problem, and the stability of these equilibrium states.

Example 1 leads us to consider the general ordinary differential equation

$$\frac{du}{dt} = f(\mu, u) \tag{4}$$

where f is a given function of u and a bifurcation parameter μ. Equilibrium solutions u of (4) satisfy the relation

$$f(\mu, u) = 0 \tag{5}$$

A graph of the locus (5) in a μu-plane is called a *bifurcation diagram*. Generally, the locus may be quite complicated, with many branches and several intersections. A heuristic argument can be developed to determine if a given point (μ, u_e) on the locus (5) represents a stable or unstable equilibrium. To this end, let $u(t) = u_e + w(t)$, where $w(t)$ represents a small perturbation from the constant equilibrium state u_e. Note that u_e depends on μ, but to keep the notation uncluttered we shall surpress this dependence in the symbolism. Substituting into (4) and expanding by a Taylor series give

$$\begin{aligned}\frac{dw}{dt} &= f(\mu, u_e + w) \\ &= f(\mu, u_e) + f_u(\mu, u_e)w + O(w^2) \\ &= f_u(\mu, u_e)w + O(w^2)\end{aligned}$$

If we linearize by ignoring the higher order term $O(w^2)$, the perturbation w must satisfy the equation

$$\frac{dw}{dt} = aw, \qquad a \equiv f_u(\mu, u_e)$$

which has solution

$$w(t) = \text{const} \cdot e^{at}$$

Therefore, if $a > 0$, the perturbation w will grow and u_e will be unstable; if $a < 0$, the perturbation w will decay to zero and u_e will be stable. This heuristic argument contains the seeds of a rigorous proof that the sign of the quantity $f_u(\mu, u_e)$ determines the nature of the stability of an equilibrium state u_e for fairly general conditions on the function f. We refer the reader to Logan [1987], for example, for a precise formulation and proof of a general result in the scalar case.

Example 2. Example 1 considers equation (2), where f is given by $f(\mu, u) = u(1-u) - \mu$. In this case, $f_u(\mu, u) = 1 - 2u$. Therefore, if u is an equilibrium solution with $u < \frac{1}{2}$, then $f_u > 0$ and u is unstable; if $u > \frac{1}{2}$, then $f_u < 0$ and u is stable. This is consistent with our earlier comment that any solution on the upper branch of the locus shown in the bifurcation diagram in Figure 8.1 is stable, while any equilibrium state on the lower branch is unstable.

These ideas easily extend to systems of ordinary differential equations. We refer the reader to Hale and Kocak [1991] for a general study.

Partial Differential Equations

In roughly the last three decades the subject of bifurcation and stability theory in the domain of nonlinear partial differential equations has attracted tremendous attention in the applied analysis community. A thorough discussion of this topic would take us beyond our scope of an introductory treatment, and therefore we consider only some elementary concepts and examples.

The idea is this. Let us consider the archetypical reaction diffusion equation

$$u_t - \Delta u = f(\mu, u), \qquad x \in D, \quad t > 0 \tag{6}$$

with boundary and initial conditions given by

$$u(x,t) = 0, \quad t > 0, \quad x \in \partial D; \qquad u(x,0) = u_0(x), \quad x \in D \tag{7}$$

where f is a given function and μ is a real parameter. Equilibrium, or time-independent, solutions are given by solutions of the elliptic problem

$$-\Delta u = f(\mu, u), \qquad x \in D \tag{8}$$

$$u = 0, \qquad x \in \partial D \tag{9}$$

From our discussion in Section 8.3, where we considered eigenvalue problems and f had the form $f(\mu, u) = \mu g(u)$, it is clear that this elliptic problem may have solutions for some values of μ and yet no solution for other values of μ. Suppose that there is an equilibrium solution $u_e(x)$ of (8)–(9) for some value of the parameter μ. In what sense is this equilibrium stable or unstable? There are several definitions of stability for reaction diffusion equations, but here we focus on one version of what is called Liapunov stability. The fundamental question is whether a time-dependent solution of the reaction diffusion problem (6)–(7) converges to an equilibrium solution of (8)–(9) as t approaches infinity. Whether or not we get convergence may depend on the set of initial functions $u_0(x)$ in the initial condition in (7). Evidently, if we start at $t = 0$ too far away from the equilibrium solution, then we may not obtain convergence. Thus stability of u_e may be local, requiring that u_0 be sufficiently close to u_e. We formulate the following definition, which should remind the reader of a similar definition probably encountered in ordinary differential equations.

Definition 1. An equilibrium solution $u_e(x)$ of (8)–(9) is said to be *stable* if for any $\varepsilon > 0$ there is a $\delta > 0$ such that

$$|u_0(x) - u_e(x)| < \delta \qquad \text{for } x \in D$$

implies that

$$|u(x,t) - u_e(x)| < \varepsilon \qquad \text{for } x \in D,\ t > 0$$

In addition, if

$$\lim_{t \to \infty} |u(x,t) - u_e(x)| = 0 \qquad \text{for } x \in D$$

then u_e is said to be *asymptotically stable*. If u_e is not stable, it is said to be *unstable*.

The set of initial conditions u_0 for which u_e is stable is called the *domain of attraction* of u_e; if the domain of attraction contains all initial functions, then u_e is called *globally stable*, or a *global attractor*.

Example 3. Consider the elliptic problem

$$-\Delta u = \mu u - u^3, \qquad x \in D \tag{10}$$

$$u = 0, \qquad x \in \partial D \tag{11}$$

which we can regard as arising in a time-asymptotic limit of a reaction–diffusion system. It is natural to ask if this problem has solutions. We shall show that if $0 < \mu < \lambda_1$, where λ_1 is the principal eigenvalue of the operator-Δ on D, no nontrivial solutions exist. By way of contradiction, assume that a nontrivial solution u of (10)–(11) exists. Then there is a domain $D' \subseteq D$ such that u satisfies the boundary value problem

$$-\Delta u = \mu u - u^3, \quad x \in D'; \qquad u = 0, \quad x \in \partial D' \tag{12}$$

and u is of one sign on D'. Let λ'_1 be the principal eigenvalue of $-\Delta$ on D' with its corresponding eigenfunction $\phi(x)$ strictly positive on D'. Thus

$$-\Delta \phi = \lambda'_1 \phi, \quad x \in D'; \qquad \phi = 0, \quad x \in \partial D' \tag{13}$$

Note that $\lambda'_1 \geq \lambda_1$ because $D' \subseteq D$. Then, using Green's identity and equations (12) and (13),

$$0 = \int_{D'} (u\,\Delta\phi - \phi\,\Delta u)\,dx$$

$$= \int_{D'} \left[-u\lambda'_1\phi + \phi(\mu u - u^3)\right] dx$$

$$= \int_{D'} u\phi(\mu - \lambda'_1 - u^2)\,dx$$

But the last integral is strictly of one sign since u is of one sign, ϕ is positive, and $\mu < \lambda_1 \le \lambda'_1$. Therefore, we have a contradiction and no nontrivial solution can exist.

Example 4. Now consider the reaction-diffusion problem

$$u_t - \Delta u = \mu u - u^3, \qquad x \in D, \quad t > 0 \tag{14}$$

$$u(x,0) = u_0(x), \qquad x \in D \tag{15}$$

$$u(x,t) = 0, \qquad x \in \partial D, \quad t > 0 \tag{16}$$

where u_0 is a bounded, continuous function. By Example 3, this problem has only on equilibrium solution, namely $u_e = 0$, for $0 < \mu < \lambda_1$, where λ_1 is the principal eigenvalue of the negative Laplacian on D. We shall now show that $u_e = 0$ is asymptotically stable in the sense that there exist positive constants a and C such that

$$|u(x,t)| \le Ce^{-at} \sup_{x \in D} |u_0(x)| \tag{17}$$

for all $t > 0$ and $x \in D$ (see Definition 1 above). To begin, pick a region $D' \supseteq D$ such that the principal eigenvalue λ'_1 of $-\Delta$ on D' satisfies the inequality $\mu < \lambda'_1 < \lambda_1$. Choose the eigenfunction $\psi(x)$ corresponding to λ'_1 to be positive and normalized to satisfy the condition $\sup_{D'} |\psi(x)| = 1$. Thus we have

$$-\Delta\psi = \lambda'_1\psi, \quad x \in D' \quad \text{with} \quad \psi = 0, \quad x \in \partial D' \tag{18}$$

Now set

$$u(x,t) = w(x,t)\psi(x)e^{-at}$$

for $x \in D$, $t > 0$, where a is to be selected later. Note that ψ is positive on D. From (14) and (18) we can easily show that w satisfies the partial differential equation

$$\Delta w + \frac{2\,\text{grad}\,\psi \cdot \text{grad}\,w}{\psi} + \left(a + \mu - \lambda'_1 - w^2\psi^2 e^{-2at}\right)w - w_t = 0, \qquad x \in D, \quad t > 0 \tag{19}$$

This equation for w has the form

$$\Delta w + \sum b_j(x) w_{x_j} + cw - w_t = 0$$

where the quantity c is given by

$$c = a + \mu - \lambda'_1 - w^2\psi^2 e^{-2at}$$

Thus equation (19) has the form for which the maximum principle for parabolic operators applies, provided that $c < 0$ (see Section 6.4). But we can ensure that $c < 0$ by choosing a sufficiently small, since $\mu < \lambda'_1$. The auxiliary conditions on w are

$$w(x,t) = 0, \quad x \in \partial D, \quad t > 0; \qquad w(x,0) = \frac{u_0(x)}{\psi(x)}, \quad x \in D$$

Therefore, by the maximum principle,

$$|w(x,t)| \le \sup_{x \in D} \left| \frac{u_0(x)}{\psi(x)} \right|, \qquad x \in D, \quad t > 0 \tag{20}$$

Selecting the constant C to be the quantity on the right side of (20) then gives the result (17), thereby showing that the zero equilibrium solution of (14)–(16) is asymptotically stable, provided that $\mu < \lambda_1$, for arbitrary initial conditions u_0.

We are left to wonder about what happens in the reaction–diffusion system (14)–(16) when the parameter μ exceeds λ_1. Clearly, $u = 0$ is still an equilibrium solution. In the following example, we show that two additional nontrivial equilibria appear. The method relies on the fundamental existence theorem on upper and lower solutions stated in Section 8.2.

Example 5. Consider (10)–(11) where $\mu > \lambda_1$, where λ_1 is the principal eigenvalue of $-\Delta$ on the domain D. We show that there exist at last two nontrivial solutions, one positive and one negative. Let $u = \alpha\phi$, where α is a constant to be selected later and ϕ is the positive eigenfunction associated with the eigenvalue λ_1. Then

$$-\Delta\phi = \lambda_1\phi, \quad x \in D; \qquad \phi = 0, \quad x \in \partial D$$

Therefore,

$$\begin{aligned} -\Delta u - \mu u + u^3 &= -\alpha\,\Delta\phi - \mu\alpha\phi + \alpha^3\phi^3 \\ &= \alpha\phi\left(\lambda_1 - \mu + \alpha^2\phi^2\right) \end{aligned}$$

Now choose $|\alpha|$ sufficiently small to force $\lambda_1 - \mu + \alpha^2\phi^2 < 0$. Consequently, if $\alpha > 0$, then $\alpha\phi$ is a positive lower solution, and if $\alpha < 0$, then $\alpha\phi$ is a negative upper solution of (10)–(11). Now it remains to determine a positive upper solution and a negative lower solution. Consider a domain D' containing D, and let $\psi(x)$ be the positive eigenfunction of the principal eigenvalue λ'_1 of the negative Laplacian on D'. That is,

$$-\Delta\psi = \lambda'_1\psi, \quad x \in D'; \qquad \phi = 0, \quad x \in \partial D'$$

We then have $\lambda'_1 \le \lambda_1 < \mu$. Now take $u = \beta\psi$, which yields

$$-\Delta u - \mu u + u^3 = \beta\psi\left(\lambda'_1 - \mu + \beta^2\psi^2\right)$$

and consider this equation on D, where $\psi > 0$. If $|\beta|$ is chosen large enough and $\beta > 0$, then $\beta\psi$ is a positive upper solution that exceeds $\alpha\phi$, $\alpha > 0$. Similarly, if $|\beta|$ is chosen large enough and $\beta < 0$, then $\beta\psi$ is a negative lower solution that does not exceed $\alpha\phi$, $\alpha < 0$. We conclude that there must be a positive solution of (10)–(11) and a negative solution of (10)–(11).

We may think of (10)–(11) as a bifurcation problem in the following sense. As the bifurcation parameter μ increases from small positive values, there is a critical value $\mu = \lambda_1$ where the trivial, equilibrium solution $u = 0$ branches, or bifurcates, into three equilibrium solutions ($u = 0$, and the positive and negative solution). We represent this phenomenon schematically on a bifurcation diagram as shown in Figure 8.3. This type of bifurcation is called a pitchfork bifurcation because of the shape of the locus. It can be shown that for $\mu > \lambda_1$ there is only one positive and one negative solution, so there are no additional branches to the diagram. See, for example, Sattinger [1973], from where these examples are adapted. Stakgold [1979] also contains an elementary discussion of some of these ideas.

Example 6. In this example we address the question of stability of the zero solution of the equation discussed above, in one spatial dimension, by another standard method called linearized stability analysis. The question is: Do small perturbations at the initial time, near the zero solution distribution, decay to zero or grow without bound as t gets large? Let us consider the

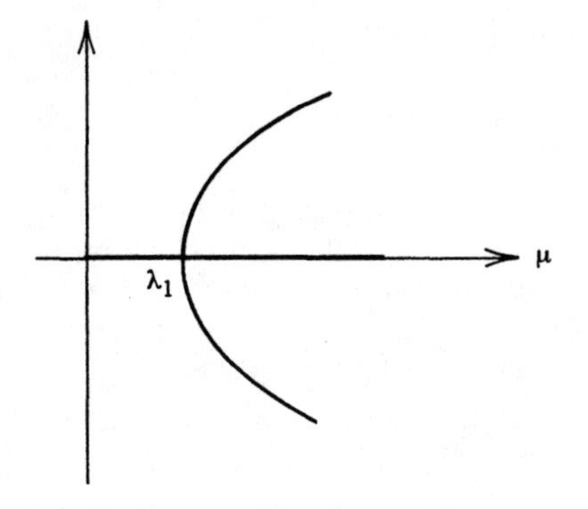

Figure 8.3. Bifurcation diagram indicating the existence of an equilibrium solution for values of the parameter μ. For $\mu < \lambda_1$ only a trivial solution exists; for $\mu > \lambda_1$ there are three equilibrium solutions, two nontrivial. The vertical axis could represent any feature of an equilibrium solution, for example, its maximum value (upper branch) or its minimum value (lower branch).

problem

$$u_t - u_{xx} = \mu u - u^3, \qquad 0 < x < \pi, \quad t > 0 \tag{21}$$

$$u(0,t) = u(\pi,t) = 0, \qquad t > 0 \tag{22}$$

$$u(x,0) = \varepsilon u_0(x), \qquad 0 < x < \pi \tag{23}$$

We note that $u = 0$ is an asymptotic solution to the problem for all $\mu > 0$. We have introduced an initial condition of the form (23), where ε is a small parameter, to study the evolution of a small initial perturbation in the problem. From our prior results, the zero solution is stable for $\mu < \lambda_1$, where λ_1 is the principal eigenvalue of $-d^2/dx^2$ on the interval $0 < x < \pi$. It is easy to determine the eigenvalues and eigenfunctions of the problem

$$-u''(x) = \lambda u(x), \quad 0 < x < \pi; \qquad u(0) = u(\pi) = 0$$

to be

$$\lambda_n = n^2, \qquad u_n(x) = \sin nx, \quad n = 1, 2, 3, \ldots$$

Therefore, the principal eigenvalue is $\lambda_1 = 1$. The standard stability argument proceeds by assuming a solution of (21)–(23) of the form $u(x,t) = \varepsilon v(x,t)$, where v is an order 1 function. One can regard this form as the zero solution plus a small perturbation represented by εv, $\varepsilon \ll 1$. If this form of u is substituted into the problem (21)–(23), we have

$$v_t - v_{xx} = \mu v - \varepsilon^2 v^3, \qquad 0 < x < \pi, \quad t > 0 \tag{24}$$

$$v(0,t) = v(\pi,t) = 0, \qquad t > 0 \tag{25}$$

$$v(x,0) = u_0(x), \qquad 0 < x < \pi \tag{26}$$

Because we are unable to solve this nonlinear equation exactly, we use an approximation and carry out a linear stability analysis. That is, we assume that the nonlinear term in (24) is small and negligible, which gives the linearized equation

$$v_t - v_{xx} = \mu v \tag{27}$$

If v stays bounded, our assumption that εv is small remains correct and it is consistent to drop the nonlinear term in (24). In this case we say that the zero solution is (linearly) stable to small perturbations. If v grows without bound, the assumption that the nonlinear term in (24) is small, with the subsequent linearization, is invalid. In this case we say that the zero solution is (linearly) unstable to small perturbations. In the latter case the full nonlinear equations must be analyzed to predict the correct evolutionary behavior of the small, initial perturbation.

Equation (27), along with the boundary and initial conditions (25)–(26) may be solved by the method of separation of variables. Let $v(x,t) = T(t)X(x)$, and substitute into (27) and the boundary conditions (25). In the usual way the PDE separates and we obtain a boundary value problem (a Sturm–Liouville problem) for X, namely

$$X'' + \lambda X = 0, \quad 0 < x < \pi, \qquad X(0) = X(\pi) = 0 \tag{28}$$

and an equation for T, namely

$$T' = (\mu - \lambda)T \tag{29}$$

Here λ is the separation constant. The boundary value problem (28) has eigenvalues and eigenfunctions given by

$$\lambda_n = n^2, \qquad X_n(x) = \sin nx, \quad n = 1, 2, 3, \ldots$$

and the T-equation therefore has solutions

$$T_n(t) = e^{(\mu - n^2)t}, \qquad n = 1, 2, 3, \ldots$$

Therefore, we have determined infinitely many solutions of the PDE(27) and the boundary conditions (25) of the form

$$v_n(x,t) = T_n(t)X_n(x) = e^{(\mu - n^2)t} \sin nx, \qquad n = 1, 2, 3, \ldots$$

These are the Fourier modes, and we can now superimpose them to determine a linear combination that meets the initial condition. Hence we take

$$v(x,t) = \sum_n a_n e^{(\mu - n^2)t} \sin nx \tag{30}$$

Applying the initial condition (26) yields

$$u_0(x) = \sum_n a_n \sin nx$$

which is the Fourier sine series for the initial function $u_0(x)$. Thus the Fourier coefficients are given by

$$a_n = \frac{2}{\pi}\int_0^{\pi} u_0(x) \sin nx \, dx \tag{31}$$

Consequently, the solution of the boundary value problem (25)–(27) is given by (30), with the a_n given by (31), which we write out as

$$v(x,t) = a_1 e^{(\mu-1)t} \sin x + a_2 e^{(\mu-4)t} \sin 2x + a_3 e^{(\mu-9)t} \sin 3x + \cdots \tag{32}$$

It is clear from (32) that if $\mu < 1$, all the exponential terms decay and $v(x,t)$ approaches zero for large t. Thus the zero solution is asymptotically stable to small perturbations. If $\mu > 1$, at least one of the terms (Fourier modes) in (32) grows exponentially and the perturbation grows without bound; the zero solution is (linearly) unstable in this case.

The stability analysis in Example 6 was based on linearization. In the case $\mu > 1$ the linearized equation (27) is no longer an accurate approximation of (24) because, as the solution (32) shows, the assumption that v^3 is small is no longer valid. Consequently, the solution (32) is no longer an accurate approximation of the solution to the nonlinear problem in this case, and we have inherent inconsistencies. Therefore, let us ask if we can deal with the nonlinear problem directly in order to get more information in the unstable regime.

Example 7. Consider the nonlinear boundary value problem (24) with auxiliary conditions (25) and (26), where the bifurcation parameter μ is just slightly larger than unity. We now introduce another standard technique by performing a normal-mode analysis for the nonlinear equation. We assume that the boundary value problem (24), (25), (26) has a solution of the form

$$v(x,t) = \sum_n y_n(t) \sin nx \tag{33}$$

where the $y_n(t)$ are to be determined. Here we are motivated by the solution form of the linearized problem and the fact that the eigenfunctions of the spatial problem are $\sin nx$. By Fourier analysis (and orthogonality) we know that the $y_n(t)$ must be given by the Fourier coefficients

$$y_n(t) = \frac{2}{\pi}\int_0^{\pi} v(x,t)\sin nx\,dx \tag{34}$$

Now multiply equation (24) by $\sin nx$ and integrate from 0 to π to obtain

$$\int_0^{\pi} v_t \sin nx\,dx - \int_0^{\pi} v_{xx}\sin nx\,dx$$
$$= \mu\int_0^{\pi} v\sin nx\,dx - \varepsilon^2\int_0^{\pi} v^3\sin nx\,dx \tag{35}$$

Then, using (33) we get

$$\int_0^{\pi}\left(\sum_m y_m'\sin mx\right)\sin nx\,dx + \int_0^{\pi}\left(\sum_m m^2 y_m\sin mx\right)\sin nx\,dx$$
$$= \mu\int_0^{\pi}\sum_m (y_m\sin mx)\sin nx\,dx - \varepsilon^2\int_0^{\pi} v^3\sin nx\,dx \tag{36}$$

Using the fact that the eigenfunctions $\sin nx$ are orthogonal on $[0,\pi]$, that is,

$$\int_0^{\pi}\sin mx\sin nx\,dx = \begin{cases} 0 & \text{if } n \neq m \\ \dfrac{\pi}{2} & \text{if } n = m \end{cases}$$

equation (36) can be written as

$$y_n' + (n^2 - \mu)y_n = -\frac{2\varepsilon^2}{\pi}\int_0^{\pi} v^3\sin nx\,dx, \qquad n = 1,2,3,\ldots \tag{37}$$

Now let us manipulate the nonlinear term on the right side of (37) by substituting for v. We have

$$\int_0^{\pi} v^3\sin nx\,dx = \int_0^{\pi}\left(\sum_k y_k\sin kx\right)^3\sin nx\,dx$$
$$= \int_0^{\pi}\left(\sum_{j,k,m} y_j y_k y_m\sin jx\sin kx\sin mx\right)\sin nx\,dx$$
$$= \sum_{j,k,m} a_{jkmn}y_j y_k y_m$$

where

$$a_{jkmn} = \int_0^{\pi} \sin jx \sin kx \sin mx \sin nx \, dx$$

Therefore, (37) becomes

$$y_n' + (n^2 - \mu) y_n = -\frac{2\varepsilon^2}{\pi} \sum a_{jkmn} y_j y_k y_m, \qquad n = 1, 2, 3, \ldots \quad (38)$$

and we have obtained a coupled system of ordinary differential equations for the coefficients $y_n(t)$. One observes that this system, because of the nonlinearity, is badly coupled since each equation contains all of the y_n. Initial conditions supplement (38) in the form

$$y_n(0) = \frac{2}{\pi} \int_0^{\pi} u_0(x) \sin nx \, dx \quad (39)$$

which comes from (34) and (26). We cannot solve (38)–(39), but we can use with advantage the assumption that μ is slightly larger than 1 to obtain the qualitative behavior of the solution in the asymptotic limit of large t. First we remark that if the terms of order ε^2 are neglected in (38), the equations decouple and the only growing solution is $y_1(t)$; the remaining y_n, for $n = 2, 3, \ldots$, still decay since $n^2 - \mu > 0$. Therefore, for a first approximation of the solution let us take

$$y_n(t) = y_n(0) e^{-(n^2 - \mu)t} \qquad \text{for } n = 2, 3, \ldots$$

Then, for $n = 1$, in equation (38) we retain only the term involving y_1 on the right side to obtain

$$y_1' + (1 - \mu) y_1 = -\frac{2\varepsilon^2 a_{1111}}{\pi} y_1^3 = -\frac{3\varepsilon^2}{4} y_1^3 \quad (40)$$

where we have used the fact that $a_{1111} = \int_0^{\pi} \sin^4 x \, dx = 3\pi/8$. We recognize equation (40) as a Bernoulli equation, which can be reduced to a linear equation via the transformation $y_1 = w^{-1/2}$. An easy calculation shows that w satisfies

$$w' - 2(1 - \mu) w = \frac{3\varepsilon^2}{2}$$

which has solution

$$w(t) = Ce^{2(1-\mu)t} - \frac{3\varepsilon^2}{4(1-\mu)}$$

Thus y_1 is given to leading order by

$$y_1(t) = \frac{1}{\sqrt{w(t)}} \tag{41}$$

and the constant C can be determined from the initial condition to be

$$C = \frac{3\varepsilon^2}{4(1-\mu)} + y_1^{-2}(0)$$

Now for large t we observe that $w(t) \to 3\varepsilon^2/[4(\mu-1)]$, and therefore

$$y_1(t) \sim \sqrt{\frac{4(\mu-1)}{3\varepsilon^2}}, \qquad t \gg 1 \tag{42}$$

Consequently, we have

$$u(x,t) \sim \sqrt{\frac{4(\mu-1)}{3}} \sin x, \qquad t \gg 1 \tag{43}$$

as the leading order approximation to (21)–(23) for large t. In conclusion, we remark that the linearized analysis in the foregoing example predicted, incorrectly, that $u(x,t)$ grows exponentially without bound if $\mu > 1$; the refined analysis shows, in fact, that u approaches a nonzero steady state given by (43) in the case when μ is only slightly larger than unity.

The nonlinear stability analysis in Example 7, which is based on normal modes, applies to many types of nonlinear problems.

The general issue of stability of equilibrium solutions to reaction–diffusion equations is an intriguing and well-developed area of study; a complete discussion of this topic, however, would take us too far afield, and therefore we limit our discussion to the example presented above. Additional sources are given in the references.

EXERCISES

1. Consider the nonlinear elliptic problem

$$-\Delta u = f(u), \qquad x \in D$$
$$u = 0, \qquad x \in \partial D$$

where the function $f(u) - \lambda_1 u$ does not vanish identically on any open interval $(0, u^*)$, where λ_1 is the principal eigenvalue of $-\Delta$ on the domain D.

(a) Prove that if this problem has a positive solution, then necessarily $f(u) - \lambda_1 u$ changes sign for $u < 0$. *Hint*: Compute the integral $\int_D (f(u) - \lambda_1 u)\phi \, dx$, using Green's identity, where ϕ is the positive eigenfunction corresponding to λ_1.

(b) Prove that if $f(u) - \lambda_1 u < 0$ for $u > 0$, a solution cannot be positive at any point in D.

2. Consider the nonlinear elliptic problem

$$-\Delta u + f(u) = \mu u, \qquad x \in D$$
$$u = 0, \qquad x \in \partial D$$

where $f \in C^2(R)$, $f(0) = f'(0) = 0$, where $f(u)/u$ is strictly increasing (decreasing) for $u > 0$ $(u < 0)$, and where $\lim_{|u| \to \infty} f(u)/u = +\infty$. Let λ_1 be as in Exercise 1. Prove the following:

(a) If $0 < \mu < \lambda_1$, the problem has only the trivial solution.

(b) If $\mu > \lambda_1$, there exists at least one positive nontrivial solution.

(c) If $\mu > \lambda_1$, the problem has a unique positive solution.

(See Hernandez [1986] or Stakgold and Payne [1973].)

3. Consider the nonlinear initial–boundary value problem

$$u_t - u_{xx} = \mu u(1 - u), \qquad 0 < x < \pi, \quad t > 0$$
$$u(0, t) = u(\pi, t) = 0, \qquad t > 0$$
$$u(x, 0) = \varepsilon u_0(x), \qquad 0 < x < \pi$$

Use a linear stability analysis to investigate the stability of the zero solution, and determine a critical value μ_c of the parameter μ above which the zero solution is unstable. For μ only slightly larger than μ_c, perform a nonlinear stability analysis as in Example 7 to determine the long-time behavior of the solution.

4. Consider the pure initial value problem for the nonlinear PDE

$$u_t - u_{xx} = \mu u(1 - u), \qquad x \in R, \quad t > 0$$

on an infinite domain. Here both $u = 0$ and $u = 1$ are asymptotic solutions. Use linear stability analysis to discuss the stability of each of these solutions (First take $u = \varepsilon v$; then take $u = 1 + \varepsilon v$. Use the superposition principle and the Fourier integral theorem as in Section 1.5 to solve the linearized problems.) Solve the nonlinear problem exactly when the initial condition is given by $u(x, 0) = \varepsilon$, $x \in R$.

REFERENCES

Elliptic equations may be the ones most studied in the literature on partial differential equations, and there are many outstanding general treatments, as well as books treating special topics. For the general study of elliptic equations one can consult Gilbarg and Trudinger [1983], Courant and Hilbert [1953, 1962], and Smoller [1983]. Linear eigenvalue problems are discussed in Courant and Hilbert [1953] and Garabedian [1986]. The classic book by Protter and Weinberger [1967] has a complete treatment of the maximum principle for elliptic and parabolic equations; see also Smoller [1983]. The recent book by Pao [1992] contains a thorough treatment of nonlinear elliptic problems and monotone iteration methods; in this regard, see also Ladde et al. [1985]. Bifurcation problems are discussed in many places; we cite the monograph by Sattinger [1973] and the review article by Stakgold [1971]. The review by Hernandez [1986] on elliptic problems is also recommended, as well as the account in the monograph by Bebernes and Eberly [1989]. Zauderer [1989] can be consulted for a treatment of the nonlinear stability analysis based on normal modes.

Bebernes, J., and D. Eberly, 1989. *Mathematical Problems from Combustion Theory*, Springer-Verlag, New York.

Courant, R. and D. Hilbert, 1953. *Methods of Mathematical Physics*, Vol. 1, Wiley-Interscience, New York.

Courant, R., and D. Hilbert, 1962. *Methods of Mathematical Physics*, Vol. 2, Wiley-Interscience, New York.

Garabedian, P. R., 1986. *Partial Differential Equations*, 2nd ed., Chelsea, New York.

Gilbarg, D., and N. S. Trudinger, 1983. *Elliptic Partial Differential Equations of Second Order*, 2nd ed., Springer-Verlag, Berlin.

Hale, J., and H. Kocak, 1991. *Dynamics and Bifurcations*, Springer-Verlag, New York.

Hernandez, J., 1986. Qualitative methods for nonlinear differential equations, in *Nonlinear Diffusion Problems*, Lecture Notes in Mathematics, No. 1224, Springer-Verlag, New York, pp. 47–118.

Ladde, G. S., V. Lakshmikantham, and A. S. Vatsala, 1985. *Monotone Iterative Techniques for Nonlinear Differential Equations*, Pitman, Boston.

Logan, J. D., 1987. *Applied Mathematics: A Contemporary Approach*, Wiley-Interscience, New York.

Pao, C. V., 1992. *Nonlinear Parabolic and Elliptic Equations*, Plenum Press, New York.

Protter, M. H., and H. Weinberger, 1967. *Maximum Principles in Differential Equations*, Prentice-Hall, Englewood Cliffs, N.J.

Sattinger, D., 1973. *Topics in Stability and Bifurcation Theory*, Lecture Notes in Mathematics, No. 309, Springer-Verlag, New York.

Smoller, J., 1983. *Shock Waves and Reaction–Diffusion Equations*, Springer-Verlag, New York.

Stakgold, I., 1971. Branching of solutions on nonlinear equations, *SIAM Rev.* **13**, 289–332.

Stakgold, I., 1979. *Green's Functions and Boundary Value Problems*, Wiley-Interscience, New York.

Stakgold, I., and L. E. Payne, 1973. Nonlinear problems in nuclear reactor analysis, in *Nonlinear Problems in the Physical Sciences and Biology*, Lecture Notes in Mathematics, No. 322, Springer-Verlag, New York, pp. 298–307.

Zauderer, E., 1989. *Partial Differential Equations of Applied Mathematics*, 2nd ed., Wiley-Interscience, New York.

INDEX

PURE AND APPLIED MATHEMATICS

A Wiley-Interscience Series of Texts, Monographs, and Tracts

ADÁMEK, HERRLICH, and STRECKER—Abstract and Concrete Categories
*ARTIN—Geometric Algebra
AZIZOV—Linear Operators in Spaces with an Indefinite Metric
BERMAN, NEUMANN, and STERN—Nonnegative Matrices in Dynamic Systems
BOYARINTSEV—Methods of Solving Singular Systems of Ordinary Differential Equations
*CARTER—Finite Groups of Lie Type
CHATELIN—Eigenvalues of Matrices
CLARK—Mathematical Bioeconomics: The Optimal Management of Renewable Resources, 2nd Edition
*CURTIS and REINER—Representation Theory of Finite Groups and Associative Algebras
*CURTIS and REINER—Methods of Representation Theory: With Applications to Finite Groups and Orders, Vol. I
CURTIS and REINER—Methods of Representation Theory: With Applications to Finite Groups and Orders, Vol. II
*DUNFORD and SCHWARTZ—Linear Operators
Part 1—General Theory
Part 2—Spectral Theory, Self Adjoint Operators in Hilbert Space
Part 3—Spectral Operators
FOLLAND—Real Analysis: Modern Techniques and Their Applications
FRÖLICHER and KRIEGL—Linear Spaces and Differentiation Theory
GARDINER—Teichmüller Theory and Quadratic Differentials
GRIFFITHS and HARRIS—Principles of Algebraic Geometry
HANNA and ROWLAND—Fourier Series, Transforms, and Boundary Value Problems, 2nd Edition
HARRIS—A Grammar of English on Mathematical Principles
*HENRICI—Applied and Computational Complex Analysis
Vol. 1, Power Series—Integration—Conformal Mapping—Location of Zeros
Vol. 2, Special Functions—Integral Transforms—Asymptotics—Continued Fractions
Vol. 3, Discrete Fourier Analysis, Cauchy Integrals, Construction of Conformal Maps, Univalent Functions
*HILTON and WU—A Course in Modern Algebra
*HOCHSTADT—Integral Equations
JOST—Two-Dimensional Geometric Variational Procedures
KOBAYASHI and NOMIZU—Foundations of Differential Geometry, Vol. I
KOBAYASHI and NOMIZU—Foundations of Differential Geometry, Vol. II
LOGAN—An Introduction to Nonlinear Partial Differential Equations
McCONNELL and ROBSON—Noncommutative Noetherian Rings
NAYFEH—Perturbation Methods
NAYFEH and MOOK—Nonlinear Oscillations
PETKOV—Geometry of Reflecting Rays and Inverse Spectral Problems
*PRENTER—Splines and Variational Methods
RAO—Measure Theory and Integration
RENELT—Elliptic Systems and Quasiconformal Mappings
RIVLIN—Chebyshev Polynomials: From Approximation Theory to Algebra and Number Theory, 2nd Edition

ROCKAFELLAR—Network Flows and Monotropic Optimization
ROITMAN—Introduction to Modern Set Theory
*RUDIN—Fourier Analysis on Groups
SENDOV—The Averaged Moduli of Smoothness: Applications in Numerical Methods and Approximations
SENDOV and POPOV—The Averaged Moduli of Smoothness
*SIEGEL—Topics in Complex Function Theory
Volume 1—Elliptic Functions and Uniformization Theory
Volume 2—Automorphic Functions and Abelian Integrals
Volume 3—Abelian Functions and Modular Functions of Several Variables
STAKGOLD—Green's Functions and Boundary Value Problems
*STOKER—Differential Geometry
*STOKER—Nonlinear Vibrations in Mechanical and Electrical Systems
*STOKER—Water Waves: The Mathematical Theory with Applications
WHITHAM—Linear and Nonlinear Waves
ZAUDERER—Partial Differential Equations of Applied Mathematics, 2nd Edition

*Now available in a lower priced paperback edition in the Wiley Classics Library.